Lecture Notes in Bioinformatics 4774

Edited by S. Istrail, P. Pevzner, and M. Waterman

Subseries of Lecture Notes in Computer Science

Jagath C. Rajapakse   Bertil Schmidt
Gwenn Volkert (Eds.)

Pattern Recognition in Bioinformatics

Second IAPR International Workshop, PRIB 2007
Singapore, October 1-2, 2007
Proceedings

 Springer

Series Editors

Sorin Istrail, Brown University, Providence, RI, USA
Pavel Pevzner, University of California, San Diego, CA, USA
Michael Waterman, University of Southern California, Los Angeles, CA, USA

Volume Editors

Jagath C. Rajapakse
Nanyang Technolocial University, Singapore
E-mail: asjagath@ntu.edu.sg

Bertil Schmidt
University of New South Wales Asia, Singapore
E-mail: bertil.schmidt@unswasia.edu.sg

Gwenn Volkert
Kent State University, USA
E-mail: volkert@cs.kent.edu

Library of Congress Control Number: Applied for

CR Subject Classification (1998): H.2.8, I.5, I.4, J.3, I.2, H.3, F.1-2

LNCS Sublibrary: SL 8 – Bioinformatics

ISSN	0302-9743
ISBN-10	3-540-75285-4 Springer Berlin Heidelberg New York
ISBN-13	978-3-540-75285-1 Springer Berlin Heidelberg New York

Springer is a part of Springer Science+Business Media

springer.com

© Springer-Verlag Berlin Heidelberg 2007
Printed in Germany

Typesetting: Camera-ready by author, data conversion by Scientific Publishing Services, Chennai, India
Printed on acid-free paper   SPIN: 12166909   06/3180   5 4 3 2 1 0

Preface

The advancements of computational and informational techniques have enabled in silico testing of many lab-based experiments in life sciences before performing them in in vitro or in vivo. Though computational techniques are not capable of mimicking all wet-lab experiments, bioinformatics will inevitably play a major role in future medical practice. For example, in the pursuit of new drugs it can reduce the costs and complexity involved in expensive wet-lab experiments. It is expected that by 2010, sequencing of individual genomes will be affordable generating an unprecedented increase of life sciences data, in the form of sequences, expressions, networks, images, literature. Pattern recognition techniques lie at the heart of discovery of new insights into biological knowledge, as the presence of particular patterns or structure is often an indication of its function.

The aim of the workshop series Pattern Recognition in Bioinformatics (PRIB) is to bring pattern recognition scientists and life scientists together to promote pattern recognition applications to solve life sciences problems. This volume presents the proceedings of the 2nd IAPR Workshop PRIB 2007 held in Singapore, October 1–2, 2007. It includes 38 technical contributions that were selected by the International Program Committee from 125 submissions. Each of these rigorously reviewed papers was presented orally at the workshop. The proceedings consists of six parts. Part 1: Sequence Analysis; Part 2: Prediction of Protein Structure, Interaction, and Localization; Part 3: Gene Expression Analysis; Part 4: Pathway Analysis; Part 5: Medical Informatics; and Part 6: Bioimaging.

Part 1 of the proceedings contains seven chapters on sequence analysis. Tang et al. propose a new design of BLAST-based gene ontology (GO) term annotator which incorporates data mining techniques and rough sets to deduce biological functions from DNA sequences. A design of ClustalW, using field programmable gate arrays (FPGA) is developed by Aung et al. to perform sequence alignment in real-time applications. Stepanova, Lin, and Lin develop a two-phase artificial neural network, and present its FPGA implementation, for genome-wide detection of response elements in steroid hormone receptors. Greene, Bill, and Moore propose an expert knowledge-guided mutation operator for the detection of genome-wide variations of DNA, using genetic programming. Luthra et al. find a conserved motif PMNYM of the transmembrane TM5 domain involved in dimerization of the A2a receptor, with a PROSITE search. Deng, Deng, and Havukkala find a strong GC and AT skew correlation in the chicken genome, using a novel visualization technique. Pearson et al. compare interval mapping to a hierarchical Bayesian method for quantitative trait loci analysis on *Arabidopsis thaliana*.

Part 2 of the proceedings contains nine chapters on the prediction of protein structure, interaction, and localization. Shi et al. propose multiple support vector machines (SVM) to handle different features and then decision templates to combine predictions so as to detect protein subcellular localization. Hoque, Chetty, and Dooley propose a generalized schemata theorem incorporating twin removal for genetic algorithms (GA) to predict protein structure. Zhang, Wei, and Ding use a fuzzy SVM to improve the prediction of structural classes of low-homology proteins. Singh and Ramani demonstrate a method to predict right-handed β-helix fold from protein sequences using SVM and report improved performance measures.

Taguchi and Gromiha investigate several amino acid features and find amino acid occurrences improve the recognition of protein fold recognition significantly over the other features. Ou, Shao, and Chen propose an efficient RBF network to identify interface residues of interacting proteins, based on PSSM profiles and biochemical properties. Ahmad presents dynamic outlier exclusion training algorithm for neural networks to enhance sequence-based predictions in residue level protein properties. Gromiha analyzes amino acid sequences of transmembrane β barrel proteins (TMBs) and finds a significantly higher occurrence of Ser, Asn and Gln in TMBs than in globular proteins. Ahmed estimates the evolutionary average hydrophobicity profile from a family of protein sequences.

Part 3 of the proceedings contains nine chapters on gene expression analysis. Yuriy et al. develop an online database for Affymetrix probe mapping and annotation (APMA) for interactive access, search, and visualization of target sequences mapping and annotation. Blanco, Martin-Merino, and Rivas combine different kinds of dissimilarity-based classifiers for the identification of cancerous samples from microarray data and illustrate its efficacy over existing classifiers. Stiglic, Khan, and Kokol propose small ensemble classifiers to visually interpret microarray data for easy comprehension of their functionality. The method is illustrated in a case-study of leukemia samples. Zhou et al. propose ant-MST, an ant-based minimum spanning tree for gene expression data clustering. McGarry, Sarfraz, and McIntyre integrate GO measures to SOM classification of gene expression data to obtain biologically meaningful clusters of genes.

Teng and Chan find order preserving clusters in gene expression data by converting each gene vector into an ordered label sequence. A method is then proposed by finding the frequent orders by iteratively combining the most frequent prefixes and suffixes in a statistical way. Mao and Tang propose correlation-based relevancy and redundancy measures for efficient gene selection and show promising results in six gene expression problems. Mundra and Rajapakse present relevancy and redundancy criteria for gene selection with an SVM-recursive feature elimination (RFE) method which selects gene subsets with better classification accuracy and generalization capability compared to the SVM-RFE method. Oja obtains digital expression profiles of human endogenous retroviruses.

Part 4 of the proceedings contains four chapters on pathway analysis. Ram and Chetty propose a framework for path analysis in gene regulatory networks by first finding the network structure by causal modeling and then enhancing the network by post-processing. Sehgal et al. reconstruct transcriptional gene regulatory network reconstruction through cross-platform fusion of gene networks. Ling et al. reconstruct protein–protein interaction pathways by mining subject-verb-objects intermediates in biological texts. Chaturvedi, Sakharkar, and Rajapakse propose a validation technique for gene regulatory networks with protein–protein interaction data by using a GA.

They demonstrate the potential of the method in an application to cell-cycle regulation.

Part 5 of the proceedings contains four chapters in medical informatics. Kurzynski and Zolnierek introduce and compare rough set- and fuzzy set-based methods for sequential medical diagnostic problems. Perumal, Lim and Sakharkar propose a comparative genomic approach for metabolic pathway analysis for in silico identification of putative drug targets in *Pseudomonas aeruginosa*. You et al. compare four methods of affinity prediction models for HLA-binding peptides and T-cell epitope identification, and find that non-liner models perform better than linear predictors. Rajapakse and Feng propose a method to identify peptides binding to MHC molecules by simultaneously optimizing entropy and evolutionary distance. Further, the binding motifs are determined by the optimal alignment of binding sites.

Part 6 of the proceedings contains five chapters on bioimaging. Dufour et al. develop a automated nuclear morphometric analysis of 3D fluorescence microscopy images by using active meshes. They also propose shape descriptors and evaluate their robustness and independence on fluorescent beads and on two cell lines. Kumar and Rajapakse propose a time-frequency-based method for detection of activation in functional MRI time-series and discuss the advantages over earlier methods. Dehzangi, Zolghadri, and Boostani develop a weighted distance neural network for high-performance classification of two imagery tasks in the cue-based brain computer interface. Zheng and Rajapakse tract the anatomical connectivity of the brain, using sequential sampling and resampling of diffusion tensor MR images. The method does not adopt fractional anisotropy as the stopping criteria and regularizes the fiber-tracking process by assigning high confidence values at low curvature points. Gong et al. develop an automated pipeline for classification of CT brain images of different head trauma, which is useful for building a content-based medical image retrieval system.

We would like to sincerely thank all authors who spent their time and effort to make important contributions to this book. Many thanks go to the reviewers whose comments have enhanced the quality of the chapters. Our gratitude also goes to the *LNBI editors* and the *managing editor* for their most kind support and help in editing this book.

We would also like to thank all individuals and institutions that contributed to the success of PRIB 2007, especially the authors for submitting the papers and all the sponsors for generously providing financial support for the workshop. We are very grateful to IAPR for the sponsorship and the IAPR Technical Committee (TC-20) on Pattern Recognition for Bioinformatics for their support and advice. Our gratitude goes to the School of Computer Engineering, Nanyang Technological University, Singapore, for supporting the workshop in many ways.

We would like to express our gratitude to all PRIB 2007 International Program Committee members and other invited reviewers for their objective and thorough reviews of the submitted papers. We fully appreciate the PRIB 2007 Organizing Committee for their time and excellent work. We thank Publicity Co-chairs, Feng Lin and Sy Loi Ho, for their hard work in getting the proceedings ready on time. We are grateful to Norhana Ahmad, PRIB 2007 secretary, for coordinating all the logistics of the workshop. Our thanks also go to Ang Linda for maintaining the workshop Web site, Tan Sing Yau for the technical support, and Jean Tan for his help in graphics design.

Last but not least, we wish to convey our sincere thanks to Springer for providing excellent professional support in preparing this volume.

October 2007

Jagath C. Rajapakse
Raj Acharya
Bertil Schmidt
Gwenn Volkert

Organization

IAPR Technical Committee (TC-20) on Pattern Recognition for Bioinformatics

Raj Acharya (Vice-chair)	Pennsylvania State University, USA
Fransisco Azuaje	University of Ulster, UK
Vladimir Brusic	University of Queensland, Australia
Phoebe Chen	Deakin University, Australia
David Corne	Heriot-Watt University, UK
Elena Marchiori	Vrije University of Amsterdam, The Netherlands
Mariofanna Milanova	University of Arkansas at Little Rock, USA
Gary B. Fogel	Natural Selection, Inc., USA
Saman K. Halgamuge	University of Melbourne, Australia
Visakan Kadirkamanathan	University of Sheffield, UK
Nik Kasabov	Auckland University of Technology, New Zealand
Irwin King	Chinese University of Hong Kong, Hong Kong
Alex V. Kochetov	Russian Academy of Sciences, Russia
Graham Leedham	Nanyang Tech. University, Singapore
Ajit Narayanan	University of Exeter, UK
Marimuthu Palaniswami	University of Melbourne, Australia
Jagath C. Rajapakse (Chair)	Nanyang Tech. University, Singapore
Gwenn Volkert	Kent State University, USA
Roy E. Welsch	Massachusetts Inst. of Technology, USA
Kay C. Wiese	Simon Fraser University, Canada
Limsoon Wong	National University of Singapore, Singapore
Jiahua (Jerry) Wu	Wellcome Trust Sanger Inst., UK
Yanqing Zhang	Georgia State University, USA
Qiang Yang	Hong Kong University of Science and Technology, Hong Kong

PRIB 2007 Organization

General Chair

Jagath C. Rajapakse (Co-chair)   Nanyang Technological University, Singapore

General Co-chair

Raj Acharya   Pennsylvania State University, USA

Program Chairs

Bertil Schmidt   University of New South Wales Asia, Singapore
Gwenn Volkert   Kent State University, USA

Special Session Chairs

Shandar Ahmad   National Institute of Biomedical Innovation,
                 Japan
Madhu Chetty   Monash University, Australia
Elena Marchiori Vrije University of Amsterdam, The Netherlands

Publicity Chairs

Saman K. Halgamuge   University of Melbourne, Australia
Roberto Tagliaferri   Università Di Salerno, Italy
Wei Wang   Fudan University, China
Yanqing Zhang   Georgia State University, USA

Publication Chairs

Sy-Loi Ho   Nanyang Technological University, Singapore
Feng Lin   Nanyang Technological University, Singapore

Local Chair

Graham Leedham   University of New South Wales Asia, Singapore

Local Organization Committee

Byron Koon Kau Choi   Nanyang Technological University, Singapore
Yulan He   Nanyang Technological University, Singapore
Hwee Kuan Lee   Bioinformatics Institute, Singapore
Jinming Li   Nanyang Technological University, Singapore

Secretariat

Norhana Binte Ahmad   Nanyang Technological University, Singapore

System Administration

Linda Ang Ah Giat   Nanyang Technological University, Singapore

Program Committee

Tatsuya Akutsu   Kyoto University, Japan
Guillaume Bourque   Genome Institute of Singapore, Singapore
Timo Rolf Bretschneider   Nanyang Technological University, Singapore
Zehra Cataltepe   Istanbul Technical University, Turkey
Phoebe Chen   Deakin University, Australia
Francis Y.L. Chin   University of Hong Kong, Hong Kong
Peter Clote   Boston College, USA
David Corne   Heriot-Watt University, UK
Carlos Cotta   University of Malaga, Spain
Antoine Danchin   Institut Pasteur, France
Joaquín Dopazo   Centro de Investigación Príncipe Felipe, Spain
James G. Evans   Massachusetts Institute of Technology, USA
Alexandru Floares   Oncological Institute Cluj-Napoca, Romania
Mikhail S. Gelfand   Institute for Information Transmission Problems,
    Russia
Ilkka Havukkala   Auckland University of Technology, New Zealand
Jaap Heringa   Vrije Universiteit, The Netherlands
Lisa Holm   University of Helsinki, Finland
Ming-Jing Hwang   Academia Sinica, Taiwan
Visakan Kadirkamanathan   University of Sheffield, UK
Nikola Kasabov   Auckland University of Technology, New Zealand
Irwin King   The Chinese University of Hong Kong,
    Hong Kong

Alex V. Kochetov	Russian Academy of Sciences, Russia
Vladimir A. Kuznetsov	Genome Institute of Singapore, Singapore
Chee Keong Kwoh	Nanyang Technological University, Singapore
Wing-Ning Li	University of Arkansas, USA
Alan Wee-Chung Liew	Chinese University of Hong Kong, Hong Kong
Frederique Lisacek	Swiss Institute of Bioinformatics, Switzerland
Hiroshi Matsuno	Yamaguchi University, Japan
Martin Middendorf	Universität Leipzig, Germany
Mariofanna Milanova	University of Arkansas at Little Rock, USA
Aleksandar Milosavljevi	Baylor College of Medicine, USA
Satoru Miyano	University of Tokyo, Japan
Jason H. Moore	Dartmouth Medical School, USA
Parvin Mousavi	Queen's University, Canada
See-Kiong Ng	Institute for Infocomm Research, Singapore
Yanay Ofran	Columbia University, USA
Christos Ouzounis	European Bioinformatics Institute, UK
Zoran Obradovic	Temple University, USA
Nikhil R. Pal	Indian Statistical Institute, India
Laxmi Parida	IBM T.J. Watson Research Center, USA
Mihail Popescu	University of Missouri, USA
Predrag Radivojac	Indiana University, USA
Nikolaus Rajewsky	Max Delbruck Center for Molecular Medicine, Germany
Jem Rowland	University of Wales Aberystwyth, UK
Meena Kishore Sakharkar	Nanyang Technological University, Singapore
Akinori Sarai	Kyushu Institute of Technology, Japan
Alexander Schliep	Max Planck Institute for Molecular Genetics, Germany
Christian Schoenbach	Nanyang Technological University, Singapore
N.Srinivasan	Indian Institute of Science, India
P. N. Suganthan	Nanyang Technological University, Singapore
Wing Kin Sung	National University of Singapore, Singapore
Anna Tramontano	University of Rome "La Sapienza", Italy
Michael Wagner	Cincinnati Children's Hospital Research Foundation, USA
Haiying Wang	University of Ulster at Jordanstown, UK
Lusheng Wang	City University of Hong Kong, Hong Kong
Michael Q. Zhang	Cold Spring Harbor Laboratory, USA

Reviewers

Konagaya Akihiko	RIKEN, Genomic Sciences Centre, Japan
Mundra Piyushkumar Arjunlal	Nanyang Technological University, Singapore
Wendy Ashlock	University of Guelph, Canada
Sansanee Auephanwiriyakul	Chiangmai University, Thailand
Jung-Hsien Chiang	National Cheng Kung University, Taiwan
Kai-Bo Duan	Center for Drug Discovery, Singapore
Julien Epps	University of New South Wales Asia, Singapore
Margaret J. Eppstein	University of Vermont, Canada
Bruno Gaeta	University of New South Wales, Australia
Shinn-Ying Ho	National Chiao Tung University, Taiwan
Masoud Jamei	Simcyp Limited, UK
Vert Jean-Philippe	Ecole des Mines de Paris, France
Vinny Just	Ohio University, USA
Marta Kasprzak	Poznan University of Technology, Poland
Kyung Joong Kim	Yonsei University, Korea
Prasanna Ratnakar Kolatkar	Genomic Institute of Singapore, Singapore
Lukasz Kurgan	University of Alberta, Canada
Weiguo Liu	Nanyang Technological University, Singapore
Pasi Luukka	Lappeenranta University of Technology, Finland
Jianmin Ma	Nanyang Technological University, Singapore
Nawar Malhis	University of British Columbia, Canada
Bernard Moret	Ecole Polytechnique Federale de Lausanne, France
Ngoc Minh Nguyen	Nanyang Technological University, Singapore
Merja Oja	University of Helsinki, Finland
Menaka Rajapakse	Institute of Infocomm Research, Singapore
Carmelina Ruggiero	University of Genoa, Italy
Muhammad Shoaib B. Sehgal	Monash University, Australia
Scott Smith	Boise State University, USA
Yuchun Tang	Georgia State University, USA
Thanos Vasilakos	University of Western Macedonia, Greece
Chandra Verma	Bioinformatics Institute, Singapore
Tiffani Williams	Texas A&M Engineering, USA
Gwan-Su Yi	Information and Communications University, Korea
Rui Xu	University of Missouri-Rolla, USA
Runxuan Zhang	Institut Pasteur, France
Shuigeng Zhou	Fudan University, China

Table of Contents

Part III: Gene Expression Analysis

Part IV: Pathway Analysis

Part V: Medical Informatics

Part VI: Bioimaging

Automated Methods of Predicting the Function of Biological Sequences Using GO and Rough Set

Xu-Ning Tang[1], Zhi-Chao Lian[2], Zhi-Li Pei[2,3], and Yan-Chun Liang[2,*]

[1] College of Software, Jilin University, Changchun 130012, China
[2] College of Computer Science and Technology, Jilin University, Key Laboratory of Symbol Computation and Knowledge Engineering of Ministry of Education, Changchun 130012, China
[3] College of Mathematics and Computer Science, Inner Mongolia University for Nationalities, Tongliao 028043, China
ycliang@jlu.edu.cn

Abstract. With the extraordinarily increase in genomic sequence data, there is a need to develop an effective and accurate method to deduce the biological functions of novel sequences with high accuracy. As the use of experiments to validate the function of biological sequence is too expensive and hardly to be applied to large-scale data, the use of computer for prediction of gene function has become an economical and effective substitute. This paper proposes a new design of BLAST-based GO term annotator which incorporates data mining techniques and utilizes rough set theory. Moreover, this method is an evolution against the traditional methods which only base on BLAST or characters of GO Terms. Finally, experimental results prove the validity of the proposed rough set-based method.

Keywords: GO BLAST Rough Set Theory.

1 Introduction

Along with the development of modern sequencing technology, the number of gene sequence is increasing everyday. A report coming from GenBank, a major repository of genomic data, shows an exponential increase in sequence data, during the last decade. As a result, biologists have to waste amount of time in finding out some useful information within specific domain. Even worse, different biological database might use different nomenclatures, which like some dialects, making information search, especially for computer-based information search, unavailable. So, how to store and take advantage of the information has become many biologists' common concern.

1.1 Gene Ontology

The emergence of Gene Ontology (GO) project has been used to solve the nomenclature problem. Gene Ontology project provides a set of unified, standard and hierarchical terms to note the functional characters of gene products [1]. People can use

---

* Corresponding author.

J.C. Rajapakse, B. Schmidt, and G. Volkert (Eds.): PRIB 2007, LNBI 4774, pp. 1–10, 2007.

nomenclature provided by GO project to annotate the biological functions of biological sequences.

Each item in GO database is composed with three key parts: gene product ID, GO terms and evidence code. Among them, gene product ID uniquely identifies the sequence of a gene product. Moreover, as sequence data alone is of limited use to biologists, GO project annotates the functions of gene products from three points of view. They are biological process, cellular component and molecular function. At last, evidence code indicates how annotation to a particular term is supported.

Essentially, each of these three types of terms can be separated into more detailed sub-categories, so that those terms construct a DAG (directed acyclic hierarchical graph), shown in Figure 1. Generally speaking, GO is a unified biological tool which can annotate gene product's function with a set of dynamic controlled vocabulary and it can keep on upgrading with the development of biology.

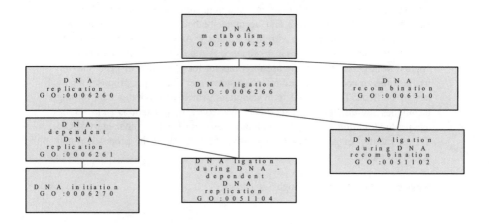

Fig. 1. Directed acyclic hierarchical graph of GO term

1.2 Basic Theory About Rough Set

Rough set has been introduced as a mathematical tool for dealing with fuzzy and uncertain knowledge in artificial intelligence application.

For convenience, we will introduce some basic concepts of rough set at first [2].

Definition 1. Given a knowledge system K= (U, R), for each subset $X \subseteq U$ and an equivalence relation $R \in ind(K)$, define two subsets:

Lower approximation: $\underline{R}X = \bigcup\{Y \in U / R \mid Y \subseteq X\}$

Upper approximation: $\overline{R}X = \bigcup\{Y \in U / R \mid Y \cap X \neq \varnothing\}$

Any subset defined by its lower and upper approximation is called a rough set.

Definition 2. Positive region: Let P and Q be equivalence relations within U, $pos_P(Q)$ is called the P-positive region of Q, such that $pos_p(Q) = \bigcup_{X \in U/Q} \underline{P}X$.

Definition 3. Let $DT = <U, C \cup D, V, f>$ be a Decision table, where C and D stand for conditional and decision attributes subsets, $C \cap D = \varnothing$, U is a non-empty, finite set called universe, V is called the value set, f stand for information function.

Definition 4. Let $\varnothing \subseteq X \subseteq C$, $\varnothing \subseteq Y \subseteq D$, $U/Y \neq \{U\}$, given $x \in X$, define significance of x with X (comparing with Y):

$$sig^Y_{X-\{x\}}(x) = (|S_X(Y)| - |S_{X-\{x\}}(Y)|)/|U|.$$

2 Relative Work and Background

Although the emergence of GO project has been used to solve the problem of unification of nomenclature successfully, there is another remarkable problem about how to apply these nomenclatures on large-scale data effectively.

At present, a number of automated BLAST-based GO term prediction applications have been published. BLAST is the most widely used sequence alignment tool [3, 4]. It permits the user to find similar sequence according to high degrees of local similarity. Normally, it is very likely that similar sequences might be homological; therefore, the similar sequences may have the same or similar functions. For these reasons BLAST has been employed to assign GO terms to a novel sequence. Nowadays, there are several methods with the idea of predicting the function of gene product using BLAST and GO, such as TOP BLAST, GOtach, GOFigue, Goblet and some others [5-10]. These approaches can be roughly divided into several main kinds: graph-based, discriminant function-based and term distance concordance-based and so on. Among them the TOP BLAST is the most commonly used approach. However, TOP BLAST is not so accurate and convincing. As a result, this paper recommends a new design of BLAST-based GO term annotator which incorporates data mining techniques and utilizes rough set theory. Under the strict criterion, the new approach provides higher quality and more accurate functional prediction for a novel sequences than TOP BLAST can.

3 Rough Set-Based Method

3.1 Data Collection

The Gene Ontology data were downloaded and divided into three parts: training set, test set and BLAST-able database. This data consist of protein sequence data and their GO term associations. UniPort annotations, proteins and their GO term associations are submitted by UniPort, is referred to as BLAST-able database. This data, consisting of 107,632 proteins, have high quality annotation. Non UniPort annotations, consisting of 3,537 proteins and their GO term associations are submitted by other sources, are referred to as training set and test set. In order to examine our method's validity, we employ cross-validation method. Each time we randomly select 1,200 proteins as test set and the other 2,337 proteins as training set.

Evidence code indicates how annotation to a particular term is supported. Some are supported by experiments, some are supported by literature and some are supported by computation method. According to different evidence codes, for training set and test set respectively we constructed 2 different experimental sets: one experimental set, called 7-evidence set, includes GO terms supporting by evidence codes such as: TAS, IDA, IC, IMP, IGI, IPI and IEP. Another experimental set, called NoIEA set, includes GO terms supporting by all evidence codes except IEA. For the reason that all GO terms within 7-evidence set are supported by evidence code which have high reliability, meanwhile the GO terms within NoIEA set just preclude those supported by evidence code of IEA, there is no doubt that GO terms in 7-evidenc are more reliable and accurate than those in NoIEA.

3.2 Accuracy Metrics

As we employ the strict evaluation method, precision and recall rate are defined as:

$$\text{Precision: } P = \frac{c}{p}$$

Where c is the number of correct predicted term assignments and p is the total number of predicted assignments.

$$\text{Recall rate: } R = \frac{c}{t}$$

Where c is the number of correct predicted term assignments and t is the total number of correct term.

$$\text{Harmonic Mean: } H = \frac{2}{1/P + 1/R}$$

Only if the predicted term is the right term which the source sequence indeed has, we count it as a correct prediction. Otherwise, prediction hit on either its parent term or its children term is considered as a false prediction.

3.3 Preparation

Before deducing rules from decision table, there are some preparation works to do.

3.3.1 Basic Concept
(1) Source sequence: we define those protein sequences which need prediction of function in training set as source sequence.

(2) Target sequence: we define those protein sequences returned by BLAST from BLAST-able database as target sequences which are similar to the source sequence.

(3) Unit: For each source sequence in training set, we returned 5 most similar sequences (in sort of ascending E-value) by BLAST from BLAST-able database. And these 5 most similar sequences construct a unit.

(4) Each GO term of those sequences belonging to the unit has 5 attributes described below: GO ID (which can uniquely identify the GO term), Rank (the ascending rank value of the highest matching result the term is found in), Times (the number of annotations using the term), E-value (a parameter returned by BLAST and stand for the similarity between source sequence and target sequence, the smaller the similar), and Score (another parameter returned by BLAST similar to E-value).

3.3.2 Calculate the Probability of Different Values of Each Attribute Within All Units

(1) For each source sequence in training set, we return a unit by BLAST and calculate those 5 attributes of the unit.

(2) For all units obtained, we calculate the probability of different values for each attribute in these units (P(Times=X) X=1,2,3,4,5; P(Rank=X) X=1,2,3,4,5; P(Score=X) X>0; P(E-Value=X) 0<X<1).

3.3.3 Calculate the Conditional Probability

When source sequence indeed has this GO term (K=1), calculate the conditional probability of 4 of these attributes: P(Times=X|K=1); P(Rank=X|K=1); P(Score=X|K=1); P(E-Value=X|K=1)).

For the reason that a particular GO term may occur in many different units, and those 4 attributes (excluding GO ID) of this GO term may have different values when they appear in different units, so it's very likely that we can judge whether the source sequence has this GO term by those 4 attributes' value. This preparation step will help us to make a discretization of rough set later.

3.4 Algorithm

Because of the difference among GO terms, it is very likely that we can predict whether the sequence has a particular GO term or not by checking those 4 attributes of the GO term within the unit. As a result, we treat all units as a whole set and generate a set of rules for each GO term in this set. With these rules, we can predict the terms of a sequence within the unit.

For each GO term, once it has occurred in at least one unit, we will deduce a set of rules about it by decision table. For example, GO: 000019 has emerged in 7 different units of training set, so we construct a raw decision table based on this GO term's situation, as shown in Table 1. After that, according to those conditional probabilities obtained from preparation step, we make discretization of the raw decision table and get the discrete decision table, as shown in Table 2.

Table 1. Raw decision table of GO: 000019, GO NO identify the GO term which needs deduction rules; Sequence stands for the unit which contains this GO term and is returned by BLAST according to a source sequence; K=1 means that the source sequence indeed contains this GO term

GO NO	Sequence	Rank	Times	E-Values	Score	K
GO:0000119	DDB|DDB019126	5	1	1.00E-82	305	0
GO:0000119	DDB|DDB021490	1	2	2.00E-11	72	0
GO:0000119	SGD|S000000397	5	1	4.00E-15	80	1
GO:0000119	SGD|S000001100	3	1	9.00E-22	102	1
GO:0000119	SGD|S000002382	5	3	8.00E-09	62	0
GO:0000119	SGD|S000002716	5	1	1.00E-10	64	1
GO:0000119	SGD|S000003095	1	1	2.00E-12	69	0

Table 2. Discrete Decision Table of GO: 000019, According to the result of preparation step, we divide: Times=4 or 5 as high times, Times=2 or 3 as mid times, Times=1 as low times; Rank=1 as high rank, Rank=2 or 3 as mid rank, Rank=4 or 5 as low rank; E-Value>1.00E-30 as high e-value, E-Value<1.E-100 as low e-value, others as mid e-value; Score<200 as low score, Score>800 as high score, others as mid score

GO NO	Sequence	Rank	Times	E-Value	Score	K
GO:0000119	DDB|DDB019126	low rank	low times	mid e-value	mid score	0
GO:0000119	DDB|DDB021490	high rank	mid times	high e-value	low score	0
GO:0000119	SGD|S000000397	low rank	low times	high e-value	low score	1
GO:0000119	SGD|S000001100	mid rank	low times	high e-value	low score	1
GO:0000119	SGD|S000002382	low rank	mid times	high e-value	low score	0
GO:0000119	SGD|S000002716	low rank	low times	high e-value	low score	1
GO:0000119	SGD|S000003095	high rank	low times	high e-value	low score	0

```
GO:000019
high rank >>>0
mid rank >>>1
mid times >>>0
mid e-value >>>0
mid score >>>0
low rank mid times >>>0
low rank mid e-value >>>0
low rank mid score >>>0
low rank mid times low score >>>0
low rank mid times high e-value >>>0
low rank low times high e-value >>>1
low rank low times mid e-value >>>0
low rank low times mid score >>>0
low rank low times low score >>>1
```

Fig. 2. Rules Deducing From Decision Table

At last, we run our program on the discrete decision table and obtain a set of rules, as shown in Figure 2. After that, we can understand which attribute or which combination of attributes can be used to decide whether the source sequence has the GO term or not.

Here is the specific algorithm of knowledge discovery based on the decision table: let decision table DT, where it contains n samples (a total of 11 rows), j conditional attributes ($|C|=j$), and one decision attribute ($|D|=1$). At first we calculate the significance for every single conditional attribute and describe rules of it. For each important conditional attribute, if there is decision attribute fully rely on this conditional attribute, the algorithm is over. Or else, we pick out the most important attribute (having the highest significance) among those important attributes. And then based on this most important attribute, we check each combination of two conditional attributes, with the purpose of finding out all important combinations of two conditional attributes and describing its rules. The step continues until all important knowledge is discovered.

Step1. //Algorithm first calculates significance of every single conditional attribute.

> Let $C' = \emptyset$
>
> For i=1 to j
>
>> $C' = C' \cup \{c_i\}$
>>
>> Compute $pos_{C'}(D)$
>>
>> IF ($sig_{\emptyset}^{D}(c_i) > 0$)
>>
>>> Output rules
>>>
>>> Let m=1 and FLAG=1
>>>
>>> $C' = C' - \{c_i\}$
>
> END FOR
>
> IF ($\exists pos_{C'}(D) = U$)
>
>> Algorithm Finish
>
> ELSE
>
>> Find max($sig_{\emptyset}^{D}(c_i)$) and then let $J_1 = c_i$
>>
>> $C' = C' \cup \{J_1\}$

Step2. //each time we add one conditional attribute to the combination and calculate its significance.

```
WHILE (FLAG)
    m=m+1
    For i=1 to j (c_i ∉ C')
```

$$C' = C' - \{c_i\}$$

```
        Compute pos_{C'}(D)
```

$$IF \left(sig^{D}_{C'-\{c_i\}}(c_i) > 0 \right)$$

```
            Output rules
            Let FLAG=1
```

$$C' = C' - \{c_i\}$$

```
    END FOR
```

$$IF \left(\exists pos_{C'}(D) = U \right)$$

```
            Algorithm Finish
        ELSE
```

$$Find\ max \left(sig^{D}_{C'-\{c_i\}}(c_i) \right)\ and\ then\ let\ J_m = c_i$$

$$C' = C' \cup \{J_m\}$$

```
    END WHILE
```

For each sequence in the test set, we also use BLAST to return 5 most similar sequences from BLAST-able database as a unit. And then we calculate the statistic result of those 4 attributes for each GO term contained in the unit. With those rules obtained from the training set, we can judge whether the source sequence contains the GO term.

4   Simulation Results and Analysis

The proposed method is examined by applying it on 7-evidence and NoIEA dataset. Moreover, it is compared with Top BLAST, as shown in Table 3:

Table 3. Comparison Between Top BLAST and Rough Set

Method	Precision		Recall		Harmonic Mean	
	7-evdience	NoIEA	7-evdience	NoIEA	7-evdience	NoIEA
Top BLAST	0.21	0.40	0.15	0.33	0.175	0.362
Rough Set	0.56	0.68	0.10	0.21	0.170	0.321

1.   Either on 7-evidence dataset or on NoIEA dataset, the rough set-based method significantly improves the accuracy for the prediction of gene product. Especially, when it comes to 7-evidence dataset, the improvement could range from 21% to 56%, which is more obvious. It is almost 167% increase than before. Also, on NoIEA dataset our method improves the accuracy for prediction of gene product from 40% to 68%. It is nearly 70% increase than before. The main reason that the Rough Set-based method performs better on NoIEA dataset than on 7-evidence dataset is that the NoIEA dataset contains GO terms coming from both electronic annotation and curator-assigned. Those electronic annotations rely highly on sequences returned by BLAST, which is completely based on the similarity between sequences. As you known, our method also highly relies on those sequences returned by BLAST. As a result, precision will be greatly improved within NoIEA dataset. Similarly, within NoIEA dataset, GO terms which have evidence code such as ISS are also based on similarity. So that when there are evidence codes such as ISS, RCA in dataset, the result will be better. By looking through the results of TOP BLAST, we can also find the same situation.
2.   On the contrary, Rough Set-based method is correspondingly lower than TOP BLAST in recall rate. It means that although most of its prediction is correct, Rough Set-based method covers only a small part of correct GO term.
3.   At last, we know that Rough Set-based method and TOP BLAST have the similar performance by comparing the harmonic mean.

5   Conclusions

It is demonstrated that the rough set theory has great potential in bioinformatics, especially in the predicting functions of gene products. This paper proposes a data-mining-oriented method using rough set theory and applies it to prediction of gene function. Experimental results show that rough set-based method is able to provide high quality, conservative functional prediction for novel sequences. The proposed method can be used to improve the accuracy significantly by comparing with TOP BLAST. This method not only enables the electronic annotation to be more reliable but also decreases the cost for functional prediction of novel sequences, which makes it an effective supplement of experimental method. However, we should not ignore the shortcoming of the rough set-based method, especially for the low recall rate. There are many reasons for the low recall rate: on one hand, test set contain plenty of situations never appear in the training set. On the other hand, our rules returned from the training set are too conservative to ensure the sensitive. In addition, according to the experiment we find that how to discretize the rough set is another key to improve the rough set-based method.

Acknowledgment

The authors are grateful to the support of the National Natural Science Foundation of China (60673023, 60433020), the support of the European Commission for the project of TH/Asia Link/010 (111084), and "985" project of Jilin University. The authors wish to thank Professor Jiwen Guan for guidance on rough set-based algorithm during this research project.

References

1. Ashburner, M., Ball, C.A., Blake, J.A., Botstein, D., Butler, H.J., Cherry, M., Davis, A.P., Dolinski, K., Dwight, S.S., Eppig, J.J., Harris, M.A., Hill, D.P., Issel-Tarver, L., Kasarskis, A., Lewis, S., Matese, J.C., Richardson, J.E., Ringwald, M., Rubin, G.M., Sherlock, G.: Gene Ontology: tool for the unification of biology. Nature Genetics 25, 25–29 (2000)
2. Pawlak, Z.: Rough Sets: Theoretical Aspects of Reasoning about Data. Kluwer, Dordrecht (1992)
3. Altschul, S., Gish, W., Miller, W., Myers, E., Lipman, D.: Basic Local Alignment Search Tool. Journal of Molecular Biology 215, 403–410 (1990)
4. Altschul, S.F., Madden, T.L., Schaffer, A., Zhang, J., Zhang, Z., Miller, W., Lipman, D.J.: Gapped BLAST and PSI-BLAST: a new generation of protein database search programs. Nucleic Acids Research 25, 3389–3402 (1997)
5. Hennig, S., Groth, D., Lehrach, H.: Automated Gene Ontology annotation for anonymous sequence data. Nucleic Acids Research 31, 3712–3715 (2003)
6. Groth, D., Lehrach, H., Hennig, S.: GOblet: a platform for Gene Ontology annotation of anonymous sequence data. Nucleic Acids Research 32, W313–W317 (2004)
7. Khan, S., Situ, G., Decker, K., Schmidt, C.J.: GoFigure: Automated Gene Ontology annotation. Bioinformatics 19, 2484–2485 (2003)
8. Martin, D.M.A., Berriman, M., Barton, G.J.: GOtcha: a new method for prediction of protein function assessed by the annotation of seven genomes. BMC Bioinformatics 5, 178 (2004)
9. Joslyn, C., Mniszewski, S., Fulmer, A., Heaton, G.: The Gene Ontology Categorizer. Bioinformatics 20, i169–i177 (2004)
10. Verspoor, K., Cohn, J., Mniszewski, S., Joslyn, C.: A Categorization Approach to Automated Ontological Protein Function Annotation. Protein Science 15, 1544–1549 (2006)

C-Based Design Methodology for FPGA Implementation of ClustalW MSA

Yan Lin Aung[1], Douglas L. Maskell[1], Timothy F. Oliver[1] Bertil Schmidt[2], and William Bong[3]

[1] School of Computer Engineering, NTU, Singapore
{yanl0003,asdouglas}@ntu.edu.sg, tim.oliver@computer.org
[2] Division of Engineering, Science and Technology, UNSW Asia, Singapore
Bertil.Schmidt@unswasia.edu.sg
[3] Network Storage Technology Division, A*STAR Data Storage Institute, Singapore
William_BONG@dsi.a-star.edu.sg

Abstract. Systolisation of the pairwise distance computation algorithm and mapping into field programmable gate arrays (FPGA) have proven to give superior performance at a lower cost, compared to the same algorithm running on a cluster of workstations. The primary design methodology for this approach is based on the hardware description languages such as VHDL and Verilog HDL. An alternative design methodology, however, is the use of a high level language such as C to describe the algorithms and generate equivalent hardware descriptions for implementation in FPGA so as to reduce time to market. This paper describes the design and implementation of the ClustalW first stage multiple sequence alignment based on the Smith-Waterman algorithm on a low cost FPGA development platform using a C language development tool suite. Performance evaluation results show that comparable performance could be achieved to that of Pentium 4 systems and other HDL-based solutions using even the smallest commercially available FPGA device with this design methodology.

Keywords: multiple sequence alignment, ClustalW, FPGA, sequence analysis.

1 Introduction

Multiple sequence alignment (MSA) is a generalized pairwise sequence alignment to include more than two sequences of protein or nucleic acid. The main purpose of MSA is to infer homology between sequences. But there are many other applications: finding diagnostic patterns, characterization of protein families, detection of similarity between new sequences and well-known families of sequences, and evolutionary analysis. As the time and space complexities for MSA are in the order of the product of the lengths of the sequences, many heuristic alignment methods have been developed. Among them, the progressive alignment method is a widely used heuristic. One popular MSA program, ClustalW, makes use of such a method and consists of three stages: distance matrix, guided tree

J.C. Rajapakse, B. Schmidt, and G. Volkert (Eds.): PRIB 2007, LNBI 4774, pp. 11–18, 2007.
© Springer-Verlag Berlin Heidelberg 2007

and progressive alignment along the tree. The first stage computes and tabulates the distance value between every pair of sequences using pairwise sequence alignment. Then, a phylogenetic or guided tree is constructed using the distance matrix obtained in the first stage. Finally, sequences are progressively aligned according to the order specified by the guided tree to produce the final MSA. Unfortunately, the progressive alignment method still requires long periods of time to compute the MSA. Profiling of the ClustalW program on a single processor showed that almost 96% of the time is spent in the first stage [1].

1.1 Motivation

Many parallel processing approaches have been proposed and developed to speed-up the time consuming MSA first stage. Ebedes et al. have implemented parallel version of the MSA first stage on a small cluster of six workstations using the Message Passing Interface and reported that linear speed-up and almost 100% efficiency could be achieved [1].

Alternatively, it is also possible and very promising to map the MSA first stage into FPGA since the underlying architecture of the FPGA is well-suited for parallel processing. This approach is concerned with the systolisation of a pairwise distance computation algorithm and mapping it into FPGA, and it has proven to give superior performance at lower cost compared to the same algorithm running on a cluster of workstations [2], [3], [4]. The dominant design methodology for this approach has been based on hardware description languages (HDL), such as VHDL and Verilog HDL.

On the other hand, the capacity of FPGAs, like other semiconductor chips, increases in accordance with Moore's law as do the design complexities for these chips. The support provided by electronic design automation software tool vendors to cope with the increasing design complexity is lagging behind Moore's law, with conventional HDL-based designs often becoming bottlenecks in the design cycle. This has created a productivity gap between design complexity and design capacity [5]. A number of extensions to existing HDLs and software tools based on high level languages have emerged in an attempt to bridge this productivity gap. One such tool is CoDeveloper from Impulse Accelerated Technologies. CoDeveloper is a C language development tool suite for FPGAs which allows designers to use standard ANSI C to express highly parallel applications and algorithms.

With the above as the basis for our motivation, we have designed and implemented the ClustalW first stage multiple sequence alignment based on the Smith-Waterman algorithm on a low cost FPGA development platform using the C language development tool suite. We explain the pairwise sequence distance computation of the ClustalW MSA in detail and describe the recurrence equations, which were used to efficiently map the algorithm into FPGA hardware, in the next section.

2 ClustalW MSA First Stage: Distance Matrix

The ClustalW program makes use of the following definition to compute the distance between two sequences.

Definition 1. *Given a set of n sequences $S = S_1, \ldots, S_n$. For two sequences $S_i, S_j \in S$, pairwise sequence distance $d(S_i, S_j)$ is defined as follows:*

$$d(S_i, S_j) = 1 - \frac{nid(S_i, S_j)}{min\{l_i, l_j\}} \tag{1}$$

where $nid(S_i, S_j)$ denotes the number of exact matches in the optimal local alignment of S_i and S_j (with respect to the given scoring system, i.e the substitution matrix sbt and gap penalty parameters α, β for affine gap penalties or just α for linear gap penalties) and l_i (l_j) denotes the length of $S_i (S_j)$.

The Smith-Waterman algorithm can be used to compute the optimal local alignment of two sequences [6]. The algorithm compares two sequences by computing a distance that represents the minimal cost of transforming one segment into another using two elementary operations: match/mutation and insertion/deletion (also called a gap penalty). For two sequences, S_1 and S_2 with length l_1 and l_2, the Smith-Waterman algorithm computes the similarity $H_A(i, j)$ of two sequences ending at position i and j in order to identify common subsequences. The computation of $H_A(i, j)$, for $1 \leq i \leq l_1, 1 \leq j \leq l_2$, is given by the following recurrences:

$$H_A(i, j) = \max\{0, E(i, j), F(i, j), H_A(i - 1, j - 1) + sbt(S_1[i], S_2[j])\} \tag{2}$$
$$E(i, j) = \max\{H_A(i, j - 1) - \alpha, E(i, j - 1) - \beta\} \tag{3}$$
$$F(i, j) = \max\{H_A(i - 1, j) - \alpha, F(i - 1, j) - \beta\} \tag{4}$$

where sbt refers to the character substitution table. Initial values are: $H_A(i, 0) = E(i, 0) = H_A(0, j) = F(0, j) = 0$ for $0 \leq i \leq l_1, 0 \leq j \leq l_2$. Multiple gap costs are taken into account as follows: α is the cost of the first gap; β is the cost of the following gaps. This type of gap cost is known as affine gap penalty. Some applications also use a linear gap penalty, i.e. $\alpha = \beta$. For linear gap penalties the above recurrence relations can be simplified to:

$$H_L(i, j) = \max\{0, H_L(i, j - 1) - \alpha, H_L(i - 1, j) - \alpha, H_L(i - 1, j - 1) + sbt(S_1[i], S_2[j])\} \ . \tag{5}$$

Each position of the matrix H_A (H_L) is the similarity value. The two segments of S_1 and S_2 producing this value can be determined by a traceback procedure. The value $nid(S_1, S_2)$ of the two sequences can then be computed by counting the number of exact character matches during the traceback procedure of the Smith-Waterman algorithm. Unfortunately, this procedure is not suitable for hardware implementation. Hence, we present a new recurrence relation for the nid-value computation [7] that is suitable for hardware implementation in this section. We first explain the idea for the linear gap penalty and then generalize it for affine gap penalties.

Definition 2. *Given two sequences S_1 and S_2 with length l_1 and l_2, linear gap penalty α and a substitution table sbt, the matrix $N_L(i, j)$ $(1 \leq i \leq l_1, 1 \leq j \leq l_2)$ is recursively defined as follows:*

$$N_L(i,j) = \begin{cases} 0 & \text{if } H_L(i,j) = 0 \\ N_L(i-1,j-1) + m(i,j) & \text{if } H_L(i,j) = H_L(i-1,j-1) \\ & \quad + sbt(S_1[i], S_2[j]) \quad (6) \\ N_L(i,j-1) & \text{if } H_L(i,j) = H(i,j-1) - \alpha \\ N_L(i-1,j) & \text{if } H_L(i,j) = H(i-1,j) - \alpha \end{cases}$$

where

$$m(i,j) = \begin{cases} 1 \text{ if } S_1[i] = S_2[j] \\ 0 \text{ otherwise .} \end{cases}$$

Theorem 1. *For the local alignment of two sequences S_1 and S_2 with linear gap penalty α and substitution matrix sbt,*

$$nid(S_1, S_2) = N_L(i_{max}, j_{max})$$

where (i_{max}, j_{max}) denotes the coordinates of the maximum value in the corresponding matrix H_L.

Proof. Consider the optimal alignment of all pairs of suffixes of the first i characters of $S_1(S_1[1 \ldots i])$ and the first j characters of $S_2(S_2[1 \ldots j])$. This alignment is called the optimal i, j suffix alignment (of S_1 and S_2). It can be found by computing a traceback in the matrix H_L starting from cell (i,j). We now show that for a given pair of indices i, j $(1 \leq i \leq l_1, 1 \leq j \leq l_2)$, $N_L(i,j)$ is equal to the number of exact matches in optimal i, j suffix alignment. The claim then follows from the fact that the optimal i_{max}, j_{max} suffix alignment is equal to the optimal local alignment.

Case 1: $H_L(i,j) = 0$. The corresponding alignment is empty and $N_L(i,j) = 0$.
Case 2: $H_L(i,j) = H_L(i-1,j-1) + sbt(S_1[i], S_2[j])$. The alignment ends with $S_1[i]$ aligned to $S_2[j]$, which contributes $m(S_1[i], S_2[j])$ to the number of exact matches. The remaining number is then equal to the number of exact matches found in the optimal $i-1$, $j-1$ suffix alignment. Hence, $N_L(i,j) = N_L(i-1,j-1) + m(S_1[i], S_2[j])$.
Case 3: $H_L(i,j) = H_L(i-1,j) - \alpha$. The alignment ends with $S_1[i]$ aligned to a gap, which contributes zero exact matches. The remaining number is equal to the number found in the optimal $i-1$, j suffix alignment. Hence, $N_L(i,j) = N_L(i-1,j)$.
Case 4: $H_L(i,j) = H_L(i,j-1) - \alpha$. Similar to Case 3 follows $N_L(i,j) = N_L(i,j-1)$.

Because $H_L(i,j)$ must be equal to one of these four cases, the Theorem is proven.

$\square$

For affine gap penalties our method is extended as follows.

Definition 3. *Given two sequences S_1 and S_2, affine gap penalties α, β, and substitution table sbt, matrix $N_A(i,j)(1 \leq i \leq l_1, 1 \leq j \leq l_2)$ is recursively defined as follows:*

$$N_A(i,j) = \begin{cases} 0 & \text{if } H_A(i,j) = 0 \\ N_A(i-1,j-1) + m(i,j) & \text{if } H_A(i,j) = H_A(i-1,j-1) \\ & \qquad + sbt(S_1[i], S_2[j]) \\ N_E(i,j) & \text{if } H_A(i,j) = E(i,j) \\ N_F(i,j) & \text{if } H_A(i,j) = F(i,j) \end{cases} \quad (7)$$

where

$$m(i,j) = \begin{cases} 1 \text{ if } S_1[i] = S_2[j] \\ 0 \text{ otherwise} \end{cases}$$

$$N_E(i,j) = \begin{cases} 0 & \text{if } j = 1 \\ N_A(i,j-1) & \text{if } E(i,j) = H_A(i,j-1) - \alpha \\ N_E(i,j-1) & \text{if } E(i,j) = E(i,j-1) - \beta \end{cases}$$

$$N_F(i,j) = \begin{cases} 0 & \text{if } i = 1 \\ N_A(i-1,j) & \text{if } F(i,j) = H_A(i-1,j) - \alpha \\ N_F(i-1,j) & \text{if } F(i,j) = F(i-1,j) - \beta \end{cases}.$$

Theorem 2. *For the local alignment of the sequences S_1 and S_2, affine gap penalty α, β and substitution matrix sbt,*

$$nid(S_1, S_2) = N_A(i_{max}, j_{max})$$

where (i_{max}, j_{max}) denotes the coordinates of the maximum value in the corresponding matrix H_A.

Proof. Similar to the proof of Theorem 1 we show that for a given pair of indices i, j $(1 \leq i \leq l_1, 1 \leq j \leq l_2)$, $N_A(i,j)$ is equal to the number of exact matches in optimal i, j suffix alignment.

Case 1: $H_A(i,j) = 0$. The corresponding alignment is empty and $N_A(i,j) = 0$.
Case 2: $H_A(i,j) = H_A(i-1,j-1) + sbt(S_1[i], S_2[j])$. Similar to Case 2 in the previous proof follows $N_A(i,j) = N_A(i-1,j-1) + m(S_1[i], S_2[j])$.
Case 3: $H_A(i,j) = F(i,j)$. The alignment ends with $S_1[i]$ aligned to a gap, which contributes zero exact matches. The remaining number is equal to the number found in the optimal $i-1$, j suffix alignment. Depending on whether this alignment ends with a gap this number is either $N_A(i-1,j)$ or $N_F(i-1,j)$. Hence, $N_A(i,j) = N_F(i,j)$.
Case 4 $H_A(i,j) = E(i,j)$. Similar to Case 3 follows $N_A(i,j) = N_E(i,j)$.

Because $H_A(i,j)$ must be equal to one of these four cases, Theorem 2 is proven.
□

3 Design and Implementation

We first implemented the recurrence equations in C, and ascertained that it produced equivalent results to the original ClustalW pairwise alignment stage. Hardware/software partitioning of the algorithm was carried out subsequently using the CoDeveloper tool suite. Equivalent hardware descriptions were generated from the hardware portion of the C program for implementation in FPGA. We decided to create a compact microprocessor system built around the MicroBlaze soft processor from Xilinx in order to test and measure the performance of the generated design in contrast to conventional design verification techniques such as functional and timing simulation of the design. The high performance fast simplex link (FSL) of the processor system is used to connect the processor system and the MSA hardware processing element (PE).

In such a system, sequences, substitution matrix and gap penalties are stored in the processor system memory. They are then sent out to the hardware and pairwise alignment scores are read back from the hardware via the FSLs. We have targeted the system to the Xilinx ML403 development platform, which consists of a Virtex-4 FX12 FPGA. Our hardware PE can handle sequences with maximum length of 550 and supports both linear and affine gap penalties. Once the functionality of the design has been verified in hardware, design optimization techniques are employed to achieve high performance. In CoDeveloper, designers can implicitly or explicitly create the C code in such a way that the optimizer is able to detect the parallelism and generate highly parallel hardware. We mainly use the pipelining optimization technique in CoDeveloper to boost the design performance.

An initial optimized implementation of the system with a single MSA PE, which utilizes just 21% of the available hardware resources (slices), is able to achieve ~ 16 MCUPS. As we can increase the device utilization by instantiating additional MSA PEs conveniently with CoDeveloper, we added two additional MSA PEs into the system. Figure 1 shows the system with three MSA PEs, which are connected to the processor via FSLs. The whole system now utilizes 99% of the available resources in which three MSA PEs contribute 62%. Performance evaluation of this system along with comparisons to a Pentium 4 system and other HDL-based solutions is provided in the next section.

4   Performance Evaluation

The above-mentioned system with three MSA PEs achieves a throughput of ~ 36 MCUPS. Comparing this performance with that of a Pentium 4 2 GHz system, it is 2.4x faster. The FPGA on the ML403 platform is the smallest Virtex-4 FX family FPGA [8]. For larger FPGA devices such as Xilinx Virtex-4 LX80 and Virtex-II XC2V6000 FPGAs, 21 and 18 MSA PEs can be implemented into these devices respectively. Estimating 90% efficiency for MSA hardware, 302 and 259 MCUPS throughput can be expected. This performance is 20x and 17x faster than that of Pentium 4 2 GHz system.

Compared to the nearest FPGA-based Verilog HDL implementation, which makes use of XC2V6000 FPGA [9], our implementation is 3.9x slower since the implementation of [9] achieves a sustained performance of ~ 1 GCUPS in the Smith-Waterman dynamic programming matrix. Although our implementation is also slower than other FPGA-based implementations using VHDL [10], [3], these designs only implement edit distance, which is of theoretical interest but not used in practice [2] since it does not allow for different gap penalties and substitution tables. This is not the case for our implementation. Figure 2 shows the performance comparison of various systems discussed previously.

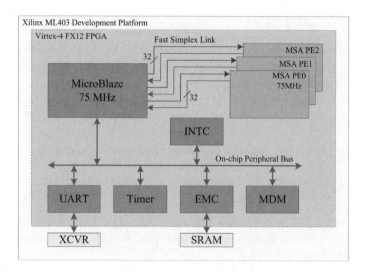

Fig. 1. MicroBlaze Processor System with Three MSA PEs

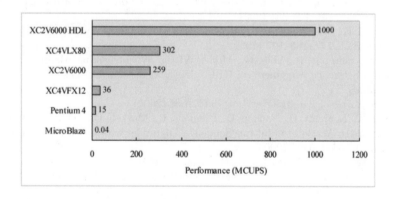

Fig. 2. ClustalW First Stage MSA Performance of Various Systems

5   Conclusion

Mapping of the first stage MSA algorithm into FPGA is solely based on the standard ANSI C language and application programming interfaces supported by the CoDeveloper tool. Neither HDL design nor simulation is done in this relatively new design methodology yet comparable high performance design can be achieved with a significant reduction in the design time. Previous FPGA-based hardware accelerators [3], [9] have been based on large and power-hungry FPGA devices. We have demonstrated that a significant performance improvement is possible using even the smallest commercially available FPGA device, opening up possibilities for using these devices as on-demand accelerators for compute-intensive applications on mobile and power constrained devices.

Acknowledgments. We would like to thank Mr. David Pellerin from Impulse Accelerated Technologies for his effort in helping us to improve the MSA hardware design performance.

References

1. Ebedes, J., Datta, A.: Multiple Sequence Alignment in Parallel on a Workstation Cluster. Bioinformatics 20(7), 1193–1195 (2004)
2. Oliver, T.F., Schmidt, B., Maskell, D.L.: Reconfigurable Architectures for Bio-sequence Database Scanning on FPGAs. IEEE Trans. Circuits Syst. II 52, 851–855 (2005)
3. Hoang, D.T.: Searching Genetic Databases on Splash 2. In: IEEE Workshop on FPGAs for Custom Computing Machines, pp. 185–191. IEEE Computer Society Press, Los Alamitos (1993)
4. Yamaguchi, Y., Maruyama, T., Konagaya, A.: High Speed Homology Search with FPGAs. In: Pacific Symposium on Biocomputing, pp. 271–282 (2002)
5. Sullivan, C., Wilson, A., Chappell, S.: Using C based Logic Synthesis to Bridge the Productivity Gap. In: Proc. of the 2004 Conference on Asia South Pacific Design Automation, pp. 349–354 (2004)
6. Smith, T.F., Waterman, M.S.: Identification of Common Molecular Subsequences. J. Mol. Biol. 147, 195–197 (1981)
7. Liu, W., Schmidt, B., Voss, G., Muller-Wittig, W.: Streaming Algorithms for Biological Sequence Alignment on GPUs. IEEE Trans. Parallel Distrib. Syst. (to be published)
8. Xilinx: Virtex-4 Family Overview. ds112.pdf (2007)
9. Oliver, T., Schmidt, B., Nathan, D., Clemens, R., Maskell, D.: Using Reconfigurable Hardware to Accelerate Multiple Sequence Alignment with ClustalW. Bioinformatics 21(16), 3431–3432 (2005)
10. Yu, C.W., Kwong, K.H., Lee, K.H., Leong, P.H.W.: A Smith-Waterman Systolic Cell. In: Proc. of 13th Int. Workshop Field Programmable Logic and Applications, pp. 375–384 (2003)

A Two-Phase ANN Method for Genome-Wide Detection of Hormone Response Elements

Maria Stepanova[1], Feng Lin[2], and Valerie C.-L. Lin[3]

[1] Bioinformatics Research Centre
[2] School of Computer Engineering
[3] School of Biological Sciences
Nanyang Technological University, Singapore 637551
{mari0004,asflin,cllin}@ntu.edu.sg

Abstract. Steroid hormone receptors compose a subgroup of regulatory proteins which tend to recognize partially symmetric response elements on DNA. Identification of the members of a gene regulatory machine conducted by steroid hormones could provide better understanding of nature and development of diseases. We present an approach based on a succession of neural networks, which can be used for highly specific detection of binding signals. It exploits the capability of a feed-forward neural network to model datasets with high confidence, while a recurrent network grants putative response elements with biologically meaningful structures. We have used a novel method to train such a two-phase artificial neural network with a set of experimentally validated response elements for steroid hormone receptors. We have demonstrated that sequence-based prediction followed by structure-based classification of putative binding sites allows to eliminate large amount of false positives. An implementation of the neural network with Field-Programmable Gate Array is also briefly described.

1 Introduction

The super-family of steroid and thyroid hormone receptor proteins includes receptors for steroid hormones, thyroid hormones, vitamin D and vitamin A (retinoic acid) [1]. Steroid hormones are involved in vital physiological processes, ranging from establishment and maintenance of pregnancy [2] to regulation in the genesis, progression and treatment of different cancers [3]. Activated by their hormone molecules, steroid hormone receptors usually bind to their target DNA – hormone response elements (HREs) – as partially symmetric homodimers [4]. HRE sequence, however, allows for a certain amount of flexibility in its nucleotide composition. Thus, recognition of these binding sites on DNA is one of the significant research topics in computational biology.

A comprehensive review on recognition of different strategies for transcription factor binding sites (TFBS) is given by Wasserman and Sandelin [5]. The most popular public resource is the P-Match tool based on a database of Position Weight Matrices (PWMs) for eukaryotic transcription factors, TRANSFAC [6];

J.C. Rajapakse, B. Schmidt, and G. Volkert (Eds.): PRIB 2007, LNBI 4774, pp. 19–29, 2007.

and another open access is the database of TFBS profiles, JASPAR [7]. Unfortunately, due to high diversity of the eukaryotic transcription factor families, PWM-based representation of general TFBS patterns has been proved not very effective [8,9]; the poor quality of datasets for particular TFs of interest in these databases is another concern. For certain vertebrate transcription factors, however, weight matrix approach used as a part of a complex prediction system may perform well, as demonstrated by Bajic et al. for estrogen response element [10]. Research has also been done for more accurate recognition of HREs or HRE-like patterns. The best result to date is implemented Hidden Markov Model (HMM) for recognition of nuclear receptor binding sites on DNA [11]. However, the reported results suffer from a dataset which included all nuclear receptors together. Another promising idea, which involves Gibbs sampling model for partially symmetric structures, was reported by Favorov et al. [12], but only preliminary results for bacterial motifs have been reported.

In this paper, we present an artificial neural network (ANN) based approach for prediction of hormone receptor binding sites on DNA. The main idea of this approach is to predict a HRE-like DNA sequence first, and then estimate its probability of being a dimeric hormone response element based on its half-site structure. For this purpose, a two-phase neural network has been developed. A feed-forward neural network returns a list of potential HREs, followed by a recurrent neural classifier which predicts a possible dimeric structure for each of them. We found this approach very promising for HRE prediction in practice, with high sensitivity and reliable specificity. Furthermore, we analyzed the efficiency and cost of implementation of such a neural network. The expensive training process is eased via use of hardware acceleration, thanks to the outstanding performance of Field-Programmable Gate Arrays (FPGA).

2   Methods

The entire system for HRE recognition has been implemented in a form of two-phase sequence analysis. The first phase is the sequence-based prediction of putative hormone response elements. It is performed by a feed-forward neural network, and used for selection of HRE-like sequences on the entire DNA sequence of interest. The output of this module is a list of sequences with posterior probability of being functional hormone response element higher than a predefined threshold. This network is trained with use of a set of experimentally verified HREs by an adapted back-propagation method. The second phase is the structure-based validation. It is performed by a recurrent neural network, and corresponds to an attempt to classify a putative HRE (predicted at the previous stage) into one of groups each representing a dimeric response element structure – direct, inverted or palindromic repeat. The rationale behind that comes from experimental observations of dimeric protein-DNA iterations: a sequence is unlikely involved in homodimeric DNA-protein binding even if it is marked as a HRE-like in case if it cannot be reliably assigned with any known HRE structure.

2.1 Data Preparation

One can easily achieve very high sensitivity and specificity in identification of HRE with just a few sequences, but that result would be unreliable. A sufficient and accurate representative dataset is essential for the model to be trustable. In our study, the HRE data was collected from more than 200 literature sources and our in-house wet-lab experiments. Such a collection of HREs has no analogs in the current public and commercial databases of TFBS profiles.

In the dataset, while a few of the regulatory elements are derived from genes of fish and birds, most of the sites are mammalian and 90% of all sites are from human or rodent genomic DNA. The collection contains response elements for androgen (218 AREs), progesterone (66 PREs) and glucocorticoid (377 GREs) hormone receptors, which have been reported to share the same DNA sequence [13]. This observation has also been confirmed with use of our dataset [9]. Thus, in our current work, we use the joint unit consisting of all the three hormone receptors of interest unless stated otherwise.

2.2 Modeling of HREs with a Feed-Forward Neural Network

For sequence-based modeling of hormone response elements on DNA, a feed-forward neural network (FFNN) has been developed. It obtains the preprocessed (with use of one-hot notations for nucleotides) DNA sequence from an encrypting module, and returns the two posterior probabilities, each corresponding to either HRE-like or non-HRE sequence. Final output of the network – whether the sequence is a HRE – is subject to adjustment by a threshold. Training of the network is performed with use of experimentally validated HRE sequences for positive patterns, and neutral DNA sequences for negative ones.

From the output of the preprocessing module, the 15bp-long DNA sequence is converted to a 60-vector. The neural network theory [14] suggests that for a reliable learning result, the number of degrees of freedom, i.e. weights to be fitted, must be at most half of the number of constrains (the inputs accompanied be desired outputs), in order to avoid over-fitting. Therefore, in the case of one hidden layer and a dataset of about 700 positive (experimentally validated) and 7000 negative (extracted from neutral DNA) HREs, we should limit the number of hidden layer neurons to about 50. In the case of two hidden layers, the maximum number of neurons on each layer is approximately 40. The amount of negative patterns can even be increased to more than ten-fold over positive ones (as we have much more neutral DNA in comparison of functional DNA), but it may increase the risk of bias of the ANN model towards the most abundant pattern.

The bipolar sigmoid function is used for activation of neuron synaptic inputs. A series of cross-validation tests allowed us to fix its coefficients (individual for each neuron).

For training of the feed-forward neural model, a back-propagation learning algorithm is implemented. The weight adjustment for each neuron is represented by the following equation:

$$w^{t+1} = w^t + \alpha^t \times \delta \times x \tag{1}$$

where w^t is a vector of weights for a particular neuron at the tth step of learning, α^t is the learning parameter at the tth step $(0 < \alpha^t < 1, \forall t > 0)$, and the delta value for each neuron of the ANN with one hidden layer and one output layer is calculated as follows:

$$\delta^{output} = \left. \frac{\partial f(u^t)}{\partial u^t} \right|_{u^t = w^t x} \times (d^t - o^t) \qquad (2)$$

$$\delta^{back\text{-}propagated} = \left. \frac{\partial f(u^t)}{\partial u^t} \right|_{u^t = w^t x} \times \sum_{k=1}^{K} w_{h \to k} \delta_k^{output} \qquad (3)$$

where d^t and o^t represent the desired and current outputs of the neuron respectively; x is the input to the layer being considered (either hidden or output), $u^t = w^t x$ is the synaptic input to the neuron, and $f(u^t)$ is the activation function of the neuron. For the back-propagated delta value, K is a number of neurons on the output layer, $w_{h \to k}$ is the weight coefficient of the connection between h^{th} neuron of hidden layer and k^{th} neuron of the output layer, δ_k^{output} is a delta value for the kth neuron of the output layer calculated as shown by formula (2). The back-propagation is terminated when

  i. Error tolerance for the accuracy of 99.99% is satisfied, or
 ii. The desired number of back-propagation cycles or the error plateau is reached.

Learning rate α is adjusted depending on whether current level of training error is decreasing (then α is increased by 10%), or it has jumped over a minimum and it is better to move back slightly slower (then α is decreased by 20%).

2.3 Classification of Respective HREs with a Recurrent Neural Network

For structure-based recognition of various HREs, we designed a recurrent neural network (RNN). It employs a quasi-heuristic approach for classification of partially overlapped datasets, and requires domain knowledge for its design.

  With a few exceptions, steroid hormone receptors bind to their response elements in a dimeric form. Thus, the structure of a response element can be treated as a repeat. However, the half-sites of a response element can occur in different orientations, and each still can interact with the zinc-fingers of a hormone receptor protein's DNA-binding domain [15,16]. The three possible structures of HREs include direct repeat, inverted repeat, and palindromic repeat of a consensus half-site. Therefore, the recurrent neural network is expected to classify a potential hormone response element into one of these HRE-like dimeric structures.

  The applicability of recurrent neural networks for biological sequence analysis has recently been reviewed [17], using examples of motif detection and prediction of subcellular localization of peptides. It shows that though the network architecture reflects the presence of bias, recurrent neural networks do provide access to biologically significant patterns.

Recurrent neural networks work in an unsupervised manner [18]. Theoretically, the goal is to design a network that stores a specific set of equilibrium points such that, when an initial condition for a neural system is provided, the network eventually comes to stability at one of these equilibrium points. Thus, the network is used for classification, and has been proven to be more flexible and powerful than distance-based clustering. The network is recursive, so its output is fed back as the input iteratively. With a properly designed RNN, each input is expected to be transformed into one of the stable states of the network in a finite number of iterations [14].

In our study, three consensuses of the respective HRE structures, and seven non-HRE sequences taken from the papers by Thackray et al. [19] and Lieberman et al. [20] are used as ten equilibrium points for the RNN. If the network comes to a stable state other than the ten desired equilibrium states, the input DNA sequence is 'unclustered'; otherwise, a certain cluster labeled 1 corresponding to a direct repeat (DR), 2 to an inverted repeat (IR), 3 to an palindromic repeat (ER) of the HRE half-site, and 7-10 to a non-HRE pattern is attributed to the input. Besides, as the sigmoid activation function returns non-integer values, the squared Euclidian distance between the final stable state and the nearest equilibrium point is estimated; for a successful classification, it must be 50% greater than that of the second nearest equilibrium point. In most cases, machine precision of the equilibrium points are reached within 100 iterations.

2.4   Hardware Acceleration

After numerous tests, we have reached quite a good accuracy of the system, but its very long processing time made it impossible to be used for genomic sizes which are often hundreds of million of base pairs. The bottleneck is identified at the operation of the recurrent ANN in the system. While for a given training set even with thousands of sequences, the back-propagation training of the feed-forward ANN is done once and forever, the recurrent neural network requires tens of iterations for each HRE-like input out of possible millions for genomic scales. We implemented the neural network on a 4-way IBM X260 server with four 3.16GHz CPUs, 3.25GB RAM, and 667MHz system bus. It took about 20 minutes to screen only 1Mb of DNA with eight parallel threads for RNN operation. Therefore, we propose hardware acceleration to make our system affordable in practice.

A full description of the acceleration with the Field-Programmable Gate Arrays is presented in another paper [21]. Briefly, referring to Fig.1, this FPGA-implemented RNN unit communicates with the rest of system implemented in a PC via a local bus. The fabricated RNN contains 60 neurons, which operate sequentially on ten identical neuron units. We used Verilog Hardware Description Language (Verilog HDL) in the circuit design for FPGA. The Alpha Data Virtex-4LX160 board was used for FPGA design. A Windows application reads DNA sequences predicted as HRE-like by a preceding feed-forward neural network from a text file, and sends the 32bit-long input vector to a configured FPGA board. It also obtains the output from the board, and proceeds to the decision making module.

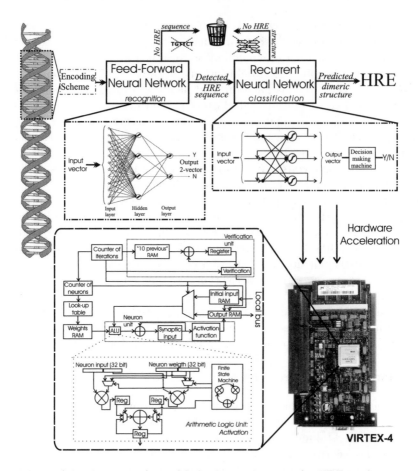

Fig. 1. A hardware-accelerated hybrid neural system for HRE prediction

Arithmetic calculations of the RNN module was implemented with use of fixed-point signed number representation, where 29 out of 32 bits were used to represent a fractional part. The sigmoid activation function was replaced by a polyline approximation, which allowed us to decrease area usage substantially with the resulting accuracy changing insignificantly.

The entire on-chip RNN system could be subdivided into two constituents: the neuron units and the control unit. The lower part of Fig. 1 shows configuration of the ANN from a point of view of arithmetic operation. Neuron units accumulate synaptic inputs of the neurons, activate them, and return current output vector into the memory.

The control unit includes the counter of iterations which defines whether an input must be put into the system for the first time (initial input). It also regulates sequence of processing of neurons by neuron units, and involves verification unit for the current output. The control unit terminates the recurrent processes if the maximum number of iterations or a stable state (no oscillations larger then 0.2% amongst outputs of the last ten iterations) is reached. The counter of neurons is reset before each iteration starts; then, it regulates the sequence of the ten neuron units involved for calculation of sixty actual neurons, and decides which values of weights must be selected from memory.

With the RNN implementation on Virtex-4 hardware, we achieved almost ten-fold speedup for a recurrent neural network. With use of more advanced digital technologies, such as flexible on-board clock management and involvement of hyper-transport techniques instead of 66MHz PCI bus, it is possible to realize an even faster hybrid neural solution.

3 Results

3.1 HRE Classification

First, the collected dataset of experimentally validated HREs was used to test the capabilities of our RNN-based classification system. The results of classification of HREs are given in the Table 1. The difference between preferred structures of the DNA response elements for the three steroid hormones of interest – progesterone, androgen, and glucocorticoid – was found to be significant (p-value 0.007), with ARE slightly different from the two other subgroups of binding sites - it is less tolerated to direct repeats, and in most cases prefers to bind to palindromic response elements.

Table 1. Results of classification of training set of HREs with RNN

HRE	PRE	GRE	ARE	U
Direct repeat	35 or 53.0%	225 or 59.7%	94 or 43.1%	354 or 53.6%
Inverted repeat	1 or 1.5%	3 or 0.8%	1 or 0.5%	5 or 0.8%
Palyndromic repeat	26 or 39.4%	114 or 30.2%	115 or 52.8%	255 or 38.6%
No dimeric structure	4 or 6.1%	35 or 9.3%	8 or 3.7%	47 or 7.1%
$\sum$	66	377	218	661

3.2 Genome-Wide HRE Recognition

The sensitivity and specificity values were tracked on each step of machine processing; for calculations, ten-fold cross-validation is used.

The first step of the machine learning is to activate the trained feed-forward neural network, which recognizes HRE-like patterns. At this stage, the sensitivity was found to be as high as 98%, that is, 15 among 661 HREs are misclassified; the specificity is 1:5.8Kb which is measured on basis of 1Mb neutral DNA sequence.

In the next step, the recurrent neural network allows to increase specificity level to 1:7.29Kb, while the sensitivity stays favorable in comparison with existing HRE prediction methods (for a review, see [9]) – it is as high as 92% (6 or 9% of PREs, 37 or 10% of GREs, and 9 or 4% AREs are now misclassified, or total 52 HREs).

Available vertebrate genomes were downloaded from the NCBI data repository, processed, and used for system performance measurements as well. To sum up, the average frequencies of prediction of HREs are as follows:

- Human genome (*Homo Sapiens*, #36.2) – 1:8.15Kb
- Chimpanzee genome (*Pan Troglodytes*, #2.1) – 1:8.13Kb
- Mouse genome (*Mus Musculus*, #36.1) – 1:7.69Kb
- Rat genome (*Rattus Norvegicus*, #4.1) – 1:7.11Kb
- Cow genome (*Bos Taurus*, #3.1) – 1:6.35Kb
- Dog genome (*Canis Familiaris*, #2.1) – 1:8.43Kb
- Opossum genome (*Monodelphis Domestica*, #2.1) – 1:7.36Kb
- Chicken genome (*Gallus Gallus*, #2.1) – 1:9.81Kb
- Zebrafish genome (*Danio Rerio*, #2.1) – 1:8.95Kb

We can see from the above that high accuracy of prediction is achieved by the combination of feed-forward and recurrent neural networks, which is crucial for the system to be used in practice. Sensitivity of 92% can be combined with random expectation level of 1 prediction per 7.29Kb of neutral DNA and 1:8.15Kb of human genomic DNA. The parameters, though, can be adjusted for a particular task.

4   Discussion and Conclusion

Due to the importance for regulation of vital processes, the group of steroid hormones has been studied extensively over decades, but very few computational methods have been available to aid experimentalists to find hormone receptor regulatory signals. In this paper, we propose an approach which combines feed-forward and recurrent neural networks for recognition of a subgroup of spaced, symmetrically structured DNA motifs, and demonstrate its performance using the examples of steroid hormone response elements. Its high level of accuracy provides an access to a powerful method for *de novo* HRE prediction, and further analysis of hormone-regulated genes as well.

The proposed approach benefits from the advantages of both feed-forward and recurrent neural networks; they employ different strategies of machine learning and allow to reveal different features of the patterns. The feed-forward neural network provides a very flexible tool to model almost any dataset of interest, but its flexibility may result in rather low specificity values. This is where the recurrent neural network may cut in; it is able eliminate most of the false positive HRE-like findings, comparing them against the symmetrically structured HREs.

On the other hand, training a highly accurate feed-forward neural network and especially elaborating all the RNN-based classifications with software implementation needs prohibitively long time. The hardware acceleration with reconfigurable FPGA computing provides us with a solution.

With use of two successive neural networks, we managed to model the HRE training set and separate it from the neutral DNA sequences quite reliably, but some outliers were detected. They were found through non-consensus binding sites for progesterone, androgen and glucocorticoid receptors in the promoters and gene regions for a number of genes: rabbit uteroglobin gene [22], chicken lysozyme gene [23], porcine uteroferrin gene [24], pro-opiomelanocortin gene [25], murine c-myc gene [26], late leader of the control region of the human poly-omavirus BK [27], gene promoter of two milk protein genes (β-casein and whey acidic protein) [28], human Na/K ATPase α1 gene promoter [29], and mouse sex-limited protein enhancer [30]. The first three are progesterone-regulated genes, the next five are glucocorticoid primary targets, and the last one is associated with androgen activity. Unless they are experimental artifacts, the possible explanation could lie in the area of complex protein-DNA interaction which is beyond DNA sequence similarity itself; it probably could be a secondary structure of DNA, or location of surrounding nucleosomes. Nevertheless, more sensitive procedures should be implemented, involving other conditions which could be related to successful formation of protein-DNA complex *in vivo*.

The dramatic progress in experimental identification of transcription factor binding sites is obvious. Thus, the availability of accurate algorithms for *in silico* binding site prediction is of great importance. Our proposed model for steroid receptor binding sites prediction can be used for determination of androgen, progesterone and glucocorticoid primary target genes with high reliability. It can also be used for detection of steroid hormone response elements *de novo*, and for evaluation of known HREs as well. Finally, the proposed model can be potentially involved for prediction of any other structured DNA motifs of interest.

Acknowledgements. This work is supported in part by two research grants, ARC02/2004 and RG50/06, from Ministry of Education, Singapore.

References

1. Larsen, P., Kronenberg, H., Melmed, S., Polonsky, K.: Williams textbook of endocrinology. Saunders, Philadelphia (2003)
2. Conneely, O.M.: Perspective: Female Steroid Hormone Action. Endocrinology 142(6), 2194–2199 (2001)
3. Ko, Y.J., Balk, S.P.: Targeting Steroid Hormone Receptor Pathways in the Treatment of Hormone Dependent Cancers. Curr. Pharm. Biotechnol. 5(5), 459–470 (2004)
4. Danielsen, M., Hinck, L., Ringold, G.M.: Two amino acids within the knuckle of the first zinc finger specify DNA response element activation by the glucocorticoid receptor. Cell 57(7), 1131–1138 (1989)
5. Wasserman, W.W., Sandelin, A.: Applied bioinformatics for the identification of regulatory elements. Nat. Rev. Genet. 5(4), 276–287 (2004)
6. Kel, A.E., Gossling, E., Reuter, I., et al.: MATCH: A tool for searching transcription factor binding sites in DNA sequences. Nucleic Acids Res. 31(13), 3576–3579 (2003)

7. Sandelin, A., Alkema, W., Engstrom, P., et al.: JASPAR: an open-access database for eukaryotic transcription factor binding profiles. Nucleic Acids Res. 32(DB), D91–D94 (2004)

8. Rahmann, S., Muller, T., Vingron, M.: On the power of profiles for transcription factor binding site detection. Stat. Appl. Genet. Mol. Biol. 2(1): Article 7 (2003)

9. Stepanova, M., Lin, F., Lin, V.: Establishing a Statistic Model for Recognition of Steroid Hormone Response Elements. Comput. Biol. Chem. 30(5), 339–347 (2006)

10. Bajic, V.B., Tan, S.L., Chong, A., et al.: Dragon ERE Finder version 2: A tool for accurate detection and analysis of estrogen response elements in vertebrate genomes. Nucleic Acids Res. 31(13), 3605–3607 (2003)

11. Sandelin, A., Wasserman, W.W.: Prediction of nuclear hormone receptor response elements. Mol. Endocrinol. 19(3), 595–606 (2005)

12. Favorov, A.V., Gelfand, M.S., Gerasimova, A.V., et al.: A Gibbs sampler for identification of symmetrically structured, spaced DNA motifs with improved estimation of the signal length. Bioinformatics 21(10), 2240–2245 (2005)

13. Khorasanizadeh, S., Rastinejad, F.: Nuclear-receptor interactions on DNA-response elements. Trends Biochem. Sci. 26(6), 384–390 (2001)

14. Hagan, M., Demuth, H., Beale, M.: Neural Network Design. PWS Publishing company, Boston (1996)

15. Evans, R.M.: The steroid and thyroid hormone receptor superfamily. Science 240(4854), 889–895 (1988)

16. Nelson, C.C., Hendy, S.C., Shukin, R.J., et al.: Determinants of DNA sequence specificity of the androgen, progesterone, and glucocorticoid receptors: evidence for differential steroid receptor response elements. Mol. Endocrinol. 13(12), 2090–2107 (1999)

17. Hawkins, J., Boden, M.: The Applicability of Recurrent Neural Networks for Biological Sequence Analysis. IEEE ACM T. Comput. BI. 2(3), 243–253 (2005)

18. Hastie, T., Tibshirani, R., Friedman, J.: The Elements of Statistical Learning: Data Mining, Inference, and Prediction. Springer, New York (2001)

19. Thackray, V.G., Lieberman, B.A., Nordeen, S.K.: Differential gene induction by glucocorticoid and progesterone receptors. J. Steroid Biochem. Mol. Biol. 66(4), 171–178 (1998)

20. Lieberman, B.A., Bona, B.J., Edwards, D.P., Nordeen, S.K.: The constitution of a progesterone response element. Mol. Endocrinol. 7(4), 515–527 (1993)

21. Stepanova, M., Lin, F., Lin, V.C.: A Hopfield Neural Classifier and Its FPGA Implementation for Identification of Symmetrically Structured DNA Motifs. J. VLSI Sig. Proc. Syst. (in press)

22. Jantzen, K., Fritton, H.P., Igo-Kemenes, T., et al.: Partial overlapping of binding sequences for steroid hormone receptors and DNaseI hypersensitive sites in the rabbit uteroglobin gene region. Nucleic Acids Res. 15(11), 4535–4552 (1987)

23. von der Ahe, D., Renoir, J.M., Buchou, T., et al.: Receptors for glucocorticosteroid and progesterone recognize distinct features of a DNA regulatory element. Proc. Natl. Acad. Sci. USA. 83(9), 2817–2821 (1986)

24. Lamian, V., Gonzalez, B.Y., Michel, F.J., Simmen, R.C.: Non-consensus progesterone response elements mediate the progesterone-regulated endometrial expression of the uteroferrin gene. J. Steroid Biochem. Mol. Biol. 46(4), 439–450 (1993)

25. Drouin, J., Trifiro, M.A., Plante, R.K., et al.: Glucocorticoid receptor binding to a specific DNA sequence is required for hormone-dependent repression of pro-opiomelanocortin gene transcription. Mol. Cell Biol. 9(12), 5305–5314 (1989)

26. Ma, T., Copland, J.A., Brasier, A.R., Thompson, E.A.: A novel glucocorticoid receptor binding element within the murine c-myc promoter. Mol. Endocrinol. 14(9), 1377–1386 (2000)
27. Moens, U., Subramaniam, N., Johansen, B., et al.: A steroid hormone response unit in the late leader of the noncoding control region of the human polyomavirus BK confers enhanced host cell permissivity. J. Virol. 68(4), 2398–2408 (1994)
28. Welte, T., Philipp, S., Cairns, C., et al.: Glucocorticoid receptor binding sites in the promoter region of milk protein genes. J. Steroid Biochem. Mol. Biol. 47(1-6), 75–81 (1993)
29. Kolla, V., Robertson, N.M., Litwack, G.: Identification of a mineralocorticoid/glucocorticoid response element in the human Na/K ATPase alpha1 gene promoter. Biochem. Biophys. Res. Commun. 266(1), 5–14 (1999)
30. Verrijdt, G., Schauwaers, K., Haelens, A., et al.: Functional interplay between two response elements with distinct binding characteristics dictates androgen specificity of the mouse sex-limited protein enhancer. J. Biol. Chem. 277(38), 35191–35201 (2002)

An Expert Knowledge-Guided Mutation Operator for Genome-Wide Genetic Analysis Using Genetic Programming

Casey S. Greene[1], Bill C. White[1], and Jason H. Moore[1]

Dartmouth College, Hanover, NH 03755, USA
{Casey.S.Greene,Bill.C.White,Jason.H.Moore}@dartmouth.edu
http://www.epistasis.org

Abstract. Human genetics is undergoing a data explosion. Methods are available to measure DNA sequence variation throughout the human genome. Given current knowledge it seems likely that common human diseases are best predicted by interactions between biological components, which can be examined as interacting DNA sequence variations. The challenge is thus to examine these high-dimensional datasets to identify combinations of variations likely to predict common diseases. The goal of this paper was to develop and evaluate a genetic programming (GP) mutator suited to this task by exploiting expert knowledge in the form of Tuned ReliefF (TuRF) scores during mutation. We show that using expert knowledge guided mutation performs similarly to expert knowledge guided selection. This study demonstrates that in the context of an expert knowledge aware GP, mutation may be an appropriate component of the GP used to search for interacting predictors in this domain.

1 Introduction

Biological and biomedical sciences are undergoing a data explosion without a corresponding knowledge explosion. This is especially true in the domain of human genetics where it is now technically and economically feasible to measure thousands of DNA sequence variations from across the human genome. For the purposes of this paper we will focus exclusively on the single nucleotide polymorphism or SNP which is a single nucleotide or point in the DNA sequence that differs among people. It is anticipated that at least one SNP occurs approximately every 100 nucleotides across the $3x10^9$ nucleotide human genome. An important goal in human genetics is to determine which of the many thousands of SNPs are useful for predicting who is at risk for common diseases. This "genome-wide" approach is expected to revolutionize the genetic analysis of common human diseases. The charge for computer science and bioinformatics is to develop algorithms for the detection and characterization of those SNPs that are predictive of human health and disease. Success in this endeavor will be difficult due to nonlinearity in the genotype-to-phenotype mapping relationship that is due, in part, to epistasis or nonadditive gene-gene interactions. The

J.C. Rajapakse, B. Schmidt, and G. Volkert (Eds.): PRIB 2007, LNBI 4774, pp. 30–40, 2007.

implication of epistasis from a data mining point of view is that SNPs need to be considered jointly in learning algorithms rather than individually. The challenge of modeling attribute interactions has been previously described [1]. Due to the combinatorial magnitude of this problem, intelligent analysis strategies are needed.

1.1   Concept Difficulty

Combining the difficulty of modeling nonlinear attribute interactions with the challenge of attribute selection yields for this domain what Goldberg [2] calls a needle-in-a-haystack problem. That is, there may be a particular combination of SNPs that together with the right nonlinear function are a significant predictor of disease susceptibility. Considered individually they may not look any different than thousands of other noisy SNPs not involved in the disease process. Under these models, the learning algorithm is truly looking for a genetic needle in a genomic haystack. A recent report from the International HapMap Consortium [3] suggests that approximately 300,000 carefully selected SNPs may be necessary to capture all of the relevant variation across the Caucasian human genome. Assuming this is true (it is probably a lower bound), we would need to scan $4.5x10^{10}$ pairwise combinations of SNPs to find a genetic needle. The number of higher order combinations is astronomical. Is GP suitable for a problem like this? At face value the answer is no. There is no reason to expect that a GP or any other wrapper method would perform better than a random attribute selector because there are no building blocks for this problem when accuracy is used as the fitness measure. The fitness of any given classifier would look no better than any other with just one of the two correct SNPs in the model. Indeed, we have observed this in our preliminary work [4,5]. Subsequent work has shown that by integrating expert knowledge into a selection scheme, it is possible to develop a GP wrapper that is able to perform better than a random attribute selector [6]. Work here examines whether or not it is also possible to integrate expert knowledge into mutation to develop a GP which performs better than one with a random attribute mutator.

1.2   Genetic Programming and Mutation

Genetic programming (GP) is an automated computational discovery tool that is inspired by Darwinian evolution and natural selection [7,8,9,10,11,12,13]. The goal of GP is to evolve computer programs to solve problems. This is accomplished by first generating computer programs that are composed of the building blocks needed to solve or approximate a solution to a problem. Each generated program is evaulated, and the good programs are selected, recombined, and mutated to form new computer programs. This process of selection based on fitness and recombination and mutation to generate variability is repeated until a best program or set of programs is identified. Genetic programming and its many variations have been applied successfully to a wide range of different problems including data mining, knowledge discovery e.g. [14], and bioinformatics [15].

Despite the many successes, there are a large number of challenges that GP practitioners and theorists must address before this general computational discovery tool becomes a standard in the modern problem solver's toolbox. Yu et al. [16] list 22 such challenges. Several of these are addressed by the present study. Previous work has shown that by integrating expert knowledge, the GP approach can successfully pick attributes from large and high-dimensional datasets [6]. Here we will examine another method of taking advantage of pre-processing based expert knowledge. Such methods may also be used for integrating domain specific or literature based expert knowledge. This paper explores the effect of a GP mutator which integrates expert knowledge on genome-wide genetic analysis in the domain of human genetics.

Previous work has shown that for a naïve mutation operator in a wide variety of problem domains, performance for naïve mutation and naïve crossver did not greatly differ [17]. There were, however, problem domains and parameter settings where mutation was found to be more or less suitable. In general mutation was more successful with more generations and crossover more successful with larger populations but the effect differed by problem domain. Previous work on expert knowledge guided mutation for genetic programming has used collective memory which contains knowledge gained by examining the population state at earlier time points within the same GP run [18]. Here we focus on expert knowledge gained by statistical pre-processing of the input data.

The goal of the present study was to develop and evaluate a GP mutation operator appropriate for genetic analysis of genome-wide data. Given the concept difficulty, we are generally interested in using expert knowledge to facilitate the generation and exploitation of good building blocks. We specifically address whether pre-processed expert knowledge can become useful for mutating trees after recombination and reproduction. In this specific situation mutation or random chance must, in addition to seeding the population with good attributes, combine the best attributes as no expert knowledge is utilized during selection or recombination. The success of expert knowledge in selection is limited to attributes available in the current population, but the use of expert knowledge in mutation is not limited to the state of the current population as attributes not present may be added. As both operators are dependent on different factors, integration of this mutation operator into a GP utilizing expert knowledge throughout may yield benefits beyond those found in this study.

2   Genetic Programming Methods

2.1   Expression Tree Representation

Figure 1A illustrates an example GP tree for this problem. We have intentionally kept the initial solution representation simple with one function in the root node and two children to evaluate the best GP parameterization. More complex trees (e.g. Figure 1B) will be explored once we understand when and how the GP works with the simpler trees. We have selected the multifactor dimensionality reduction or MDR approach as an attribute constructor for the function set

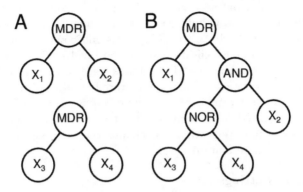

Fig. 1. Example GP trees for solutions (A). Example of a more complex tree that will be considered in future studies (B).

because it is able to capture interaction information (see Section 3). Each tree has two leaves or terminals consisting of attributes. In the case of our study, the terminal set consists of 1000 attributes.

2.2 Fitness Function

The fitness function used in this study was accuracy estimated using a naïve Bayes classifier. Here, accuracy is defined as how well the model predicts the case-control status of each simulated individual. Each tree is evaluated as a constructed attribute using the MDR function in the root node. It is this single constructed attribute with two levels that is assessed using the classifier. The classification accuracy of a tree is its fitness.

2.3 A Sensible Mutation Operator

The goal of this study was to use expert knowledge to ensure good building blocks are reintroduced into the population through mutation. We compared the random mutator with a new sensible or expert knowledge-guided mutator. The random mutator works by mutating the designated percentage of the population each generation. For each chosen individual a random attribute is picked for mutation and replaced with an attribute selected randomly from the set of all attributes. For the sensible mutator, we used pre-processed attribute quality from the Tuned ReliefF (TuRF) algorithm as our expert knowledge (see Section 4). The sensible mutation operator is modeled after the sensible selection operator from Moore and White [6] and mutates individuals from some percentage of the population with the greatest difference in TuRF scores between the attributes, chooses to mutate the attribute in that individual with the lowest TuRF score and iteratively creates trees where the mutated attribute is replaced with every attribute from the top 1% of TuRF scores. Next, trees are evaluated, and the tree with the best accuracy is retained in the population as the result of the mutation. This sensible mutator ensures that poor building blocks are replaced by good building blocks throughout evolution. This method is designed to mutate attributes which are unlikely to lead to success and combine the newly chosen attribute with attributes which are likely to be involved in successful models. In the event that no combination of attributes leads to success, a higher TuRF scored attribute than the one which was mutated is retained in the population and may participate in recombination. The percentage of the population chosen for mutation is adjusted to account for the iterative replacement attempts so that the number of mutation attempts remains the same between the random mutator and the sensible mutator.

2.4   Parameter Settings

For this study, we used a population size of 500 and ran the GP for 10 generations as previously used by Moore and White [6]. We used a crossover probability of 0.9 and varied the mutation probability in increments of 10%. Since each tree has exactly two attributes, an initial population size of 500 trees will include 1,000 total attributes. Each initial population was generated such that each of the 1,000 attributes was represented once and only once across the 500 trees. This sensible initialization ensures that all building blocks are represented. It is important to note that the probability of any one tree receiving both functional attributes (i.e. the solution) is only $0.001x0.001$ or 10^{-6}. Thus, it is unlikely that any one tree in the initial population will be the correct solution. The size of the search space is approximately 500,000 or 1000 choose 2. With a population size of 500 and 10 generations the GP is exploring at most of 1% of the search space. The GP was implemented in C++ using GAlib (http://lancet.mit.edu/ga/). The crossover operator was modified to ensure binary trees of depth one.

3   Multifactor Dimensionality Reduction (MDR) for Attribute Construction

Multifactor dimensionality reduction (MDR) was developed as a nonparametric and genetic model-free data mining strategy for identifying combination of SNPs that are predictive of a discrete clinical endpoint [19,20,21,22]. The MDR method has been successfully applied to detecting gene-gene interactions for a variety of common human diseases including adverse drug reactions [23]. At the heart of the MDR approach is an attribute construction algorithm that creates a new attribute by pooling genotypes from multiple SNPs. Constructive induction using the MDR kernel is accomplished in the following way. Given a threshold T, a multilocus genotype combination is considered high-risk if the ratio of cases (subjects with disease) to controls (healthy subjects) exceeds or equals T, otherwise it is considered low-risk. Genotype combinations considered to be high-risk are labeled G1 while those considered low-risk are labeled G0. This process constructs a new one-dimensional attribute with levels G0 and G1. It is this new single variable that is returned by the MDR function in the GP function set. The MDR method is described in more detail by Moore et al. [21]. Open-source MDR software is freely available from www.epistasis.org.

4 Expert Knowledge from Tuned ReliefF (TuRF)

Our goal was to provide an external measure of attribute quality that could be used as expert knowledge by the GP. Here this external measure used was statistical, but it could just as easily be biological. There are many statistical and computational methods for determining the quality of attributes. Our goal was to identify a method that is capable of identifying attributes that predict class primarily through dependencies or interactions with other attributes. Kira and Rendell [24] developed an algorithm called Relief that is capable of detecting attribute dependencies. Relief estimates the quality of attributes through a nearest neighbor algorithm that selects neighbors (instances) from the same class and from the different class based on the vector of values across attributes. Weights (W) or quality estimates for each attribute (A) are estimated based on whether the nearest neighbor (nearest hit, H) of a randomly selected instance (R) from the same class and the nearest neighbor from the other class (nearest miss, M) have the same or different values. This process of adjusting weights is repeated for m instances. The algorithm produces weights for each attribute ranging from -1 (worst) to +1 (best). Kononenko [25] improved upon Relief by choosing n nearest neighbors instead of just one. This new ReliefF algorithm has been shown to be more robust to noisy attributes and missing data [26] and is widely used in data mining applications.

We have developed a modified ReliefF algorithm for the domain of human genetics called Tuned ReliefF (TuRF). We have previously shown that TuRF is significantly better than ReliefF in this domain [27]. The TuRF algorithm systematically removes attributes that have low quality estimates so that the ReliefF values if the remaining attributes can be re-estimated. We applied TuRF as described by Moore and White [27] to each dataset.

5 Data Simulation and Analysis

The goal of the simulation study is to generate artificial datasets with high concept difficulty to evaluate the power of GP in the domain of human genetics.

Table 1. Penetrance values for an example epistasis model

	AA (0.36)	Aa (0.48)	aa (0.16)
BB (0.36)	0.077	0.656	0.880
Bb (0.48)	0.892	0.235	0.312
bb (0.16)	0.174	0.842	0.106

We first developed 60 different penetrance functions (i.e. genetic models) that on genotypes from two SNPs in the absence of any independent effects. The 60 penetrance functions include groups of five with heritabilities of 0.025, 0.05, 0.1, 0.2, 0.3, or 0.4. These heritabilities range from a very small to a large genetic effect size. In half of the cases, each functional SNP had two alleles with frequencies of 0.4 and 0.6. In the other half of the cases, each functional SNP had two alleles with frequencies of 0.2 and 0.8. Table 1 summarizes the penetrance values to three significant digits for one of the 60 models. The values in parentheses are the genotype frequencies. Each of the 60 models was used to generate 100 replicate datasets with a sample size of 1600. Each dataset consisted of an equal number of case (disease) and control (no disease) subjects. Each pair of functional SNPs was combined within a genome-wide set of 998 randomly generated SNPs for a total of 1000 attributes. A total of 6,000 datasets were generated and analyzed.

For each set of 100 datasets we counted the number of times the correct two functional attributes were selected as the best model by the GP. This count, expressed as a percentage, is an estimate of the power of the method. This percentage represents how often the GP finds the answer that we know is present.

6   Experimental Results

Figure 2 summarizes the average power for each method for the models with 0.6 major allele frequency. Results for 0.8 major allele frequency were similar and are available upon request. Each point represents the power averaged over 500 datasets (5 models with 100 datasets each). Power represents the number of times out of 100 that the GP found the right two attributes. The unfilled circles represent the average power for a GP using the random mutation operator with otherwise identical parameters. The filled circles represent the average power for a GP using our sensible mutation operator with TuRF pre-processing scores as expert knowledge. Within each parameter setting mutation was varied such that 0 to 100 percent of the individuals were mutated during each generation for the random mutator in increments of 10 percent. For the sensible mutation operator, this corresponds to 0 to 10 percent of the population being mutated during each generation in increments of 1 percent to retain identical numbers of attempted mutations. These results clearly show the value of using TuRF scores in the mutation operator.

Comparing the TuRF sensible mutator to TuRF sensible recombination shows that the TuRF mutator performs similarly to the TuRF selector [6]. The previously examined TuRF selector showed slightly higher power, though that is not unexpected. Without an expert knowledge guided selector, there was no way to utilize the strength of attributes already in the population for recombination.

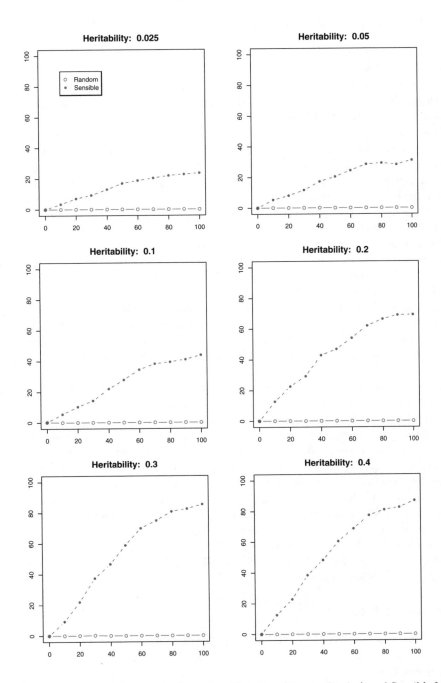

Fig. 2. Summary of the power of Random Mutation (Empty Circles) and Sensible Mutation (Filled Circles) using sensible initialization with accuracy for fitness under with random mutation equivalents (RMEs) from 0 to 100 under models with heritabilities from 0.025 to 0.4 and a major allele frequency of 0.6

7   Discussion and Conclusion

There are several conclusions to draw from this study. Again it has been shown that expert knowledge can provide building blocks necessary to find the genetic needle in the genome-wide haystack. Secondly, sensible mutation seems to perform similarly to using expert knowledge in selection and much better than random mutation. Sensible mutation itself is deterministic given a current population and set of TuRF scores. For the purposes of studying sensible mutation, we have separated random and sensible mutation. To retain the ability to reach the entire search space, in practical situations the pairing of sensible mutation and a non-deterministic mutation operator would probably be the best analysis strategy. In addition, combining sensible mutation with similar knowledge guided strategies in selection and recombination may provide additional benefits.

Previous work has focused on the use of expert knowledge in initialization, selection, and fitness [6,5]. We have focused on sensible mutation for this study. Future work will examine how different operators that integrate expert knowledge may be combined. Does using expert knowledge to guide both mutation and recombination work better than either alone?

In this work the building blocks of outside knowledge were obtained by pre-processing data with TuRF. For the realm of genetic studies, outside knowledge could also be obtained from the numerous public databases available to geneticists. Tools are being developed which integrate knowledge across these public databases and generate information about relationships between genes and disease in the context of protein interactions [28]. Future work will also focus on integrating multiple distinct expert knowledge types and sources.

We have again found that, given domain specific building blocks and operators which use these building blocks it is possible for a GP to outperform a random search, even for a needle-in-a-haystack problem. This indicates that GP may be a useful wrapper for genome wide analysis of common human diseases with a complex genetic architecture. Moore et al. have recently shown that Symbolic Discriminant Analysis (SDA), which uses a GP approach to generate models, was able to successfully model predictors of atrial fibrillation in a well characterized dataset which included a two-way epistatic interaction [29]. Integrating expert knowledge into the SDA approach should increase the efficiency of the search and assist SDA in finding higher order nonlinear interactions and allow SDA to be applied to larger genome-wide datasets.

Acknowledgement

This work was supported by NIH grants LM009012 and AI59694.

References

1. Freitas, A.A.: Understanding the crucial role of attribute interaction in data mining. Artif. Intell. Rev. 16(3), 177–199 (2001)
2. Goldberg, D.E.: The Design of Innovation: Lessons from and for Competent Genetic Algorithms. Kluwer Academic Publishers, Norwell, MA, USA (2002)

3. Consortium, T.I.H.: A haplotype map of the human genome. Nature 437(7063), 1299–1320 (2005)
4. White, B.C., Gilbert, J.C., Reif, D.M., Moore, J.H.: A statistical comparison of grammatical evolution strategies in the domain of human genetics. In: Proceedings of the IEEE Congress on Evolutionary Computing, pp. 676–682. IEEE Computer Society Press, Los Alamitos (2005)
5. Moore, J.H., White, B.C.: Genome-wide genetic analysis using genetic programming: The critical need for expert knowledge. In: Genetic Programming Theory and Practice IV, Springer, Heidelberg (2006)
6. Moore, J., White, B.: Exploiting expert knowledge in genetic programming for genome-wide genetic analysis. In: Runarsson, T.P., Beyer, H.-G., Burke, E., Merelo-Guervós, J.J., Whitley, L.D., Yao, X. (eds.) Parallel Problem Solving from Nature - PPSN IX. LNCS, vol. 4193, pp. 969–977. Springer, Heidelberg (2006)
7. Koza, J.R.: Genetic programming: on the programming of computers by means of natural selection. MIT Press, Cambridge, MA, USA (1992)
8. Koza, J.R.: Genetic programming II: automatic discovery of reusable programs. MIT Press, Cambridge, MA, USA (1994)
9. Koza, J.R., Andre, D., Bennett, F.H., Keane, M.A.: Genetic Programming III: Darwinian Invention & Problem Solving. Morgan Kaufmann Publishers Inc., San Francisco, CA, USA (1999)
10. Koza, J.R.: Genetic Programming IV: Routine Human-Competitive Machine Intelligence. Kluwer Academic Publishers, Norwell, MA, USA (2003)
11. Banzhaf, W., Nordin, P., Keller, R.E., Francone, F.D.: Genetic programming: an introduction: on the automatic evolution of computer programs and its applications. Morgan Kaufmann Publishers Inc., San Francisco, CA, USA (1998)
12. Langdon, W.B., Koza, J.R.: Genetic Programming and Data Structures: Genetic Programming + Data Structures = Automatic Programming! Kluwer Academic Publishers, Norwell, MA, USA (1998)
13. Langdon, W.B., Poli, R.: Foundations of Genetic Programming. Springer, Heidelberg (2002)
14. Freitas, A.A.: Data Mining and Knowledge Discovery with Evolutionary Algorithms, Secaucus, NJ, USA. Springer, New York (2002)
15. Fogel, G., Corne, D.: Evolutionary Computation in Bioinformatics. Morgan Kaufmann, San Francisco (2003)
16. Yu, T., Riolo, R., Worzel, B.: Genetic Programming: Theory and Practice (2006) 10.1007/0-387-28111-8_1
17. Luke, S., Spector, L.: A revised comparison of crossover and mutation in genetic programming. In: Koza, J.R., Banzhaf, W., Chellapilla, K., Deb, K., Dorigo, M., Fogel, D.B., Garzon, M.H., Goldberg, D.E., Iba, H., Riolo, R. (eds.) Genetic Programming 1998: Proceedings of the Third Annual Conference, University of Wisconsin, Madison, Wisconsin, USA, pp. 208–213. Morgan Kaufmann, San Francisco (1998)
18. Bearpark, K., Keane, A.: The use of collective memory in genetic programming. In: Jin, Y. (ed.) Knowledge Incorporation in Evolutionary Computation. Studies in Fuzziness and Soft Computing, pp. 15–36. Springer, Heidelberg (2005)
19. Ritchie, M.D., Hahn, L.W., Roodi, N., Bailey, L.R., Dupont, W.D., Parl, F.F., Moore, J.H.: Multifactor dimensionality reduction reveals high-order interactions among estrogen metabolism genes in sporadic breast cancer. American Journal of Human Genetics 69, 138–147 (2001)

20. Moore, J.H.: Computational analysis of gene-gene interactions using multifactor dimensionality reduction. Expert Review of Molecular Diagnostics 4(6), 795–803 (2004)
21. Moore, J.H., Gilbert, J.C., Tsai, C.T., Chiang, F.T., Holden, T., Barney, N., White, B.C.: A flexible computational framework for detecting, characterizing, and interpreting statistical patterns of epistasis in genetic studies of human disease susceptibility. Journal of Theoretical Biology 241(2), 252–261 (2006)
22. Moore, J.H.: Genome-wide analysis of epistasis using multifactor dimensionality reduction: feature selection and construction in the domain of human genetics. In: Knowledge Discovery and Data Mining: Challenges and Realities with Real World Data. IGI (2007)
23. Wilke, R.A., Reif, D.M., Moore, J.H.: Combinatorial pharmacogenetics. Nature Reviews Drug Discovery 4, 911–918 (2005)
24. Kira, K., Rendell, L.A.: A practical approach to feature selection. In: Machine Learning: Proceedings of the AAAI'92 (1992)
25. Kononenko, I.: Estimating attributes: Analysis and extension of relief. In: Bergadano, F., De Raedt, L. (eds.) ECML 1994. LNCS, vol. 784, pp. 171–182. Springer, Heidelberg (1994)
26. Robnik-Sikonja, M., Kononenko, I.: Theoretical and empirical analysis of relieff and rrelieff. Mach. Learn. 53(1-2), 23–69 (2003)
27. Moore, J.H., White, B.C.: Tuning relieff for genome-wide genetic analysis. LNCS, vol. 4447, pp. 166–175 (2007)
28. Gonzalez, G., Uribe, J.C., Tari, L., Brophy, C., Baral, C.: Mining gene-disease relationships from biomedical literature: Weighting protein-protein interactions and connectivity measures. In: Pacific Symposium on Biocomputing, vol. 12, pp. 28–39 (2007)
29. Moore, J.H., Barney, N., Tsai, C.T., Chiang, F.T., Gui, J., White, B.C.: Symbolic modeling of epistasis. Hum. Hered. 63(2), 120–133 (2007)

cDNA-Derived Amino Acid Sequence from Rat Brain $A_{2a}R$ Possesses Conserved Motifs PMNYM of TM 5 Domain, Which May Be Involved in Dimerization of $A_{2a}R$

Pratibha Mehta Luthra[1], Sandeep Kumar Barodia[1], Amresh Prakash[1], and Ramraghubir[2]

[1] Dr. B.R. Ambedkar Center for Biomedical Research, University of Delhi, P.O. Box No. 2148, Delhi-110007, India
[2] Department of Pharmacology, Central Drug Research Institute, Lucknow 226001, India
pmlsci@yahoo.com

Abstract. The human adenosine A_{2a} receptor ($A_{2a}R$) belongs to the family of G-protein coupled receptors (GPCRs), characterized by seven transmembrane (TM) helices. TMs are involved in various cellular processes including dimerization-mediated recognition of ligand. TM5 has been suggested to self associate and may be involved in the dimerization of $A_{2a}R$. However the role of dimerization and the motifs involved in dimerization of TM 5 have not been revealed. To study the folding and assembly of $A_{2a}R$, the cDNA of the adenosine $A_{2a}R$ from rat brain was isolated and sequenced (DQ098650). The computational analysis (gil70727927lgblAAZ07991.1l) showed that the protein of 42 amino acid residues aligned in TM 5 domain region of AA2AR_RAT (P30543). PROSITE search illustrated that the motif PMNYM was conserved in $A_{2a}R$ and the motif PMSYM was present in $A_{2b}R$ respectively. The minimal dimerization motif in the TM 5 domain of the rat A_{2a} receptor sequence DQ098650 has found to be the motif PXXXM/Y.

Keywords: Adenosine A_{2a} receptor, cDNA, GPCRs, RT-PCR, TM 5.

1 Introduction

Adenosine A_{2a} receptors have a localized distribution and have emerged as a promising drug target for treating many neurological and psychiatric disorders such as Parkinson's disease [4,5,6,7], schizophrenia and affective disorders [8,9,10]. The adenosine A_{2a} receptors ($A_{2a}R$) belong to the G-protein coupled receptor (GPCR) super family characterized by seven transmembrane (TM) helices arbitrating a surfeit of signals across the plasma membrane in the cell modulating many physiological processes [11,12,13]. Assembly of transmembrane (TM) domains is a critical step in the function of membrane proteins. The determinants of transmembrane receptors structure, folding, assembly, activation mechanism and oligomeric states to function as monomers, dimers, or larger oligomers are wrapped. The existence of GPCRs as homodimers, heterodimers, or even as higher order oligomers [14, 15, 16, 17] assist in

J.C. Rajapakse, B. Schmidt, and G. Volkert (Eds.): PRIB 2007, LNBI 4774, pp. 41–50, 2007.
© Springer-Verlag Berlin Heidelberg 2007

GPCRs' functions, including ligand binding, receptor activation, desensitization, and trafficking, as well as receptor signaling [18, 19, 20]. The essential residues required for recognition of adenosine receptor agonists and/or antagonists binding within the transmembrane helical domains (TMs) 3, 5, 6, and 7, coincide largely with the corresponding amino acids of the binding site of cis-retinal in rhodopsin, although there are additional interaction sites within TMs 6 and 7 of the ARs in comparison with the binding site of rhodopsin [21]. Essentially, the constitutive and ligand-induced oligomerization has been established in other receptors [22]. However mechanism of intermolecular interaction remains unclear. A five residue motif (GxxxG) responsible for specific homodimerization for TM helices of a bioptic membrane protein have been reported. The GxxxG motif present in TM 1 of yeast α-factor is essential in oligomerization. Polar clamps and serine zipper motifs have also been identified. Since GPCRs may not have similar structures due to differences in helix-orientation, helix-helix interactions different mechanism of folding may exist in GPCR [23]. The ability of TM in oligomerization and dimerization has been reported recently and the involvement of the five residue PxxxM pattern have been suggested in dimerization of $A_{2a}R$ [24]. To study the mechanism of ligand interaction to the $A_{2a}R$, we implicated our attention to understand the determinants of $A_{2a}R$ folding and assembly.

In the present study, a primer set specific for adenosine A_{2a} receptor gene encoding TM 5 domain region have been designed. A partially amplified cDNA fragment of approx. 127bp was obtained by RT-PCR and sequenced (NCBI Gen Bank database accession no. DQ098650). Computation of sequence analysis showed that the rat brain adenosine A_{2a} receptor protein of 42 amino acid residues (gi|70727927| gb|AAZ07991.1|) aligned in TM 5 domain region of AA2AR_RAT (P30543) and exhibited 85.17% homology (36 amino acid residues) with TM 5 domain region of AA2AR_HUMAN (P29274). We carried the PROSITE search of PMNYMV residues present in the submitted sequence with the known amino acid sequences of mammalian adenosine receptors. We found that PMNYM motif was conserved in the TM 5 region of all mammalians $A_{2a}R$ and PMSYM motif was present in TM 5 region of all mammalians $A_{2b}R$ [25], thus suggesting a general role of these patterns/motifs in TM assembly. This is the first reported evidence of showing the presence of conserved motifs PMNYM in $A_{2a}R$ and PMSYM in $A_{2b}R$ respectively, which may be involved in transmembrane domain self-association, and lays ground for more apprehended analysis of adenosine receptors dimerization. The dimerization of $A_{2a}R$ involving PMNYM motif present in TM 5 is focus of our further investigations employing *in silico* and *in vitro* experimental studies.

2   Materials and Methods

2.1   Dissection and Isolation of Rat Brain Striatal Tissues

Adult wistar rats (~250g from animal house facility at Dr. B.R. Ambedkar Center for Biomedical Research, India) were sacrificed by cervical dislocation and decapitated. The freshly collected skull was cut-opened from the dorsal side. Whole brain was immediately placed in a sterilized glass petridish containing ice-cold PBS (pH 7.4) and striatal tissue was dissected out from the mid-brain region. Striatal tissue was processed immediately for the RNA extraction [26].

2.2 Total RNA Isolation

Total striatal RNA was extracted by TRIzol reagent method [27]. Briefly, tissue (100-200mg) was rinsed in cold PBS buffer (137mM NaCl, 2.7mM KCl, 4.3mM Na2HPO4, 1.4mM KH2PO4, pH 7.5), placed in 1 ml of Trizol reagent (0.8M Guanidine thiocyanate, 0.4M Ammonium thiocyanate, 0.1M Sodium acetate pH 5.0, 38% Saturated phenol and 5% Glycerol) and immediately homogenized with a glass homogenizer to quickly dissociate the tissue and incubated at room temperature for 3 minutes. About 200µl of the chloroform (Sigma-Aldrich, USA) was added to the tube, mixed gently and incubated at room temperature for 10 minutes. The sample was centrifuged at 13,000 rpm for 15 minutes at 4^0C. Clear supernatant containing total RNA and trace amount of DNA was transferred to a fresh eppendorf tube and was precipitated by adding chilled isopropanol (at -20^0C) (Sigma-Aldrich, USA). The tube was kept at -20^0C for 30 minutes and centrifuged at 13,000 rpm for 10 minutes at 4^0C. RNA pellet was washed twice with ice-cold 70% ethanol and air dried for 30 minutes. Pellet was resuspended in 20µl of Diethylpyrocarbonate (DEPC)-treated water. All solutions were prepared from DEPC-treated autoclaved, distilled water.

2.3 DNase I Treatment

To the tube containing total RNA, 2 µl of 10X DNase buffer (0.5M Tris-HCl pH 7.5, 0.5mg/ml BSA) and 2 µl of 10 U/µl DNase I (Sigma-aldrich; 20 units total) was added and incubated at 37°C for 2 hours. RNA was re-extracted by adding: 2 µl of 2M sodium acetate, pH 4.0, 22 µl water saturated phenol and 6 µl Chloroform-isoamyl alcohol. vortexed vigorously for 15 seconds. Placed on wet ice for 15 minutes and centrifuged for 10 minutes at 4°C. Upper layer was transferred to fresh tube [28].

2.4 mRNA Purification and Quantitation

Striatal mRNA was extracted by Nucleotrap mRNA mini purification kit (BD Biosciences, USA) by using manufacturer's protocol. Purified mRNA was quantitated by using UV/vis. Spectrophotometer (Schimadzu Corp, Kyoto, Japan) [29].

2.5 Primer Designing

Specific primers were designed based on the TM 5 sequence of the adenosine A_{2a} receptor DNA from *Rattus norvegicus* (30). Primer 3 (online software for primer designing) was used to evaluate secondary priming sites and inter and intra-primer complementation (31).

2.6 RT-PCR Amplification

Reverse transcriptase (RT)-PCR was performed on the purified mRNA using a Titanium One-Step RT-PCR kit (Clontech/BD Biosciences, Palo Alto, CA) according to the manufacturer's protocol. A forward oligo DNA primer with sequence (5'-CCA

TGC TGG GCT GGA ACA-3'), a reverse oligo DNA primer with sequence (5' – GAA GCG GCA GTA ACA CGA ACG-3') and an oligo (dT)17 primer were used in the RT-PCR reaction at a concentration of 45 µmol/L (primers synthesized by Microsynth, Switzerland). A mouse ß-actin primer (Clontech) served as the positive control. The RT-PCR reaction and amplification were performed under the following conditions: 1 hour (50°C), 5 minutes (94°C), 30 times (92°C 1 minute, 55°C 1 minute, 72°C 1 minute), 2 minutes (68°C). The amplified RT-PCR product was then subjected to electrophoresis on a 1.5% agarose gel (0.5 µg/ml ethidium bromide) for 1.5 hours at 80 V. The gel was visualized using Alpha Imager 1220 documentation and analysis system (Alpha Innotech Corporation, San Leandro, California). The PCR product (approx. 150bp fragment) was eluted from the low melting agarose and purified by Genei quick PCR purification kit (Bangalore genei, India). Purified fragments were lyophilized and sent to Microsynth, Switzerland for sequencing.

2.7 Sequence Analysis

Nucleotide sequence of the cDNA fragment was analyzed by Nucleotide-nucleotide BLAST (blastn) to find out cDNA fragment homology with all organism database and was submitted online to NCBI GenBank database [32]. Further, PSI-BLAST was carried on SWISS Prot Data Base for determining the positions of conserved amino acids using GAP extension 7/2 having opening penalty 7 and extension penalty 2 [33]. Multiple sequence alignment was done by Clustal-X (Version 1.83) to find out sequence homology to human and rat brain adenosine A_{2a} receptor TM 5 domain with submitted sequence (accession no. AAZ07991.1). PROSITE search was carried to find out conserved motif.

3   Results and Discussion

3.1   RT-PCR Amplification of Adenosine A_{2a} Receptor cDNA Encoding TM 5 Domain

The TM4 and TM5 have been shown to be involved in intradimeric contact in oligomeric molecular model of rhodopsin [34, 35, 36] while TM1, TM2, and the cytoplasmic loop connecting TM5 and TM6, facilitate the formation of rhodopsin dimer rows. Two conserved serine residues in TM 5 postulated to be part of a ligand-binding site in the adrenergic receptor [37]. The amino acid sequence encoded by cDNA fragment from rat brain contained conserved sequences which had characteristics of the G-linked class of receptors and displayed sequence homology in TM 5 domains with the human A_{2a} receptors (85%). We designed highly specific primer set by considering the exon sequence of adenosine A_{2a} receptor gene to isolate the cDNA fragment encoding TM 5 domain of rat brain adenosine A_{2a} receptor using Primer3 primer design tool (http://frodo.wi.mit.edu/cgi-bin/primer3/primer3_www.cgi) [38].). Rat brain striatal RNA was isolated using TRIzol reagent method and was depicted by 1% agarose gel electrophoresis (Fig.1). A desired cDNA fragment of approximately 150bp was obtained by RT-PCR

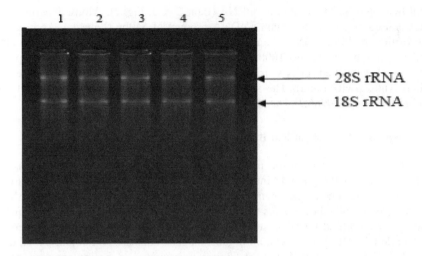

Fig. 1. Total rat brain striatal RNA. TRIzol reagent was used to isolate the total RNA . Lane 1-5 represents two distinct bands of 28S rRNA and 18S rRNA.

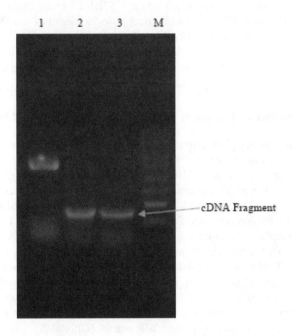

Fig. 2. RT-PCR products. Single stranded cDNA was synthesized from 1μg of total poly (A)⁺ striatal RNA using MMuLV-RTase. Lane 1. Control (mouse β-actin cDNA product, 540bp), Lane 2 & 3. Amplified cDNA fragment of approx. 150bp, Lane 4. 100bp DNA marker, Light faint bands below 100bp represents non-specific cDNA fragments.

amplification of purified rat brain mRNA (Lane 2 & 3, Fig. 2). Mouse β-actin primer set was used as control to obtain a cDNA product of 540bp (Lane 1.). Approximate size of the amplified cDNA fragment was illustrated with the aid of a 100bp DNA ladder. Faint bands below 100bp represent the non-specific fragments or primer dimers. The purified cDNA fragments possessed 127bp length on sequencing (Microsynth, Switzerland). The sequence was submitted to the GenBank database (Accession No. DQ098650, Dated: 19 Jun, 2005).

3.2 Sequence Analysis of Rat Brain Adenosine A_{2a} Receptor cDNA

In order to enable mechanistic understanding of $A_{2a}R$, computational studies were carried to predict the possible interaction interfaces. We conceded the evolutionary relation existing among the adenosine receptors in terms of their sequences that is measurable in common elements of their structural and functional features. Sequence analysis of amplified rat adenosine A_{2a} receptor cDNA was carried by Nucleotide-nucleotide BLAST (blastn) with all organisms gene database. We found a total of 69 blast hits and it was found that rat (*Rattus norvevicus*) adenosine receptor mRNA exhibited 90% homology (Score = 178 bits (90), Expect = 3e-42 Identities = 96/98 (97%), Gaps = 0/98 (0%) with the isolated cDNA sequence. Besides, it has also been observed that mouse (*Mus musculus*) strain C57BL/6J clone rp23-288i20 showed 68% homology (Score = 135 bits (68), Expect = 4e-29 Identities = 89/96 (92%), Gaps = 0/96 (0%)), Homo sapiens ADORA2A, mRNA (cDNA clone IMAGE:100000002) showed 34% homology (Score = 67.9 bits (34), Expect = 7e-09 Identities = 67/78 (85%), Gaps = 0/78 (0%)) with the RT-PCR amplified cDNA fragment of rat brain adenosine A_{2a} receptor.

3.3 Multiple Sequence Alignment of cDNA Derived Amino Acid Sequence

In the application to over 700 aligned GPCR sequences from classes A (rhodopsin-like), B (secretin-like) and C (metabotropicglutamate-like), an enhanced evolutionary trace method using Monte-Carlo techniques [39] suggested a potential functional-site on the lipid exposed faces of TM5 and TM6 is common to each family or subfamily of receptors [40] and therefore may be engaged in dimerization interface of GPCRs having specific detectable patterns/motifs. To identify such pattern/motif, single code amino acid sequence (gi|70727927|gb|AAZ07991.1|) of the submitted cDNA sequence of *Rattus norvegicus* (accession no. DQ098650) was retrieved from NCBI protein database. The rat A_{2a} receptor sequence (accession no. DQ098650) was input as a query in PSI-BLAST with inclusion threshold 0.005 (Calculation matrix BLOSUM 62 and E-value 0-1) to find the occurrence of specified pattern in the sequence. Sequences showing identities more than 80% were selected for multiple sequence alignment (Clustal-X) [41]. The results showed that rat brain adenosine A_{2a} receptor protein of 42 residues (gi|70727927|gb|AAZ07991.1|) aligned in TM 5 domain region of AA2AR_RAT (P30543) and exhibited 85.17% homology (36 residues) with TM 5 domain region of AA2AR_HUMAN (P29274) (Fig. 3).

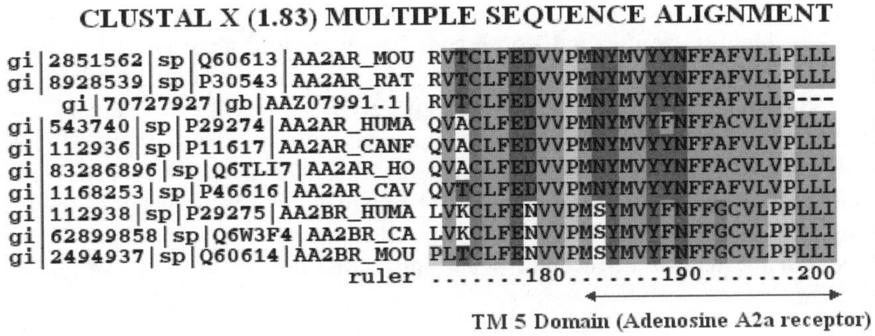

CLUSTAL X (1.83) MULTIPLE SEQUENCE ALIGNMENT

```
gi|2851562|sp|Q60613|AA2AR_MOU    RVTCLFEDVVPMNYMVYYNFFAFVLLPLLL    187
gi|8928539|sp|P30543|AA2AR_RAT    RVTCLFEDVVPMNYMVYYNFFAFVLLPLLL    187
   gi|70727927|gb|AAZ07991.1|     RVTCLFEDVVPMNYMVYYNFFAFVLLP---    31
gi|543740|sp|P29274|AA2AR_HUMA    QVACLFEDVVPMNYMVYFNFFACVLVPLLL   192
gi|112936|sp|P11617|AA2AR_CANF    QVACLFEDVVPMNYMVYYNFFAFVLVPLLL   192
gi|83286896|sp|Q6TLI7|AA2AR_HO    QVACLFEDVVPMNYMVYYNFFACVLVPLLL   192
gi|1168253|sp|P46616|AA2AR_CAV    QVTCLFEDVVPMNYMVYYNFFAFVLVPLLL   189
gi|112938|sp|P29275|AA2BR_HUMA    LVKCLFENVVPMSYMVYFNFFGCVLPPLLI   197
gi|62899858|sp|Q6W3F4|AA2BR_CA    LVKCLFENVVPMSYMVYFNFFGCVLPPLLI   196
gi|2494937|sp|Q60614|AA2BR_MOU    PLTCLFENVVPMSYMVYFNFFGCVLPPLLI   197
                         ruler    .......180.......190.......200
```

TM 5 Domain (Adenosine A2a receptor)

Fig. 3. Homology alignments cDNA sequence of adenosine A$_{2a}$ receptor *Rattus norvegicus* (gi|70727927|gb|AAZ07991.1|) and known mammalian A$_{2a}$ receptor sequences lie in TM5 domain

Previously, it has been reported that TM helix dimerization motifs GxxxG, AxxxA, SxxSSxxT, polar clamps, serine zipper, leucine zipper do not appear in TM5. However, statistical analysis of amino acid patterns in TM helices revealed that PM4 pair (PxxxM) was the most overpresented doublet pattern from any combination of PxxxY doublet pattern suggesting its role in the adenosine A$_{2a}$ receptor dimerization [24].

AA2A_HUMAN	(166) CLFEDVV**PMNYM**VYFNFFAC	(185)
AA2A_CAVPO	(163) CLFEDVV**PMNYM**VYYNFFAF	(182)
AA2A_MOUSE	(161) CLFEDVV**PMNYM**VYYNFFAF	(180)
AA2A_RAT	(161) CLFEDVV**PMNYM**VYYNFFAF	(180)
AA2A_CANFA	(166) CLFEDVV**PMNYM**VYYNFFAF	(185)
Q4F987_RAT	(8) CLFEDVV**PMNYM**VYYNFFAF	(27)
AA2B_CHICK	(171) CLFENVV**TMSYM**VYFNFFGC	(190)
AA2B_MOUSE	(171) CLFENVV**PMSYM**VYFNFFGC	(190)
AA2B_RAT	(171) CLFENVV**PMSYM**VYFNFFGC	(190)
AA2B_HUMAN	(171) CLFENVV**PMSYM**VYFNFFGC	(190)

Fig. 4. UniProtKB/Swiss-Prot Hits for USERPAT1 {PMNYMV} motif on all (release 51.0), UniProtKB/TrEMBL (release 34.0), PDB (2-Nov-2006) databases sequences

The PROSITE search was carried using PMNYMV residues present in the submitted sequence to examine the occurrence of this pattern in known amino acid sequences of mammalian adenosine receptor. We found that PMNYM motif was conserved in the TM 5 region of all mammalians A$_{2a}$R and PMSYM motif was present in TM 5 region of all mammalians A$_{2b}$R. From the detailed domain sequence descriptions employing UniProtKB/Swiss-Prot (Fig. 4 www.expasy.org), we found that 42-residue sequence contains only one PM4 pattern (PMNYM) that lies in the TM 5 domain, the analysis revealed a significant difference in the distribution of conserved TM 5 domain motif at subtype level of adenosine receptors. In adenosine A_{2a} receptors, the motif PMNYM is highly specific and conserved, however, in adenosine A_{2b} receptor asparagine (N) residue is replaced by serine (S) generating the motif PMSYM [42, 43] thus differentiating the two isoforms of receptors functionally. Finally, we interpret in the context of transmembrane dimerization motifs that conserved motif (PxxxM) may play a role in the dimerization of adenosine receptors .The motif PMNYM of $A_{2a}R$ and PMSYM of $A_{2b}R$ may be involved in TM assembly of the two isoforms of the receptors respectively. The information may provide an insight into the molecular mechanism of receptor-ligand interaction leading to design of tailored compounds.

Acknowledgements

Work carried out in the author's laboratory was supported by the Council of Scientific and Industrial Research (CSIR), New Delhi, India. We are thankful to CSIR, New Delhi, India and University Grant Commission (UGC), Delhi India for financial assistance.

References

1. Zhou, N.E., Kay, C.M., Hodges, R.S.: Synthetic model proteins. Positional effects of interchain hydrophobic interactions on stability of two-stranded alpha-helical coiled-coils. J. Biol. Chem. 267, 2664–2670 (1992)
2. Lazarova, T., Brewin, K.A., Stoeber, K., Robinson, C.R.: Characterization of peptides corresponding to the seven transmembrane domains of human adenosine A2a receptor. Biochemistry 43, 12945–12954 (2004)
3. Lau, S.Y., Taneja, A.K., Hodges, R.S.: Synthesis of a model protein of defined secondary and quaternary structure. Effect of chain length on the stabilization and formation of two-stranded alpha-helical coiled-coils. J. Biol. Chem. 259, 13253–13261 (1984)
4. Chen, J.F., Xu, K., Petzer, J.P., Staal, R., Xu, Y.H., Beilstein, M., Sonsalla, P.K., Castagnoli, K., Castagnoli Jr., N., Schwarzschild, M.A.: Neuroprotection by caffeine and A(2A) adenosine receptor inactivation in a model of Parkinson's disease. J. Neurosci. 21, RC143 (2001)
5. Ferre, S., Popoli, P., Gimenez-Llort, L., Rimondini, R., Muller, C.E., Stromberg, I., Ogren, S.O., Fuxe, K.: Adenosine/dopamine interaction: implications for the treatment of Parkinson's disease. Parkinsonism Relat. Disord. 7, 235–241 (2001)
6. Richardson, P.J., Gubitz, A.K., Freeman, T.C., Dixon, A.K.: Adenosine receptor antagonists and Parkinson's disease: actions of the A2A receptor in the striatum. Adv. Neurol. 80, 111–119 (1999)
7. Schwarzschild, M.A., Chen, J.F., Ascherio, A.: Caffeinated clues and the promise of adenosine A(2A) antagonists in PD. Neurology 58, 1154–1160 (2002)
8. Ferre, S.: Adenosine– dopamine interactions in the ventral striatum. Implications for the treatment of schizophrenia. Psychopharmacology (Berl.) 133, 107–120 (1997)
9. Ferre, S., Fredholm, B.B., Morelli, M., Popoli, P., Fuxe, K.: Adenosine– dopamine receptor– receptor interactions as an integrative mechanism in the basal ganglia. Trends Neurosci. 20, 482–487 (1997)

10. Rimondini, R., Ferre, S., Ogren, S.O., Fuxe, K.: Adenosine A2A agonists: a potential new type of atypical antipsychotic. Neuropsychopharmacology 17, 82–91 (1997)
11. Shichida, Y., Imai, H.: Visual pigment: G-protein-coupled receptor for light signals. Cell. Mol. Life Sci. 54, 1299–1315 (1998)
12. Gether, U.: Uncovering molecular mechanisms involved in activation of G protein-coupled receptors. Endocr. Rev. 21, 90–113 (2000)
13. Gurrath, M.: Peptide-binding G protein-coupled receptors: New opportunities for drug design. Curr. Med. Chem. 8, 1605–1648 (2001)
14. Jones, K.A., Borowsky, B., Tamm, J.A., Craig, D.A., Durkin, M.M., Dai, M., Yao, W.J., Johnson, M., Gunwaldsen, C., Huang, L.Y., et al.: GABA (B) receptors function as a heteromeric assembly of the subunits GABA(B)R1 and GABA(B)R2. Nature 396, 674–679 (1998)
15. Jordan, B.A., Devi, L.A.: G-protein-coupled receptor heterodimerization modulates receptor function. Nature 399, 697–700 (1999)
16. Bai, M.: Dimerization of G-protein-coupled receptors: Roles in signal transduction. Cell Signal 16, 175–186 (2004)
17. Fotiadis, D., Liang, Y., Filipek, S., Saperstein, D.A., Engel, A., Palczewski, K.: The G protein-coupled receptor rhodopsin in the native membrane. FEBS Lett. 564, 281–288 (2004)
18. Rios, C.D., Jordan, B.A., Gomes, I., Devi, L.A.: G protein-coupled receptor dimerization: Modulation of receptor function. Pharmacol. Ther. 92, 71–87 (2001)
19. George, S.R., O'Dowd, B.F., Lee, S.P.: G-protein-coupled receptor oligomerization and its potential for drug discovery. Nat. Rev. Drug Discov. 1, 808–820 (2002)
20. Breitwieser, G.E.: G protein-coupled receptor oligomerization: Implications for G-protein activation and cell signaling. Circ. Res. 94, 17–27 (2004)
21. Fredholm, B.B., IJzerman, A.P., Jacobson, K.A., Klotz, K.N., Linden, J.: International Union of Pharmacology. XXV. Nomenclature and Classification of Adenosine Receptors. Pharmacol. Rev. 53, 527–552 (2001)
22. Schlessinger, J.: Ligand-induced, receptor-mediated dimerization and activation of EGF receptor. Cell 110, 669–672 (2002)
23. Palczewski, K., Kumasaka, T., Hori, T., Behnke, C.A., Motoshima, H., Fox, B.A., Le Trong, I., Teller, D.C., Okada, T., Stenkamp, T.E., Yamamoto, M., Miyano, M.: Crystal Structure of Rhodopsin: A G Protein-Coupled Receptor. Science 289, 739–745 (2000)
24. Thevenin, D., Lazarova, T., Roberts, M.F., Robinson, C.R.: Oligomerization of the fifth transmembrane domain from the adenosine A2a receptor. Protein Science 14, 2177–2186 (2005)
25. Hulo, N., Sigrist, C.J., Le Saux, V., Langendijk-Genevaux, P.S., Bordoli, L., Gattiker, A., De Castro, E., Bucher, P., Bairoch, A.: Recent improvements to the PROSITE database. Nucleic Acids Res. 32, D134–D137 (2004)
26. Haobam, R., Sindhu, K.M., Chandra, G., Mohanakumar, K.P.: Swim-test as a function of motor impairment in MPTP model of Parkinson's disease: a comparative study in two mouse strains. Behav. Brain Res. 163, 159–167 (2005)
27. Chomczynski, P., Sacchi, N.: Single-step method of RNA isolation by acid guanidinium thiocyanate-phenol-chloroform extraction. Anal. Biochem. 162, 156–159 (1987)
28. Sutton, D.H., Conn, G.L., Brown, T., Lane, A.N.: The dependence of DNase I activity on the conformation of oligodeoxynucleotides. Biochem. J. 321, 481–486 (1997)
29. Hooper-McGrevy, K.E., MacDonald, B., Whitcombe, L.: Quick, simple, and sensitive RNA quantitation. Analytical Biochemistry 318, 318–320 (2003)

30. Wu, D.Y., Ugozzoli, L., Pal, B.K., Qian, J., Wallace, R.B.: The effect of temperature and oligonucleotide primer length on the specificity and efficiency of amplification by the polymerase chain reaction. DNA and Cell Biology 10, 233–238 (1991)
31. Rozen, S., Skaletsky, H.J.: Bioinformatics Methods and Protocols: Methods in Molecular Biology, pp. 365–386. Humana Press, Totowa, NJ (2000)
32. Altschul, S.F., Gish, W., Miller, W., Myers, E.W., Lipman, D.J.: Basic local alignment search tool. J. Mol. Biol. 215, 403–410 (1990)
33. Altschul, S.F., Madden, T.L., Schaffer, A.A., Zhang, J., Zhang, Z., Miller, W., Lipman, D.J.: A new generation of protein database search programs. Nucl. Acids Res. 25, 3389–3402 (1997)
34. Fotiadis, D., Liang, Y., Filipek, S., Saperstein, D.A., Engel, A., Palczewski, K.: Atomic-force microscopy: Rhodopsin dimers in native disc membranes. Nature 421, 127–128 (2003)
35. Liang, Y., Fotiadis, D., Filipek, S., Saperstein, D.A., Engel, A., Palczewski, K.: Organization of the G Protein-coupled Receptors Rhodopsin and Opsin in Native Membranes. J. Biol. Chem. 278, 21655–21662 (2003)
36. Fotiadis, D., Liang, Y., Filipek, S., Saperstein, D.A., Engel, A., Palczewski, K.: The G protein-coupled receptor rhodopsin in the native membrane. FEBS Lett. 564, 281–288 (2004)
37. O'Dowd, B.F., Nguyen, T., Tirpak, A., Jarvie, K.R., Israel, Y., Seeman, P., Niznik, H.B.: Cloning of two additional catecholamine receptors from rat brain. FEBS Lett. 262, 8–12 (1990)
38. Yu, L., Frith, M.C., Suzuki, Y., Peterfreund, R.A., Gearan, T., Sugano, S., Schwarzschild, M.A., Weng, Z., Fink, J.S., Chen, J.F.: Characterization of genomic organization of the adenosine A2A receptor gene by molecular and bioinformatics analyses. Brain Res. 1000, 156–173 (2004)
39. Upton, G., Fingleton, B.: Spatial data analysis by example. Wiley, Chichester (1985)
40. Dean, M.K., Higgs, C., Smith, R.E., Bywater, R.P., Snell, C.R., Scott, P.D., Upton, G.J., Howe, T.J., Reynolds, C.A.: Dimerization of G-protein-coupled receptors. J. Med. Chem. 44, 4595–4614 (2001)
41. Chenna, R., Sugawara1, H., Koike, T., Lopez, R., Gibson, T.J., Higgins, D.G., Thompson, J.D.: Multiple sequence alignment with the Clustal series of programs. Nucleic Acids Research. 31, 3497–3500 (2003)
42. Bateman, A., Coin, L., Durbin, R., Finn, R.D., Hollich, V., Griffiths-Jones, S., Khanna, A., Marshall, M., Moxon, S., Sonnhammer, E.L.L., Studholme, D.J., Yeats, C., Eddy, S.R.: The Pfam protein families database. Nucleic Acids Research 32, D138–D141 (2004)
43. Castro, E., Christian, J.A., Sigrist, G.A., Bulliard, V., Langendijk-Genevaux, P.S., Gasteiger, E., Bairoch, A., Hulo, N.: ScanProsite: detection of PROSITE signature matches and ProRule-associated functional and structural residues in proteins. Nucleic Acids Research 34, W362–W365 (2006)

Strong GC and AT Skew Correlation in Chicken Genome

Xuegong Deng[1], Xuemei Deng [2], and Ilkka Havukkala[3]

[1] Northeastern university, College of Science, Shenyang, China
dengxuegong@tom.com
[2] Department of Animal Genetics and Breeding & National Key Lab, China Agricultural
University, Beijing 100094, China
deng@cau.edu.cn
[3] Auckland University of Technology, Knowledge Engineering and Discovery Research
Institute, Auckland, New Zealand
ilkka.havukkala@aut.ac.nz

Abstract. Chicken genome AT and GC skews for individual chromosomes were visualized simultaneously using a novel method of 2-dimensional color-coded pixel matrix. The visualizations were compared to those of human, mouse and possum genomes. A strikingly strong correlation of AT skew and GC from small to large scale in chicken genome was found, compared to the other vertebrates. Some local skew correlations were also found for the other vertebrates, but only in small genomic scale. Quantitative measures of correlation were developed, and confirmed the special characteristic of chicken chromosomes. Possible explanations for uniqueness of birds in this respect are discussed. The phylogenetic distribution and evolutionary pressures responsible for this previously unreported skew correlation warrant further study.

Keywords: AT/GC skew, skew correlation, chicken genome, chromosome, visualization, 2D.

1 Introduction

Division of genomes to large-scale segments of low and high GC% (isochores) has been attributed to natural selection [1], mutational biases, biased gene conversion [2] and recombination [3]. In addition, the nucleotide skews (A-T)/(A+T) and (C-G)/(G+C) can vary locally and seem to be related to gene distribution, transcription direction and the origin of replication and many other important biological properties in bacteria [4] and in mammals [5]. This suggests that genomes at a local scale need frequently to deviate from Chargaff's 2nd rule [6], which states that in the whole genome scale the frequencies of A and T are similar in single stranded DNA, as well as the frequencies of G and C [7].

Traditionally, CG-skews have been analyzed by using separate cumulative skew diagrams [8]. Recently AT and CG skews have been combined as cumulative total skew (sum of AT and CG skews) over 1 kb non-overlapping windows, and used for prediction of replication origins in mammalian genomes [5]. Bacterial genomes [9] and organellar genomes [10] show correlations between nucleotide skews but there is little data from higher eukaryote main genomes in this respect.

J.C. Rajapakse, B. Schmidt, and G. Volkert (Eds.): PRIB 2007, LNBI 4774, pp. 51–59, 2007.
© Springer-Verlag Berlin Heidelberg 2007

Visualizations of GC skew for circular bacterial genomes has been developed [11,12], but these are not suitable to display skew correlation in large eukaryotic genomes. Therefore other better methods are needed for large-scale easy visualization and analysis of AT and CG skews for whole genomes and eukaryotic chromosomes. We develop here a new method of 2-D color display to show simultaneously both AT skew and GC skew at varying scales in whole chromosomes. Correlation of the skews is easily seen as symmetry along the diagonal in the 2D matrix representation.

A new summary parameter is developed for symmetry level (correlation) of AT and CG skews. The developed new visualization method, called Base Skew Double Triangles (BSDT), and correlation quantitation method facilitate nucleotide skew comparisons between various genomes and at various genomic scales. The BSDT visualization software and related information is available at request from the authors.

The new visualization method and quantitative correlation measures discovered that chicken chromosomes have a strong skew correlation in large scale, in strong contrast to the three higher vertebrates studied (opossum, mouse, human). Possible explanations for this difference are discussed.

2 Methods

Nucleotide skews show preponderance of one nucleotide frequency over another, e.g. AT skew = (A + T) / (A – T). They are normally calculated over limited windows, or cumulatively along the sequence. In contrast, here we visualize the AT skew and CG skew of whole chromosomes in a square matrix of pixels at scale of 1024*1024 pixels. The DNA sequence is divided into 1024 equal size windows, W^i, with window length $l = L/1024$ where L is the length of the whole sequence. For any pair m,n<1024, m<n, we define the matrix of subsequences from W^m to W^n as

$$D_{m,n} = \bigcup_{i=m}^{n} W^i \tag{1}$$

and the AT and CG skews of $D_{m,n}$ can be denoted as $D_{m,n}^{AT}$ and $D_{m,n}^{GC}$. By defining a RGB color function, we draw two symmetrical pixels (m,n) and (n,m) (at the symmetry position) with color $color(D_{m,n}^{AT})$ and $color(D_{m,n}^{GC})$. When the skew value changes from high negative through zero to high positive value, the color changes from deep blue through blue, pink, black, yellow, green to deep green (see Figure 1a). For simultaneous visualization of both AT and CG skews at all scales, we us a color square consisting of two triangles, separated by the diagonal, the bottom-left for the CG skew and the top-right for AT skew. We call this image of triangles Base Skew Double Triangles (BSDT). The skews of $1024^2/2=524288$ segments are shown in one triangle. Each point in the square represents a base skew of a different subsequence, The changes in AT and GC skews are visualized clearly as color gradients at different scales and locations in the chromosome sequence. Thus, we can clearly observe changes in the skew along the sequence.

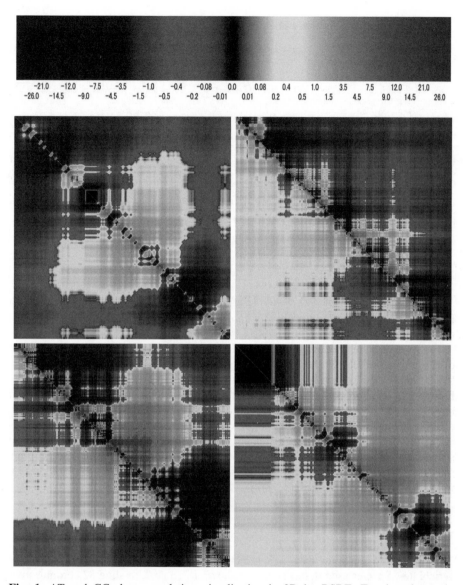

Fig. 1. AT and GC skew correlation visualization in 2D by BSDT. Top bar shows the chromatogram (color scale). Top left: chicken chromosome 8; top right: opossum chromosome 3; bottom left: mouse chromosome 1; bottom right: human chromosome 14. The lower left triangle represents the AT bias, the upper right triangle represents the GC bias. The clear correlation of AT and CG skews in chicken is evident as symmetry along the diagonal.

By visual observation, most BSDTs of chicken chromosomes are very symmetrical across the diagonal, meaning that the correlation between the AT skew and CG skew is very high for most large chromosomes of chicken at any scale. This visual symmetry phenomenon can be quantified as a summary correlation measure as follows:

Taking 12 pairs of symmetry lines parallel to the diagonal in the square,

$$m - n = \beta \times k + 1 \tag{2}$$

$$m - n = \beta \times k - 1 \tag{3}$$

where β is the distance between 2 neighboring lines. k=0,1,2,3,...,to 11 in equation (2) and k=0,-1,-2,-3, to -11 in equation (3). Since each point pair (n,m) and (m,n) in the lines corresponds two base skews $D_{n,m}^{AT}$ and $D_{n,m}^{CG}$, each line pair has a a set of paired values for each genome sequence. For the 12 correlation coefficients from the 12 paired lines denoted as C_i, we define the average C_i as the correlation level (CL) of the target sequence:

$$CL_{\beta}^{species\_chr*} = \frac{1}{12} \sum_{i=1}^{12} C_i \tag{4}$$

The species_chr* means for which DNA sequence and β means in what scale. Positive values of CL mean positively correlated AT and CG skews, negative values mean negative correlation.

Interpretation of the images is as follows: 1) The skew level of any segment displayed in the BSDT shows strong negative skews as deep blue, changing via black color at no skew to deep green for strong positive skews (see color scale in Figure 1a). 2) The closer the point to the diagonal, the smaller the sequence it represents. 3) The symmetrical color pattern across the diagonal means correlation of GC skew and AT skew of that genome area 4) By comparing the color changes along the diagonal direction we can observe local deviations in skews along the sequence. Note, that in the corners furthest from the diagonal (= largest genomic scale) the skew is less (color is often pink or yellow, that is, closer to black portion in the middle of the color scale).

3 Data

The eukaryotic genomic data was selected to compare main branches of higher vertebrates. Species used in this study are shown in Table 2, totaling 87 chromosomes and 9,794 Mb of sequence data.

Table 1. Genomic data analyzed for AT and GC skew correlation

Organism	Latin name	Chrom, # - size (Mb)	Data source
chicken	*Gallus gallus*	32— 984 Mb	UCSC(galGal3)
opossum	*Monodelphis domestica*	10—3420 Mb	UCSC(monDom4)
mouse	*Mus musculus*	21—2470 Mb	UCSC(mm6)
human	*Homo sapiens*	24—2920 Mb	UCSC (hg17)
Total		87—9794 Mb	

4 Results

4.1 AT and GC Skew Correlations

We generated 2D displays of skew correlation (BSDTs) for 87 chromosomes of 4 fully sequenced species. Representative samples of visualizations are shown in Figure 1. We then calculated the correlation level CL values for $\beta = 2,5,10,20,50$ in equations (2) and (3). The CL values of all chromosomes from the four species are listed in table 2, using $\beta = 20$. It is remarkable, that chicken chromosomes have clearly the highest symmetry level CL compared with all the other vertebrates.

Table 2. AT and GC skew correlation level CL of ranked chromosomes of four vertebrates, β=20

Rank	Chicken	CL.	Human	CL.	Mouse	CL	Opossum	CL
1	chr8	0.984	chr14	0.887	chrY	0.730	chr1	0.239
2	chr14	0.98	chr15	0.865	chr1	0.718	chrUn	0.207
3	chr10	0.98	chr7	0.814	chr12	0.685	chr6	0.050
4	chr5	0.977	chr3	0.789	chr3	0.669	chr8	-0.023
5	chr13	0.975	chr12	0.780	chr14	0.632	chr3	-0.251
6	chr6	0.975	chr16	0.777	chr18	0.610	chr2	-0.335
7	chr4	0.971	chr20	0.753	chr13	0.602	chr4	-0.381
8	chr19	0.97	chr11	0.738	chr2	0.584	chr5	-0.403
9	chr7	0.969	chr5	0.737	chr10	0.575	chr7	-0.445
10	chr12	0.967	chr8	0.701	chr9	0.554	chrX	-0.456
11	chr11	0.967	chr13	0.698	chr5	0.540		
12	chr20	0.963	chr9	0.687	chr16	0.513		
13	chr3	0.963	chr2	0.683	chrX	0.505		
14	chr9	0.957	chr6	0.675	chr6	0.498		
15	chr24	0.956	chr17	0.658	chr17	0.391		
16	chr1	0.955	chr1	0.658	chr15	0.377		
17	chr18	0.954	chr4	0.649	chr19	0.370		
18	chr15	0.94	chr22	0.590	chr7	0.321		
19	chr23	0.925	chr18	0.544	chr4	0.232		
20	chr22	0.923	chrX	0.519	chr11	0.218		
21	chrZ	0.921	chr10	0.365	chr8	0.181		
22	chr26	0.915	chr21	0.239	chrM	-0.030		
23	chr17	0.913	chr19	0.110				
24	chr2	0.906	chrY	-0.252				
25	chr21	0.89						
26	chr28	0.804						
27	chr27	0.724						
28	chr25	0.381						
29	chr32	0.247						
30	chr16	-0.2						
31	chrM	-0.33						
32	chrW	-0.38						

Since the chromosomes listed in Table 2 have quite different size, but they are represented in a 2D square with same number of pixels, chromosome size might be thought to bias the comparisons. However, we have calculated the CL for many parameter values of β for correlation level in many genomic scales, and consistently found that CL for chicken chromosomes were clearly higher than in the other vertebrates for all combinations of species, and almost all chromosomes, using parameter combinations $\beta = 2,5,10,20,50$ and n for first 25 chromosomes listed in Table 2. Thus we conclude that chicken chromosomes have higher skew correlation compared to other animals in any scale, based on both visual observation of the BSDT images and the quantitative analysis by CL measure.

One can also compare the chromosomes at the same window scale by a simple visual method. For example, if two DNA sequences have length 30Mb and 10Mb, then we can compare them by sliding a square 1024/3 in length along diagonal of the BDST of the bigger one, with step 1. If the longer sequence has the same skew correlation level CL or higher in the smaller window scale, the sliding square should on average will have the same CL or higher compared to the smaller one (see Figure 2).

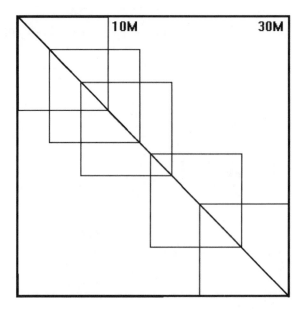

Fig. 2. Sketch map of sliding a smaller square in a bigger BSDT along the diagonal

It is very interesting that we indeed can find some local symmetry segments in other animal's chromosomes with lower CL. For example, human chromosome 2 has some high symmetry local segments found by sliding the square window. Figure 3a is the BDST of the human whole chromosome 2 and has CL 0.584. that is much smaller than the top ranked chromosomes in Table 2. We enlarged 3 non-overlapping sliding square windows at length 1024/3, shown in Figure 3d and we can see some local symmetry areas in the chromosome. In certain locations, blue and green color have symmetry along the diagonal, indicating local negative correlation of AT and CG skews. This would lead to a lower CL of the whole chromosome. Unlike birds, other animals show little symmetry in whole genome scale, although many have some symmetry locally.

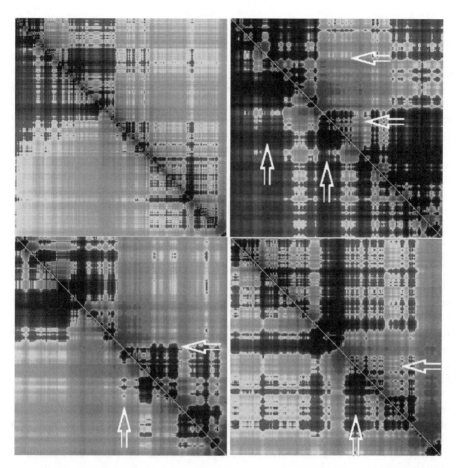

Fig. 3. The BSDT of chromosome 2 of human. Top left: whole chromosome, top right, bottom left and bottom right: successive non-overlapping 1024/3 sliding windows. The white arrows show AT skew and CG skew negatively correlated regions.

5 Discussion

In this study, we introduced a new 2D visualization method of BSDT to display and quantify correlation of AT and GC skews in the genome, and discovered a strong correlation in large-scale in the chicken genome. We predict it is prevalent in other bird genomes as well. Further study of the phylogenetic distribution and evolution of this skew correlation awaits access to more genomic data from other bird species and related taxa.

We also discovered local skew symmetries in other animals, which also warrant further investigation. Such smaller scale local skew correlation regions may be related to replication origins at protein coding regions, as shown for human genome (5), or other local features tied to transcription activity along the genome. It will be interesting to examine the role and contribution of non-coding and repetitive DNA sequences in this skew correlation. Also the relation between isochore structures and skew symmetry remains to be explored.

Chargaff's second rule is a global rule with many deviations in local genomic scales (4). It must have something to do with double-stranded genome organization, as single-stranded RNA and DNA genomes do not obey the rule (13). Isochore structure may be related to this phenomenon and for such studies the recently constructed human chromosome isochore maps (14) should be useful. It seems established that isochores are declining in current evolution of mammals (15). Webster et al (16) suggest that birds have a history of biased gene conversion, causing and maintaining the isochore structure and more increased divergence in GC-rich regions compared to mammals. As for other sauropsids, only limited reptile data is available, including GC composition in various species measures by analytical centrifugation (17). It would be interesting to see how skew correlation is related to the variation of GC composition at the same genomic scale.

The exact biochemical determinants of evolutionary selection pressures causing and maintaining the special skew correlations reported in this paper are largely unknown, as is their relation to the Chargaff's second rule and isochores. Some possible explanations for the uniqueness of birds are suggested here. Firstly, there may be some unique metabolic aspects in birds, compared to both lizards and mammals, affecting the nucleotide composition. Metabolic constraints and availability of energy in microbes have been shown to affect AT/GC composition (18), as well as environmental factors (19).

Secondly, birds have a higher body temperature and faster metabolism than mammals, due to unidirectional airflow in bird lung and more efficient blood circulation and maximized ATP production (20). Higher temperature and faster metabolism could mean faster mutation rates, affecting biased gene conversion. This could be studied by relating body temperature in various birds and other animals to skew correlation.

Thirdly, chicken genome has been reported to be a mosaic of GC and AT rich isochores (21), though comparisons to reptiles and mammals are still lacking. Isochore structure might affect the biased gene conversion, thus affecting skew correlation.

With more genomic data accumulating from a variety of birds and related taxa we expect interesting findings in this area. The new methods of visualization by BSDT and skew correlation measure CL and their variations should serve as useful tools in this endeavor.

Acknowledgments. This work was funded by the state Major Basic Research Development Program (2006CB102100), National Natural Science Foundation of China (No. 30471233), and National High Technology Research and Development Program of China (No.2001AA222191) for Deng Xuemei. The Auckland University of Technology, Knowledge Engineering and Discovery Institute in Auckland, New Zealand is thanked for facilities for Ilkka Havukkala.

References

1. Bernardi, G.: Isochores and the evolutionary genomics of vertebrates. Gene. 241, 3–17 (2000)
2. Eyre-Walker, A., Hurst, L.D.: The evolution of isochores. Nat. Rev. Genet. 2(7), 549–555 (2001)
3. Meunier, J., Duret, L.: Recombination drives the evolution of GC-content in the human genome. Mol. Biol. E. 21(6), 984–990 (2004)
4. Bell, S.J., Forsdyke, D.R.: Deviations from Chargaff's second parity rule correlate with direction of transcription. J. Theor. Biol. 197(1), 63–76 (1999)
5. Touchon, M., Nicolay, S., Audit, B., Brodie, E.-B., d'Aubenton-Carafa, Y., Arneodo, A., Thermes, C.: Replication-associated strand asymmetries in mammalian genomes: Toward detection of replication origins. PNAS 102, 9836–9841 (2005)
6. Chargaff, E.: Some recent studies on the composition and structure of nucleic acids. J. Cell Physiol. 38, 41–59 (1951)
7. Forsdyke, D.R., Mortimer, J.R.: Chargaff's legacy. Gene. 261(1), 127–137 (2000)
8. Grigoriev, A.: Analyzing genomes with cumulative skew diagrams. Nucleic Acids Research 26(10), 2286–2290 (1998)
9. Nikolaou, C., Almirantis, Y.: A study on the correlation of nucleotide skews and the positioning of the origin of replication: different modes of replication in bacterial species. Nucleic Acids Res. 33(21), 6816–6822 (2005)
10. Nikolaou, C., Almirantis, Y.: Deviations from Chargaff's second parity rule in organellar DNA Insights into the evolution of organellar genomes. Gene. 381, 34–41 (2006)
11. Ghai, R., Hain, T., Chakraborty, T.: GenomeViz: visualizing microbial genomes. BMC Bioinformatics 5, 198 (2004)
12. Pritchard, L., White, J.A., Birch, P.R.J., Toth, I.K.: GenomeDiagram: a python package for the visualization of large-scale genomic data. Bioinformatics 22(5), 616–617 (2000)
13. Mitchell, D., Bridge, R.: A test of Chargaff's second rule. Biochem. Biophys. Res. Commun. 340(1), 90–94 (2006)
14. Constantini, M., Clay, O., Auletta, F., Bernardi, G.: An isochore map of human chromosomes. Genome Res. 16(4), 536–541 (2006)
15. Duret, L., Eyre-Walker, A., Galtier, N.: A new perspective on isochore evolution. Gene. 385, 71–74 (2006)
16. Webster, M.T., Axelsson, E., Ellegren, H.: Strong regional biases in nucleotide substitution in the chicken genome. Mol. Biol. Evol. 23(6), 1203–1216 (2006)
17. Hughes, S., Clay, O., Bernardi, G.: Compositional patterns in reptilian genomes. Gene. 295(2), 323–329 (2002)
18. Rocha, E.P., Danchin, A.: Base composition bias might result from competition for metabolic resources. Trends Genet. 18(6), 291–294 (2002)
19. Foerstner, K.U., von Mering, C., Hooper, S.D., Bork, P.: Environments shape the nucleotide composition of genomes. EMBO Rep. 6(12), 1208–1213 (2005)
20. Gao, F., Zhang, C.T.: Isochore structures in the chicken genome. FEBS J. 273(8), 1637–1648 (2006)

Comparative Analysis of a Hierarchical Bayesian Method for Quantitative Trait Loci Analysis for the Arabidopsis Thaliana

Caroline Pearson[1], Susan J. Simmons[1], Karl Ricanek Jr.[2], and Edward L. Boone[3]

[1] University of North Carolina Wilmington, Department of Mathematics and Statistics,
[2] University of North Carolina Wilmington, Department of Computer Science,
601 South College Road, Wilmington, North Carolina, 28403, USA
[3] Virginia Commonwealth University,
Department of Statistical Sciences and Operations Research,
Richmond, Virginia, 23284, USA
caroline.pearson310@gmail.com, {simmonssj,ricanekk}@uncw.edu,
elboone@vcu.edu

Abstract. This work performs an analysis on two, quite different, techniques for Quantitative Trait Loci (QTL) Analysis. Interval Mapping (IM) as described by Karl Broman is compared to a Hierarchical Bayesian Model (HBM) technique that reduces the problem of QTL analysis down to one of model selection. Simulations were generated for the flowering plant of the *Arabidopsis thaliana* for evaluation of the techniques. It is shown that the HBM technique was much more successful at determining the appropriate loci/markers and corresponding chromosomes than the IM technique given a single loci. It was further elucidated through simulation runs that the HBM was robust against two loci/markers, whereas IM completely failed. The contribution of this work is in the comparison and analysis of the IM method to that of the HBM; hence, demonstrating through simulations that the HBM technique is superior to that of the IM for the Arabidopsis simulated data.

Keywords: Quantitative Trait Loci (QTL) Analysis, Arabidopsis, Interval Mapping, Hierarchical Bayesian Model.

1 Introduction

The commencement of the genomic era has witnessed an increased interest in identifying locations on a genome responsible for a quantitative trait, referred to as quantitative trait loci (QTL). Mapping a quantitative trait to a location on a genome is made possible through a genetic map, which illustrates the relative distance between known markers or genes on an organism's genome. Alfred Henry Sturtevant constructed the first genetic map in 1913, and the first analysis relating genes to quantitative traits was done in 1923 by Sax [1]. Since these initial works and a few others, not much else has been done in the area of QTLs until the 1980's.

J.C. Rajapakse, B. Schmidt, and G. Volkert (Eds.): PRIB 2007, LNBI 4774, pp. 60–70, 2007.
© Springer-Verlag Berlin Heidelberg 2007

The last two decades has seen an explosion of algorithms proposed for the identification of QTLs. A few of the more notable initial methods include Lander and Botstein [2]; Jensen [3]; Zeng [4]; Wright and Mowers [5]; Kearsey and Hyne [6]; Wu and Li [7]; and, Sen and Churchill [8]. In the last decade a technique known as interval mapping [9], in which pseudo-markers were placed in the interval between two known markers to evaluate the possibility of a QTL in the interval, has dominated the research landscape. Variations on the Interval Mapping (IM) algorithm such as Composite Interval Mapping [4, 10] and Multiple Interval Mapping [11] have also gained much attention in the 1990's. Yet, still other influential methods have been proposed to analyze QTLs; for example, generalized estimating equations [2], partial least squares [12], and Bayesian approaches [13, 14, 15, 16, 17]. Each method has its own advantages and weaknesses; however, all of these approaches assume only one observation per genotype.

A single trait is usually determined by many genes; as a result, many QTLs are usually associated with a single trait. The number of QTLs associated with each phenotypic trait tells us the genetic makeup and the variation of this trait. For instance, a small effect can be determined if there are many QTLs correlated with a single trait and a large effect can be determined if there are only a few QTLs correlated with a single trait. The information gleamed from the QTL can help us better understand the chemical structure of these traits and better understand the evolution of these traits over a period of time. Furthermore, QTLs can ultimately enable the alteration of the chemical structure of these traits. One potential benefit of understanding the QTLs for plants is the ability to alter the chemical structure of a plant to make it more tolerant of ultraviolet (UV) radiation, which may help agriculturalists deal with the depleting ozone layer; this layer filters much of the UV radiation before it can enter the atmosphere and ultimately the terra firma.

As mentioned above, most QTL methods have been developed mainly for human and animal genotypes. Since all humans and animals have unique genetic composition, each observation per genotype (or line) consists of a single observation value. However, plants are often cloned and their genetic makeup reproduced, generating multiple observations per genotype (or line). The clones within each line consist of the same marker information on their genetic makeup; therefore the question must be asked, *Are the methods for human and animal genotypes appropriate for plant genotypes?*

Current state-of-the-art algorithms for QTL detection are limited to the use of a single value for the quantitative trait per genotype; therefore, plant biologists will take the mean or median value of the quantitative trait within each line to perform their analysis. Although, plants have the unique capability of being cloned, important information, like variability within a line, may be lost due to the suppression of data into a single value for methods such as interval mapping. The hierarchical Bayesian model (HBM) proposed in [18] and further elucidated for QTL analysis in [19] over comes the limitation of a single value for the quantitative trait. The following sections will detail the development of the hierarchical Bayesian model used for this work (section 2 Methods), the simulations that were generated to evaluate the IM against HBM (section 3 Simulations), the results (section 4) and conclusions drawn (section 5).

1.1 Arabidopsis Thaliana

The *Arabidopsis thaliana*, which is illustrated in figure 1, is an angiosperm, a dicot from the mustard (*Brassicaceae*) family. This little plant has become to plant biology what *Drosophila melanogaster* and *Caenorhabditis elegans* are to animal biology. Although this plant has no commercial viability, it has proved to be an ideal organism for the study of plant development.

Fig. 1. Arabidopsis thaliana member of the mustard (*Brassicaceae*) family, which is shown here a grow tray

The main attractions of this plant as a model organism for cellular and molecular biology of flowering plants are, short germination to seed maturation (6 weeks); seed production is prolific and the plant is easily cultivated; large number of mutant lines and genomic resources are available; extensive genetic and physical maps of all 5 chromosomes are readily available; one of the smallest genomes in the plant kingdom with little junk DNA; mutations can be easily generated; and it is self-pollinated so recessive mutations become quickly homozygous.

2 Methods

Hierarchical models have proven to be invaluable in many instances (e.g. Boone *et al.* [18] and Simmons *et al.* [20]). In the case of plant QTL experiments, hierarchical models make the most sense and are flexible enough to adequately model the data. The response, y_{ij} is the numerical value of the quantitative trait for $i = 1,...,L$ and $j = 1,..., n_i$ where L is the number of lines and n_i represents the number of replicates within each line. Each y_{ij} is assumed to be linearly dependent on the genetic composition of the plant, in other words,

$$y_{ij} = \beta_0 + \beta_1 x_{i1} + \beta_2 x_{i2} + ... \beta_M x_{iM} \qquad (1)$$

where $x_{ij} = \begin{cases} 1 & \text{if marker from parent A} \\ 0 & \text{if marker from parent B} \end{cases}$

and M is the number of markers. Most models for QTL experiments involve similar forms of a linear association between the quantitative trait and the markers. However, most QTL models assume that there is only one observation per genotype, or per line, and that the variance is the same within each line [2, 20]. The hierarchical model does not make this assumption of homogeneity of variance and is able to incorporate the replicate information within each line.

The hierarchical model assumes that each y_{ij} is normally distributed with a mean of θ_i and a variance of σ_i^2. The next level of the hierarchy assumes that the θ_i's are normally distributed with a mean of $\beta_0 + \beta_1 x_{i1} + \beta_2 x_{i2} + \dots \beta_M x_{iM}$ and a variance of τ^2. This model allows different variances within each line (σ_i^2) and a variance between the lines (τ^2). The Bayesian paradigm is extremely flexible and easy to incorporate hierarchical structures evident in these experiments. The following prior distributions will be assumed

$$\beta_m \sim N(0,100) \tag{2}$$

$$\tau^2 \sim \text{Inverse-}\chi^2 (2) \tag{3}$$

$$\sigma_i^2 \sim \text{Inverse-}\chi^2 (2). \tag{4}$$

The Inverse-χ^2 (2) is a natural choice for the variances, since it has an infinite variance (Boone *et al.* [22]). The prior distribution for the βs assume that no markers have an effect on the quantitative trait, and forces the data to dictate which markers are most important with respect to the quantitative trait.

Combining this information into a hierarchical model creates a full joint posterior distribution of the form

$$p(\theta,\beta,\sigma^2,\tau^2 \mid y) \propto (\tau^{\tau_0+2+L} \prod_i (\sigma_i^{n_i+\sigma_{0i}+2}))^{-1} \cdot$$
$$\exp\left[-\sum_i \frac{1}{2\sigma_i^2} - \frac{1}{2\tau^2} - \frac{1}{200}\beta'\beta - \frac{1}{2\tau^2}(\theta - X\beta)'(\theta - X\beta) - \sum_i \sum_j \frac{1}{2\sigma_i^2}(y_{ij} - \theta_i)^2\right]. \tag{5}$$

The Gibbs Sampler, a Markov Chain Monte Carlo technique, can generate samples from the full joint posterior distribution in (5) by using the following conditional posterior distributions

$$p(\tau^2 \mid \theta,\beta,\sigma^2,y) \sim Inv-Gamma\left(\frac{L+\tau_0^2}{2}, \frac{(\theta - X\beta)'(\theta - X\beta)+1}{2}\right) \tag{6}$$

$$p(\sigma_i^2 \mid \theta,\beta,\tau^2,y) \sim Inv-Gamma\left(\frac{n_i+\sigma_0^2}{2}, \frac{\sum_j (y_{ij} - \theta_i)^2 +1}{2}\right) \tag{7}$$

$$p(\beta \mid \tau^2, \theta, \sigma^2, y) \sim N\left(\left(\frac{I}{100} + \frac{X'X}{\tau^2}\right)\frac{X'\theta}{\tau^2}, \left(\frac{I}{100} + \frac{X'X}{\tau^2}\right)^{-1}\right) \tag{8}$$

$$p(\theta \mid \tau^2, \beta, \sigma^2, y) \sim N\left(\frac{\dfrac{X_i\beta}{\tau^2} + \dfrac{n_i\overline{y}_i}{\sigma_i^2}}{\dfrac{1}{\tau^2} + \dfrac{n_i}{\sigma_i^2}}, \frac{1}{\dfrac{1}{\tau^2} + \dfrac{n_i}{\sigma_i^2}}\right). \tag{9}$$

This information can be used to find posterior probabilities and ultimately determine which markers are important in controlling the quantitative trait. For example, the posterior probability of interest is the probability of model η_i given the data

$$P(\eta_i \mid D) = \frac{P(D \mid \eta_i)P(\eta_i)}{\displaystyle\sum_{j=1}^{|H|} P(D \mid \eta_i)P(\eta_i)}, \tag{10}$$

where $P(\eta_i)$ is the prior probability of model η_i and $|H|$ is the set of all possible models. Assuming no prior information is known about the markers, equal probability is assigned to each η_i. The quantity $P(D \mid \eta_i)$ in equation (10) is calculated by

$$P(D \mid \eta_i) = \int P(D \mid \omega_i, \eta_i)P(\omega_i \mid \eta_i)d\omega_i, \tag{11}$$

where ω_i is the vector of unknown parameters under model η_i. This integration is computationally intensive, but can be estimated by

$$\int P(D \mid \omega_i, \eta_i)P(\omega_i \mid \eta_i)d\omega_i \approx \frac{1}{t}\sum_{j=1}^{t} P(D \mid \omega_i^{(j)}, \eta_i), \tag{12}$$

where $\omega_i^{(j)}$ are samples from the full posterior distribution under model η_i. Since there are many unknown parameters in this model, a large number of samples from the posterior are recommended. In this research, t was chosen to be 100,000 with a burn-in period of 2,000. The posterior probability of the model given the data can be used to find the activation probability of a marker or region, $P(\beta_j \neq 0 \mid D)$. The activation probability is defined as,

$$P(\beta_j \neq 0 \mid D) = \sum_{|H|} P(\beta_j \neq 0 \mid \eta_i, D)P(\eta_i \mid D). \tag{13}$$

However, to calculate the activation probability for each marker means that 2^M models need to be fit. This can become computationally intensive since most genetic maps have more than 100 markers (M is generally larger than 100). Therefore, a search strategy is needed. We define a conditional search strategy that continues to break the genome into smaller and smaller regions and retains only those regions of importance.

The search strategy first breaks the genome into chromosome regions. In this instance, there are 2^K number of models that need to be evaluated where K is the number of chromosomes. The activation probability for each region is evaluated by

$$P(\kappa_j \neq 0 \mid D) = \sum_{|H|} P(\kappa_j \neq 0 \mid \eta_i, D)P(\eta_i \mid D). \tag{14}$$

Regions with posterior probability larger than 0.5 are regarded as potential QTLs and retained in the model. Once all potential regions are identified, those regions retained are divided in half. For example, in a hypothetical example with 7 chromosomes, the search algorithm would first find which chromosomes make a significant contribution to the QTL by searching through all $2^7 = 128$ possible models and calculating the activation probability for each chromosome. For this example, the following activation probabilities were obtained $C_1 = 0.01$, $C_2 = 0.03$, $C_3 = 0.67$, $C_4 = 0.33$, $C_5 = 0.90$, $C_6 = 0.21$ and $C_7 = 0.84$. Chromosomes 3, 5 and 7 have activation probabilities higher than 0.5 and are kept for further analysis. Dividing these chromosomes in half, there are now six regions to explore (i.e. $2^6 = 64$ models). These regions are defined as C_{31}, C_{32}, C_{51}, C_{52}, C_{71} and C_{72}. The algorithm is rerun and activation probabilities for each of these six regions are calculated. Only those regions with activation probability higher than 0.5 are retained and then divided in half. This algorithm is repeated until the activation probabilities are calculated on individual markers.

3 Simulations

The dataset that was utilized for the X matrix (165 lines x 38 markers) is from a real marker structure, the Bay-0 x Shahdara population created by Oliver Loudet and Sylvian Chaillou [23] which is depicted below in Figure 1.

A genetic map is a map based on the frequencies of recombination between markers during crossover of homologous chromosomes. The greater the frequency of recombination (segregation) between two genetic markers, the farther apart they are assumed to be. Conversely, the higher the frequency of association between the markers, the smaller the physical distance between them [24]. Distance on a genetic map is measured in centiMorgans (cM) which is a relative distance between two markers.

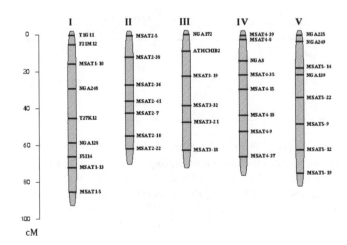

Fig. 2. The genetic map of the *Arabidopsis Thaliana* Bay-O by Shahdara

The genetic map of the Bay-0 x Shahdara population is made up of five chromosomes consisting of approximately 6 to 9 markers each. Marker values (x_i) were set to $x_i = 0$, 0.5 or 1. The marker value $x_i = 0$ came from parent A and the marker $x_i = 1$ came from parent B, whereas, the marker value $x_i = 0.5$ is a missing or unknown value. Ten response values (y_{ij}) were simulated per line around approximately a 26 unit mean (μ) which depended on the QTL location and the actual genotypic trait. Bimodal standard deviations $(\sigma_1$ and $\sigma_2)$ were created around the mean value (μ) for the observations in each line. The response values (y_{ij}) for the one QTL simulation were drawn from a random normal distribution. The marker values of the QTL were simulated as follows,

$$y_{ij} = \mu + 2^*a_i {}^*x_i + \varepsilon_{ij}, \tag{15}$$

where, μ = the underlying true mean and a_i = QTL affect, x_i = 0, 0.5 or 1 and ε_{ij} is random error noise. The random error noise ε_{ij} has standard deviation σ_1 or σ_2 depending on a random draw from a Bernoulli with probability of success 0.5.

In the case of two QTLs, the following equation was used to simulate the QTL value,

$$y_{ij} = \mu + a_1{}^*x_{1i} + a_2{}^*x_{2i} + \varepsilon_{ij} \tag{16}$$

where, μ = the underlying true mean and a_1, a_2 = QTL affects and ε_{ij} is random error noise, as defined previously.

One or two QTLs were arbitrarily placed on or around random markers with two different effects as shown in Table 1 and Table 2. The Bay-0 x Shahdara X-matrix and the simulated response values were run through the two methods introduced in section one to see which one performed better given the different variances created.

Table 1. Simulations for one QTL that were used for this work. The chromosome were the marker of choice is located (ground truth) given the effect and number of QTL's.

Effects	Standard Deviations	
	$\sigma_1 = 2.0, \sigma_2 = 4.5$	$\sigma_1 = 4.2, \sigma_2 = 9.1$
a = 2	Chrom 3 M 18	Chrom 5 M 36
a = 12	Chrom 3 M 15	Chrom 2 M 12

The two methods being evaluated in this sensitivity study have different criteria associated with their respective selection mechanism for QTL markers. The Interval Mapping method uses the Logarithm of the Odds (LOD), shown in equation (17), such that any loci with a score greater than a threshold value is said to be a potential QTL. From [20] it has been stated that a LOD score of 11 or greater are deemed significant; hence a marker selection can be determined.

$$\text{LOD} = -2\ln\left[\frac{\max(\text{likelihood assuming no QTLs})}{\max(\text{likelihood assuming QTL at location})}\right]. \tag{17}$$

Table 2. Simulations for two QTLs that were used for this work. The chromosome were the marker of choice is located (ground truth) given the effect and number of QTL's.

Effects	Standard Deviations	
	$\sigma_1= 1.5, \sigma_2 =2.5$	$\sigma_1= 2.0, \sigma_2 =4.5$
$a_1 = 1$ $a_2 = 2$	Chrom 1 M 5	Chrom 2 M 11
	Chrom 4 M 24	Chrom 5 M 36
$a_1 = 1$ $a_2 = 2$	Chrom 2 M 14	Chrom 1 M 8
	Chrom 3 M 21	Chrom 3 M 18
$a_1 = 2$ $a_2 = 12$	Chrom 1 M 5	Chrom 2 M 11
	Chrom 4 M 24	Chrom 5 M 36

HBM as stated above in the Methods section (section 2) uses a conditional activation probability to determine loci of interest. It is further stated that a conditional activation probabilities of 0.50 or greater is deemed significant. Hence, probability scores of 0.50 are used to determine the loci of markers.

4 Results

The locations of the simulated QTLs illustrated in Tables 1 and 2 were arbitrarily chosen for this study. Small effect sizes (1 and 2) and large effect sizes were chosen to study the sensitivity of the methods to the size of the effect. In a similar fashion, larger and smaller variances were also chosen to study the influence of different variations.

Table 3 clearly indicates that the HBM is able to detect the simulated QTLs. In three instances in Table 3 the HBM also detects adjoining markers, which is not uncommon in QTL analysis. Markers that are close together on a genetic map tend to be highly correlated, so often more than one marker will be detected in a QTL analysis. Table 3 also indicates that the IM algorithm can detect the approximate location of the QTL; however, the IM algorithm tends to choose the general region of the QTL by selecting the marker immediately after the simulated QTL.

Table 4 illustrates the two QTL case where the HBM again is able to detect the correct locations of the QTLs and occasionally picks up adjacent markers. The IM algorithm is only able to detect the larges effect size of 12 in the study. Effect sizes of 1 and 2 are not detected in the IM algorithm.

Table 3. One QTL Summary Table. ([1]Final conditional activation probability. [2]LOD score.)

| Effects | Standard Deviations | | | |
| | Hierarchical Bayesian Method | | Interval Mapping Method | |
	$\sigma_1 = 2.0$ $\sigma_2 = 4.5$	$\sigma_1 = 4.2$ $\sigma_2 = 9.1$	$\sigma_1 = 2.0$ $\sigma_2 = 4.5$	$\sigma_1 = 4.2$ $\sigma_2 = 9.1$
2	Chrom 3 M 18 (1.000)[1] M 19 (0.987)[1]	Chrom 5 M 35 (0.586)[1] M 36 (1.000)[1]	Chrom 3 M 19 (52.49)[2]	Chrom 5 M 37 (17.53)[2]
12	Chrom 2 M 15 (1.000)[1]	Chrom 2 M 11 (0.989)[1] M 12 (1.000)[1]	Chrom 3 M 16 (117.84)[2] Loc 2.5 cM (63.05)[2]	Chrom 2 M 13 (106.39)[2]

Table 4. Two QTL Summary Table. ([1]Final conditional activation probability. [2]LOD score.)

| Effects | Standard Deviations | | | |
| | Hierarchical Bayesian Method | | Interval Mapping Method | |
	$\sigma_1 = 1.5$ $\sigma_2 = 2.5$	$\sigma_1 = 2.0$ $\sigma_2 = 4.5$	$\sigma_1 = 1.5$ $\sigma_2 = 2.5$	$\sigma_1 = 2.0$ $\sigma_2 = 4.5$
1,2	Chrom 1 M 5 (1.000)[1] Chrom 4 M 24 (1.000)[1]	Chrome 2 M 11 (0.997)[1] Chrom 5 M 36 (1.000)[1]	Chrom 4 Loc, 17.5 cM (21.73)[2]	LOD all < 11
1,2	Chrom 2 M14 (0.999)[1] Chrom 3 M 21 (1.000)[1] M 22 (0.543)[1]	Chrom 1 M 8 (0.963)[1] Chrom 3 M18 (1.000)[1] M 19 (0.535)[1]	Chrom 3 Loc, 65 cM (33.94)[2]	Chrom 3 M 19 (16.37)[2]
2,12	Chrom 1 M5 (1.000)[1] Chrom 4 M 24 (1.000)[1]	Chrom 2 M11 (1.000)[1] Chrom 5 M 36 (1.000)[1]	Chrom 4 M25 (88.08)[2]	Chrom 5 M37 (93.55)[2]

5 Conclusions

Many novel approaches have been developed for QTL analysis over the past 25 years. However, most methods assume only one observation per genotype, or equal variances within each genotype. Plant biologists have the luxury of cloning plants and creating replicates within each line. These replicates provide information about the QTL, but also on the variances within each line. Summarizing the replicates into one observation to utilize available software disregards the abundant information provided by the replicates. In these simple simulations of allowing the variances to have two values, we illustrate the importance of incorporating the different variance structures into the analysis. The additional complexity introduced by different variances within lines is easy to incorporate by using a Bayesian hierarchical model and is more appropriate in situations such as this. More extensive simulations to determine power and Type I error are needed, but these initial results are promising.

References

1. Sax, K.: The association of size differences with seed-coat pattern and pigmentation in Phaseolus vulgaris. Genetics 8, 552–560 (1923)
2. Lander, E.S., Botstein, D.: Mapping mendelian factors underlying traits using RFLP linkage maps. Genetics 121, 185–199 (1989)
3. Jansen, R.C.: A General Mixture Model for Mapping Quatitative Loci by Using Molecular Markers. Theoretical and Applied Genetics 85, 252–260 (1992)
4. Zeng, Z.B.: Theoretical basis for separation of multiple linked gene effects in mapping quantitative trait loci. Proceedings of the National Academy of Science USA 90, 10972–10976 (1993)
5. Wright, A.J., Mowers, R.P.: Multiple regression for molecular-marker: quantitative trait data from large F2 populations. Theoretical and Applied Genetics 89, 305–312 (1994)
6. Kearsey, M.J., Hyne, V.: QTL Analysis, A simple marker regression approach. Theoretical and Applied Genetic 89, 698–702 (1994)
7. Wu, W.R., Li, W.M.: A New Approach for Mapping Quantitative Trait Loci Using Complete Genetic Marker Linkage Maps. Theoretical and Applied Genetics 89, 535–539 (1994)
8. Sen, S., Churchill, G.A.: A statistical framework for quantitative trait mapping. Genetics 159, 371–387 (2001)
9. Lamon, E.C., Clyde, M.A.: Accounting for Model Uncertainty in Prediction of Cholophyll A in Lake Okeechobee. Journal of Agricultural Biological and Environmental Statistics 5, 297–322 (2000)
10. Zeng, Z.B.: Precision mapping of quantitative trait loci. Genetics 136, 1457–1468 (1994)
11. Zeng, Z.B., Kao, C.H., Basten, C.J.: Estimating the genetic architecture of quantitative traits. Genetic Research 74, 279–289 (1999)
12. Bao, H.: Bayesian Hierarchical Regression Model to Detect Quantitative Trait Loci. UNCW Thesis (2006)
13. Satagopan, J.M., Yandell, B.S., Newton, M.A., Osborn, T.C.: A Bayesian approach to detect quantitative trait loci using Markov chain Monte Carlo. Genetics 144, 805–816 (1996)
14. Sillanpaa, M.J., Arjas, E.: Bayesian mapping of multiple quantitative trait loci from incomplete inbred line cross data. Genetics 148, 1373–1388 (1998)
15. Sillanpaa, M.J., Corander, J.: Model choice in gene mapping, what and why. Trends in Genetics 18, 301–307 (2002)
16. Xu, S.: Estimating Polygenic Effects Using Markers of the Entire Genome. Genetics 163, 789–801 (2003)
17. Yi, N., Xu, S., Allison, D.B.: Bayesian model choice and search strategies for mapping interacting quantitative trait Loci. Genetics 165, 867–883 (2003)
18. Boone, E.L., Ye, K., Smith, E.P.: Evaluating the Relationship Between Ecological and Habitat Conditions Using Hierarchical Models. Journal of Agriculture, Biological, and Environmental Statistics 10(2), 1–17 (2005)
19. Bjornstad, A., Westad, F., Martens, H.: Analysis of genetic marker-phenotype relationships by jack-knifed partial least squares regression (PLSR). Hereditas 141, 149–165 (2004)

20. Broman, K.W., Wu, H., Sen, Ś., Churchill, G.A.: R/qtl, QTL mapping in experimental crosses. Bioinformatics 19, 889–890 (2003)
21. Simmons, S.J., Piegorsch, W.W., Nitcheva, D., Zeiger, E.: Combining environmental information via hierarchical modeling, an example using mutagenic potencies. Environmentrics 14, 159–168 (2003)
22. Boone, E., Ye, K., Smith, E.P.: Assessment of Two Approximation Methods for Computing Posterior Model Probabilities. Computational Statistics and Data Analysis 48, 221–234 (2005)
23. Loudet, O., Chaillou, S., Camilleri, C., Bouchez, D., Daniel-Vedele, F.: Bay-0 x Shahdara recombinant inbred lines population, a powerful tool for the genetic dissection of complex traits in Arabidopsis. Theoretical and Applied Genetics 104(6-7), 1173–1184 (2002)
24. Lynch, M., Walsh, B.: Genetics and Analysis of Quantitative Traits. Sinauer Associates, Inc., Sunderland, MA (1998)

Using Decision Templates to Predict Subcellular Localization of Protein

Jianyu Shi[1], Shaowu Zhang[2], Quan Pan[2], and Yanning Zhang[1]

[1] School of Computer Science and Engineering,
[2] School of Automation, Northwestern Polytechnical University
710072 Xi'an, China
snake5947@msn.com, {zhangsw,quanpan,ynzhang}@nwpu.edu.cn

Abstract. Theoretical and computational methods for the prediction of protein subcellular localization have been proposed and are developing continuously. Many representations of protein sequence are proposed but a new problem arises: how to organize them together to improve prediction. It is an available solution to serialize multiple representations to single bigger one, but is still hard to avoid calculation error derived from greatly different feature values and causes huge computational burden natively because of high dimensional feature vector. We present a novel method based on decision templates(DT) for such problems in this paper. First, a protein sequence is represented as three new types of feature vectors. Then, the feature vectors are further taken as the inputs of individual SVM classifiers respectively. Finally, the outputs of these classifiers are aggregated by decision templates. The results demonstrate that DT is superior to other methods of subcellular localization prediction.

Keywords: decision templates, subcellular localization prediction, multi-scale energy, moment descriptor, amino acid composition distribution, support vector machines.

1 Introduction

As one of the most important areas in post-genome era, proteomics aims to understand proteins' potential roles, elucidate their interaction in a cellular context, and further make the corresponding functional annotation. Determination of subcellular location of proteins is of essence and importance to their functional annotation. However, the biological experiment of protein subcellular localization will be hard to meet the demands due to both time-consuming and expensive cost. Therefore, to bridge this gap, there is a need to develop more effective methods.

During the last decade, many theoretical and computational methods were developed in an attempt to predict subcellular localization of protein. Originally, Nakashima and Nishikawa represented protein sequence with amino acid composition (AAC) and indicated that intracellular and extracellular proteins are significantly different in this representation[1]. The subsequent studies showed that AAC is closely related to protein subcellular localizations[2-4]. Although AAC can represent the major

J.C. Rajapakse, B. Schmidt, and G. Volkert (Eds.): PRIB 2007, LNBI 4774, pp. 71–83, 2007.
© Springer-Verlag Berlin Heidelberg 2007

information of sequence, it always ignores the sequence-order and structure information of protein. Hence two sequences, different in function and localization but similar in AAC, may be predicted as the same localization. To represent protein sequence better, some improved representations have been proposed[5-13].

So many representation methods give us lots of choices to predict subcellular localization of protein, but also push a new problem out: how to organize them together to achieve better prediction than what any single method can do.

One available solution is to serialize multiple representations to single bigger one, in other words, combined feature. In this way, it is proved that the prediction obtained can be better[5]. However, there are still two problems need to be solved. First, it is hard to avoid calculation error because feature values derived from multiple representations are always greatly different. Secondly, the combined feature vector is always of high dimension which bring out huge computational burden natively. Chou's pseudo amino acid composition (PseAA) provides a good way to the former problem[5,6,14]. However, the latter one is still the obstacle to build the online application of subcellular localization prediction which is already the tendency of this research area[15].

To solve above problems, some methods are presented by performing majority vote algorithm to the outputs of several classifiers which use different feature representations as inputs respectively. In this paper, we introduce three types of feature representation methods, and then use a multiple classifier fusion scheme, decision templates (DT), to perform the prediction of protein subcellular localization.

2 Database

In this paper, we use several databases which are presented in [5] and [15] respectively. These databases vary with the version of SWISS-PROT, the type of locations, the count of subcellular localization and total number of sequences. The similarities between sequences are all less than 80%.

Table 1. The database used in this paper

Database	CH01-J[5]	CH01-I[5]	HO06-P[15]
SWISS-PROT	Release 35.0	Release 35.0	Release 42.0
Location			
chloroplast(ch)	145	112	449
cytoplasm(cy)	571	761	1411
cytoskeleton(cs)	34	19	—
endoplasmic reticulum (er)	49	106	198
extracellular (ex)	224	95	843
Golgi apparatus (go)	25	4	150
lysosome (ly)	37	31	—
mitochondria (mi)	84	163	510
nucleus (nu)	272	418	837
peroxisome (pe)	27	23	157
plasma membrane (pm)	699	762	1238
vacuole (va)	24	—	63
Total	2191	2494	5856

Remarkably, the former database is composed of a training set and a testing set; the latter one contains three databases of which many locations and sequences overlap, therefore only the plant database are used here as a result of holding the most locations. Consequently, there are three databases used in these papers, which are shortly denoted by CH01-J, CH01-I, and HO06-P and listed in Table 1.

3 Representation Methods

Without loss of generality, we assume that there are N protein sequences in the dataset, let L_k be the length of the k th sequence p_k, and α_i be the i th element of 20 natural amino acids represented by English letters A, C, D, E, F, G, H, I, K, L, M, N, P, Q, R, S, T, V, W and Y respectively.

3.1 Multi-scale Energy

According to amino acid composition, the protein sequence p_k can be characterized as a 20-D feature vector:

$$AAC_k = \left[c_1^k, \cdots, c_i^k, \cdots, c_{20}^k \right], \quad k = 1, \cdots, N , \tag{1}$$

where $c_i^k = n^i / L_k$ is the normalized occurrence frequency of amino acid α_i, and n^i is the count of α_i appearing in sequence p_k.

However, it is not sufficient to represent a specific protein sequence only based on AAC. Consequently, there is a need to improve AAC or develop other representations of protein sequence to deal with such case.

Using discrete wavelet transform (DWT) and Mallat fast algorithm[16], we proposed the multi-scale energy (MSE) representation to improve AAC[8]. Each protein sequence can be firstly coded into digital signal by mapping all amino acid residues of protein sequence to the corresponding numerical value according to one of amino acid indices[17]. Here we choose the hydrophilicity index HOPT810101 from amino acid index database. Hence, such a coded protein sequence can be treated as a digital signal and further processed by DWT.

According to DWT and Mallat fast algorithm, the fine-scale and large-scale information of a protein hydrophilicity signal can be simultaneously investigated by projecting the mapped digital signal onto a set of wavelet basis functions with various scales. Here, the wavelet basis function used is symlet wavelet. The features extracted from the wavelet-based multi-resolution information, can discriminate different types of protein signals. Consequently, sequence p_k can be characterized as a (m+1)-D feature vector of multi-scale energy(MSE):

$$MSE_k = \left[d_1^k, \cdots, d_j^k, \cdots, d_m^k, a_m^k \right] . \tag{2}$$

Here m is the coarsest scale of decomposition, d_j^k is the root mean square energy of the wavelet detail coefficients in the corresponding j th scale, and a_m^k is the root mean square energy of the wavelet approximation coefficients in the scale m. The energy factors d_j^k and a_m^k are defined as

$$d_j^k = \sqrt{\frac{1}{N_j}\sum_{n=0}^{N_j-1}[u_j^k(n)]^2} \;,\; a_m^k = \sqrt{\frac{1}{N_m}\sum_{n=0}^{N_m-1}[v_m^k(n)]^2} \;,\; j=1,2,\cdots,m\,, \tag{3}$$

where N_j is the number of the wavelet detail coefficients, N_m is the number of the wavelet approximation coefficients, $u_j^k(n)$ is the nth detail coefficient in the corresponding jth scale, and $v_m^k(n)$ is the nth approximation coefficient in the scale m. For the protein sequence p_k with length L_k, m equals $INT\left(\log_2\left(L_k\right)\right)$.

Obviously, MSE contains the approximation and detail information of protein signal which reflect sequence-order effects. In order to get better representation, we combine MSE with AAC and construct the following (20+m+1)-D feature vector $\bar{x}_k$ to represent sequence p_k.

$$\bar{x}_k = [c_1^k,\cdots,c_i^k,\cdots,c_{20}^k,\lambda_1^k,\cdots,\lambda_j^k,\cdots,\lambda_m^k,\lambda_{m+1}^k]^T\,, \tag{4}$$

where $\lambda_j^k = d_j^k, \lambda_{m+1}^k = a_m^k, j = 1,\cdots,m$.

3.2 Moment Descriptor

Considering the order amino acid, we proposed a new feature representation method, called moment descriptor (MD)[12].

Firstly, instead of using the direct definition of AAC in (1), we calculate c_i^k by introducing position indicator $\Delta_{i,j}^k$ as follows:

$$c_i^k = \frac{1}{L_k}\sum_{j=1}^{L_k}\Delta_{i,j}^k\,, \tag{5}$$

$$\Delta_{i,j}^k = \begin{cases} 1 & \text{if } \alpha_i \text{ is present at position } j \text{ in } p_k \\ 0 & \text{if } \alpha_i \text{ is NOT present at position } j \text{ in } p_k \end{cases}. \tag{6}$$

Obviously, (5) is the sampled statistical mean (raw moment) of position indicator. Hence, we choose it as the first MD of protein sequence.

Secondly, considering the position of amino acid α_i in sequence p_k, we define another feature for amino acid α_i:

$$m_i^k = \frac{1}{L_k}\sum_{j=1}^{L_k}\left(\Delta_{i,j}^k\cdot j\right). \tag{7}$$

where m_i^k represents mean of position of α_i. We choose it as the second MD.

Thirdly, the sampled variance v_i^k of position of amino acid α_i in sequence p_k is considered:

$$v_i^k = \frac{1}{L_k} \sum_{j=1}^{L_k} \left(\Delta_{i,j}^k \cdot j - m_i^k \right)^2 . \tag{8}$$

where v_i^k represents the second-order central moment of position of amino acid α_i in sequence p_k. We choose it as the third MD of protein sequence.

Eventually, we get a combined feature vector for sequence p_k by serializing above three moment descriptors:

$$\bar{x}_k = \left[c_1^k, \cdots, c_i^k, \cdots, c_{20}^k, m_1^k, \cdots, m_i^k, \cdots, m_{20}^k, v_1^k, \cdots, v_i^k, \cdots, v_{20}^k \right]^T , k = 1, \cdots, N . \tag{9}$$

3.3 Amino Acid Composition Distribution

As we know, the tertiary structure of protein is always composed of several secondary structure units, such as α-helix or β-sheet. Considering this fact, we present a new representation, amino acid composition distribution (AACD) which divide a protein sequence p_k equally into multiple segments and then calculate AAC of each segment in series[13]. So the sequence p_k can be represented as the following formula:

$$AACD_n^k = \left\{ \begin{matrix} c_{1,1}^k & \cdots & c_{1,m}^k & \cdots & c_{1,n}^k \\ \cdots & \cdots & c_{i,m}^k & \cdots & \cdots \\ c_{20,1}^k & \cdots & c_{20,m}^k & \cdots & c_{20,n}^k \end{matrix} \right\}_{20 \times n} , \tag{10}$$

where n is the count of segments, $\left[c_{1,m}^k, \cdots, c_{i,m}^k, \cdots c_{20,m}^k \right]^T$ is the AAC of the m th segment of p_k, and $c_{i,m}^k$ is define as:

$$c_{i,m}^k = n \cdot t_{i,m}^k / L_k , \quad m = 1, \cdots, n, \quad i = 1, \cdots, 20, \tag{11}$$

where $t_{i,m}^k$ is the count of α_i appearing in the m th segment of sequence p_k.

In order to be the input of classifier, this representation of protein sequence p_k is turned to a feature vector as the following:

$$\bar{x}_k = \left[c_{1,1}^k, \cdots, c_{1,n}^k, \cdots, c_{i,1}^k, \cdots, c_{i,n}^k, \cdots, c_{20,1}^k, \cdots, c_{20,n}^k \right]^T , \quad k = 1, \cdots, N . \tag{12}$$

4 Classification and Assessment

4.1 Support Vector Machines

When the representation of protein sequence is set, next step is just to choose a classifier to perform the prediction of subcellular localization. Many types of classifiers which have been applied to such prediction, such as neural network[2], covariant discriminant algorithm[5], fuzzy KNN[18] and support vector machines(SVM)[9,10,15].

In these classifiers, SVM has been more broadly applied to such prediction due to its good performance of classification. SVM was originally designed for binary classification[19] while such prediction is M-class classification. Usually, we can construct M-class SVMs to solve such problem based on the binary class SVM. That is an ongoing research issue. Extensive experiments have shown that "One-Versus-Rest" (OVR)[19], "One-Versus-One" (OVO)[20]and "Directed Acyclic Graph" (DAG)[21]are practical[8,12,22,23]. Because of its convenient usage, OVO is used in this paper.

To perform the prediction, the SVM software, LIBSVM, is used, and can be freely downloaded from http://www.csie.ntu.edu.tw/~cjlin/libsvm/ for academic research[22]. In addition, we do the training only with the RBF kernel in all experiments.

4.2 Multiple Classifier System and Decision Template

If we have different feature sets, different training sets, different classification methods or different training sessions, then multiple classifier system (MCS) is proposed to improve classification or prediction accuracy by combining the outputs of a set of classifiers[24,25]. Various combined schemes for MCS can be grouped into three basic main categories according to their architecture: serial (cascading), hierarchical and parallel architectures[25]. Most combination schemes in the literatures belong to the parallel MCS which involves two kinds of aggregated schemes. The one is known as selection rule[26] of which clustering and selection, dynamic classifier selection with local accuracy are always used. The other is referred to as fusion rule which includes lots of related algorithms, for example, majority vote[24].

If the set of classifiers is fixed, the problem focuses on the aggregated function or rule. It is also possible to use a fixed combiner and optimize the set of input classifiers. We only consider the former one in this paper. Due to the higher efficiency and flexibility, we select parallel MCS and use the fusion rule decision templates (DT). DT is non-sensitive to poorly trained individual classifiers, and can achieve good and stable performance without strict probabilistic conditions[27].

Let $\bar{x} \in \Re^n$ be a feature vector (a representation of a protein sequence), $\{\omega_1,...,\omega_j,...,\omega_M\}$ be the label set of M classes, and $\{e_1,...,e_i,...,e_L\}$ be the set of L classifiers. We denote the output of the i-th classifier as $D_i(\bar{x}) = \left[d_{i,1}(\bar{x}),...,d_{i,j}(\bar{x}),...,d_{i,M}(\bar{x}) \right]^T$, where $d_{i,j}(\bar{x})$ is the degree of "support" given by classifier e_i to the hypothesis that $\bar{x}$ comes from class ω_j. The outputs of L classifiers can be organized in a decision profile (DP) as the matrix [27]:

$$DP(\bar{x}) = \begin{bmatrix} d_{1,1}(\bar{x}) & \cdots & d_{1,j}(\bar{x}) & \cdots & d_{1,M}(\bar{x}) \\ & & \vdots & & \\ d_{i,1}(\bar{x}) & \cdots & d_{i,j}(\bar{x}) & \cdots & d_{i,M}(\bar{x}) \\ & & \vdots & & \\ d_{L,1}(\bar{x}) & \cdots & d_{L,j}(\bar{x}) & \cdots & d_{L,M}(\bar{x}) \end{bmatrix}, \tag{13}$$

where the components $d_{i,j}(\bar{x})$ can be regarded as an estimate of the posterior probability $P_i(\omega_j \mid \bar{x})$ produced by classifier e_i for class ω_j and the given $\bar{x}$, $i = 1, 2, ..., L$, $j = 1, 2, ..., M$.

Let Z be the crisp labeled training dataset. The decision template for class ω_j denoted DT_j can be regarded as the expected $DP(\bar{x})$ for class ω_j:

$$DT_j = \frac{1}{N_j} \sum_{\substack{\bar{z}_k \in \omega_j \\ \bar{z}_k \in Z}} DP(\bar{z}_k), j = 1, \cdots M. \tag{14}$$

where $\bar{z}_k \in \Re^n$ is a feature vector of the training dataset, N_j is the number of samples of Z from class ω_j, and DT_j is an $L \times M$ matrix.

For a tested feature vector $\bar{x} \in \Re^n$, we can calculate the squared Euclidean distance between $DP(\bar{x})$ and each DT_j

$$d_E\left(DP(\bar{x}), DT_j\right) = \frac{1}{L \cdot c} \sum_{i=1}^{M} \sum_{k=1}^{L} \left(d_{k,i}(\bar{x}) - dt_j(k,i)\right)^2, \tag{15}$$

where $dt_j(k,i)$ is the k, i-th element in decision template DT_j.

Then, we can get the find predicted class label of $\bar{x}$ by:

$$j^*(\bar{x}) = \arg\max_j \left(1 - d_E\left(DP(\bar{x}), DT_j\right)\right), j = 1, \cdots, M. \tag{16}$$

4.3 Prediction of Assessment

To make a fair and full comparison, jackknife test are used to CH01-J and independent test to CH01-I and 5-fold cross validation (5CV) to HO06-P respectively.

During the process of jackknife test, each protein in training dataset is singled out in turn as a test sample, and the remaining are used as training samples. For independent test, proteins in training dataset are used as training samples, and those in independent test dataset are used as test samples. The quality of independent test indicates the ability of generalization of prediction system. To assess the quality of jackknife and independent test, the total prediction accuracy is always used and defined as:

$$Q = \frac{1}{N} \sum_{\omega=1}^{\Omega} p(\omega) .$$ (17)

According to 5CV procedure, the dataset is split randomly and equally into 5 subsets. In turn, we take each subset as the testing set to evaluate the prediction, and use the rest subsets to build classification modal, in other words, to do the training. The average and the standard deviation of the accuracies of all evaluations are used to indicate the performance of prediction, and are defined respectively as:

$$\bar{Q} = \sum_{i=1}^{k} Q_i \Big/ k, S = \sqrt{\sum_{i=1}^{k} \left(Q_i - \bar{Q} \right)^2 \Big/ (k-1)}, i = 1, \cdots, k, \quad k = 5 .$$ (18)

where Q_i is the accuracy of the i th evaluation and k is the count of cross-validation.

Besides the total accuracy or average accuracy, the sensitivity ($Sens^i$), the specificity ($Spec^i$) and the Matthews correlation coefficient (MCC^i) of each location are also used to assess the wider performance of prediction:

$$Sens^i = \frac{p_i}{\left(p_i + u_i \right)}, Spec^i = \frac{p_i}{\left(p_i + o_i \right)},$$

$$MCC^i = \frac{p_i n_i - u_i o_i}{\sqrt{\left(p_i + u_i \right)\left(p_i + o_i \right)\left(n_i + u_i \right)\left(n_i + o_i \right)}}$$ (19)

5   Experiment and Discussion

Each of protein sequences is firstly represented as three types of feature vectors which are corresponding to MSE, MD, and AACD respectively. Then, the proposed feature representations are taken as the inputs of three multi-class SVM classifiers respectively. Finally, the prediction is performed by fusing the outputs of these individual SVM classifiers with decision templates rule.

5.1   Comparison with the Former Methods

In order to validate the effectiveness of decision templates and make fair comparisons, we perform DT on all selected databases, use jackknife test to CH01-J, independent test to CH01-I, and 5CV test to HO06-P respectively. The results are shown in Table 2 and Table 3 respectively.

We can see that DT wins the best accuracies and achieves the lowest standard deviations. To dataset CH01-J, the accuracy achieved by DT is 83.75%, and 10.72%, 16.07% and 10.18% higher than those achieved by [5,6,14], respectively. To dataset CH01-I, the accuracy achieved by DT is 88.41%, and 7.54%, 14.55% and 8.62% higher than those achieved by [5,6,14], respectively. To dataset HO06-P, the accuracy

Table 2. The results of the comparison with the former methods for database Chou

Methods	CH01-J	CH01-I
	Jackknife(%)	Independent(%)
Chou[5]	73.03	80.87
Pan[6]	67.68	73.86
Xiao[14]	73.57	79.79
Our(DT)	83.75	88.41

Table 3. The results of the comparison with the former methods for database HO06-P by 5CV

Loc	PORST[28]			MultiLoc[15]			DT		
	Sens	Spec	MCC	Sens	Spec	MCC	Sens	Spec	MCC
ch	0.49	0.58	0.50	0.88	0.85	0.85	0.85	0.89	0.86
cy	0.40	0.70	0.42	0.68	0.85	0.70	0.87	0.75	0.73
er	0.21	0.11	0.11	0.72	0.54	0.61	0.61	0.86	0.71
ex	0.74	0.70	0.67	0.68	0.81	0.70	0.86	0.86	0.83
go	0.02	0.13	0.04	0.75	0.41	0.54	0.71	0.82	0.76
mi	0.65	0.53	0.54	0.85	0.81	0.81	0.74	0.77	0.73
nu	0.59	0.60	0.53	0.82	0.75	0.75	0.80	0.82	0.78
pe	0.47	0.16	0.24	0.71	0.34	0.47	0.39	0.79	0.55
pm	0.81	0.75	0.72	0.74	0.89	0.77	0.93	0.90	0.89
va	0.13	0.06	0.07	0.70	0.20	0.36	0.48	0.75	0.59
Q(%)	57.50			74.60 ± 0.80			82.68 ± 0.38		

achieved by DT is 82.68%, and 25.18% and 8.08% higher than those achieved by [28] and [15], respectively. In addition, DT achieves the lowest standard deviation.

5.2 The Performance Analysis of DT

In order to analyze the performance of DT, we make a contrast with the "single best" classifier and the "oracle". The "single best" classifier is referred to as the member classifier which achieves the best accuracy. The "oracle" is such procedure which assign the correct class label to $\bar{x}$ as long as one individual classifier produces the correct class label of $\bar{x}$ [27]. The result of "oracle" can reflect the maximum bound of the classification ability of MCS. The results of Chou-J, Chou-I and HO06-P are listed in Table 4, 5, and 6 respectively.

Firstly, these results show that DT fusion rule can achieve better total accuracy and lower standard deviation than the "single best" classifier. Secondly, the sensitivity, the specificity and MCC of each location are always improved in most cases. Especially, the performances of prediction on those locations which have minority protein sequences are improved greatly. Finally, the results of "oracle" show that the fusion of MSE, MD and AACD can represent protein sequence better than any single feature representation because the maximum bound of the classification ability of MCS for three dataset are 89.14%, 93.42 and 90.86%, respectively.

Furthermore, we also compare DT with other popular aggregated rules which are majority vote (MV) [24] and dynamic classifier selection with local accuracy (DCS_LA) [26]. The results are shown in Table 7.

Table 4. The result of multi-classifier fusion for the Chou-J dataset

Loc	Single Best			Decision Template			Oracle
	Sens	Spec	MCC	Sens	Spec	MCC	Sens
ch	0.79	0.82	0.79	0.75	0.88	0.80	0.86
cy	0.91	0.76	0.76	0.93	0.77	0.78	0.97
cs	0.44	0.94	0.64	0.47	1.00	0.68	0.56
er	0.39	0.83	0.56	0.41	0.95	0.62	0.55
ex	0.76	0.77	0.74	0.75	0.80	0.74	0.86
go	0.20	0.83	0.40	0.28	0.78	0.46	0.40
ly	0.54	0.80	0.65	0.59	0.73	0.65	0.76
mi	0.32	0.71	0.46	0.43	0.82	0.58	0.48
nu	0.88	0.72	0.76	0.88	0.74	0.78	0.93
pe	0.26	1.00	0.51	0.26	0.44	0.33	0.26
pm	0.94	0.94	0.91	0.96	0.96	0.94	0.98
va	0.29	1.00	0.54	0.29	0.88	0.50	0.46
Q(%)	82.15			83.75			89.14

Table 5. The result of multi-classifier fusion for the Chou-I dataset

Loc	Single Best			Decision Template			Oracle
	Sens	Spec	MCC	Sens	Spec	MCC	Sens
ch	0.67	0.78	0.71	0.74	0.81	0.77	0.86
cy	0.91	0.87	0.84	0.93	0.91	0.89	0.96
cs	1.00	0.95	0.97	1.00	0.95	0.97	1.00
er	0.89	0.96	0.92	0.93	0.98	0.95	0.94
ex	0.88	0.74	0.80	0.86	0.81	0.83	1.00
go	0.50	1.00	0.71	0.50	1.00	0.71	0.50
ly	0.94	0.97	0.95	1.00	0.97	0.98	1.00
mi	0.18	0.85	0.37	0.29	0.89	0.49	0.53
nu	0.84	0.85	0.82	0.87	0.87	0.84	0.94
pe	0.39	1.00	0.62	0.61	0.78	0.69	0.65
pm	0.98	0.84	0.86	0.99	0.87	0.89	0.99
va	-	-	-	-	-	-	-
Q(%)	85.49			88.41			93.42

Table 6. The result of multi-classifier fusion for the HO06-P dataset

Loc	Single Best			Decision Template			Oracle
	Sens	Spec	MCC	Sens	Spec	MCC	Sens
ch	0.84	0.83	0.82	0.85	0.89	0.86	0.90
cy	0.81	0.71	0.66	0.87	0.75	0.73	0.94
er	0.55	0.81	0.65	0.61	0.86	0.71	0.69
ex	0.84	0.79	0.77	0.86	0.86	0.83	0.95
go	0.57	0.83	0.68	0.71	0.82	0.76	0.76
mi	0.72	0.75	0.71	0.74	0.77	0.73	0.93
nu	0.75	0.79	0.72	0.80	0.82	0.78	0.89
pe	0.32	0.85	0.51	0.39	0.79	0.55	0.45
pm	0.87	0.84	0.81	0.93	0.90	0.89	0.97
va	0.40	0.81	0.56	0.48	0.75	0.59	0.48
Q(%)	78.09 ± 1.01			82.68 ± 0.38			90.86

Table 7. The results of the comparison with other aggregated rules

Methods	CH01-J	CH01-I	HO06-P
	Jackknife(%)	Independent(%)	5CV(%)
MV	83.02	88.05	80.28 ± 0.53
DCS_LA (k=10)	83.20	88.01	80.67 ± 0.49
DT	83.75	88.41	82.68 ± 0.38

These results also show that DT fusion rule can achieve better total accuracy and lower standard deviation than both MV and DCS_LA rules.

As above described, DT is an effective and robust method to subcellular localization prediction.

6 Conclusion

In this paper, we have presented three types of representation methods, MSE, MD and AACD, and then performed prediction of protein subcellular localization by using the parallel MCS. Instead of serializing the proposed representations of protein sequence to single bigger one, the presented method integrates them together by DT fusion rule. DT aggregates the outputs of three individual SVM classifiers which take those representations as the inputs respectively. Compared with other prediction methods, other aggregated rules and the single best classifier, the results show that DT achieves better prediction of subcellular localization and is more effective and robust to subcellular localization prediction. In addition, DT also avoids huge computational burden increased by high dimension derived from the serialization of multiple representations. Consequently, DT can be applied to develop the application of subcellular localization prediction.

Acknowledgments. The authors would like to thank Prof. Guo-Ping Zhou for his critical and constructive comments and suggestions. This paper was supported in part by the National Natural Science Foundation of China (No. 60372085 and 60634030) and the Technological Innovation Foundation of Northwestern Polytechnical University (No. KC02).

References

1. Nakashima, H., Nishikawa, K.: Discrimination of Intracellular and Extracellular Proteins Using Amino Acid Composition and Residue-Pair Frequencies. J. Mol. Biol. 238, 54–61 (1994)
2. Reinhardt, A., Hubbard, T.: Using Neural Networks for Prediction of the Subcellular Localization of Proteins. Nucleic Acids Research 26, 2230–2236 (1998)
3. Chou, K.C., Elrod, D.: Protein Subcellular Localization Prediction. Protein Eng. 12, 107–118 (1999)
4. Hua, S.J., Sun, Z.R.: Support Vector Machine Approach for Protein Subcellular Localization Prediction. Bioinformatics 17, 721–728 (2001)
5. Chou, K.C.: Prediction of Protein Cellular Attributes Using Pseudo-Amino Acid Composition. Proteins: Struct. Funct. Genet. 43, 246–255 (2001)

6. Pan, Y.X., Zhang, Z.Z., Guo, Z.M., Feng, G.Y., Huang, Z., He, L.: Application of Pseudo Amino Acid Composition for Predicting Protein Subcellular Location: Stochastic Signal Processing Approach. Journal of Protein Chemistry 22, 395–402 (2003)

7. Gao, Y., Shao, S.H., Xiao, X., Ding, Y.S., Huang, Y.S., Huang, Z.D., Chou, K.C.: Using Pseudo Amino Acid Composition to Predict Protein Subcellular Location: Approached with Lyapunov Index, Bessel Function, and Chebyshev Filter. Amino Acids 28, 373–376 (2005)

8. Shi, J.Y., Zhang, S.W., Pan, Q., Cheng, Y.M., Xie, J.: Prediction of Protein Subcellular Localization by Support Vector Machines Using Multi-Scale Energy and Pseudo Amino Acid Composition. Amino Acids 33, 69–74 (2007)

9. Park, K.J., Kanehisa, M.: Prediction of Protein Subcellular Locations by Support Vector Machines Using Compositions of Amino Acids and Amino Acid Pairs. Bioinformatics 19, 1656–1663 (2003)

10. Cui, Q., Jiang, T., Liu, B., Ma, S.: Esub8: A Novel Tool to Predict Protein Subcellular Localizations in Eukaryotic Organisms. BMC Bioinformatics 5, 66–72 (2004)

11. Bhasin, M., Raghava, G.P.S.: Eslpred: SVM-Based Method for Subcellular Localization of Eukaryotic Proteins Using Dipeptide Composition and Psi-Blast. Nucl. Acids Res. 32, W414–W419 (2004)

12. Shi, J.Y., Zhang, S.W., Liang, Y., Pan, Q.: Prediction of Protein Subcellular Localizations Using Moment Descriptors and Support Vector Machine. In: Rajapakse, J.C., Wong, L., Acharya, R. (eds.) PRIB 2006. LNCS (LNBI), vol. 4146, pp. 105–114. Springer, Heidelberg (2006)

13. Shi, J.Y., Zhang, S.W., Pan, Q., Zhou, G.-P.: Amino Acid Composition Distribution: A Novel Sequence Representation for Prediction of Protein Subcellular Localization. In: The 1st IEEE International Conference on Bioinformatics and Biomedical Engineering, pp. 115–118. IEEE Computer Society Press, Los Alamitos (2007)

14. Xiao, X., Shao, S.H., Ding, Y.S., Huang, Z.D., Huang, Y., Chou, K.C.: Using Complexity Measure Factor to Predict Protein Subcellular Location. Amino Acids 28, 57–61 (2005)

15. Höglund, A., Dönnes, P., Blum, T., Adolph, H.-W., Kohlbacher, O.: Multiloc: Prediction of Protein Subcellular Localization Using N-Terminal Targeting Sequences, Sequence Motifs and Amino Acid Composition. Bioinformatics 22, 1158–1165 (2006)

16. Mallat, S.: A Wavelet Tour of Signal Processing, 2nd edn. Academic Press, London (1999)

17. Kawashima, S., Ogata, H., Kanehisa, M.: AAindex: Amino Acid Index Database. Nucleic Acids Research 27, 368–369 (1999)

18. Huang, Y., Li, Y.D.: Prediction of Protein Subcellular Locations Using Fuzzy K-NN Method. Bioinformatics 20, 21–28 (2004)

19. Vapnik, V.: Statistical Learning Theory. Wiley, New York (1998)

20. Kreßel, U.H.: Pairwise Classification and Support Vector Machines. In: Schölkopf, B., Burges, C.J., Smola, A.J. (eds.) Advances in Kernel Methods: Support Vector Learning, pp. 255–268. MIT Press, Cambridge, MA (1999)

21. Platt, J., Cristianini, N., Shawe-Taylor, J.: Large Margin Dags for Multiclass Classification. Advances in Neural Information Processing Systems 12, 547–553 (2000)

22. Hsu, C., Lin, C.J.: A Comparison of Methods for Multi-Class Support Vector Machines. IEEE Transactions on Neural Networks 13, 415–425 (2002)

23. Rifin, R., Klautau, A.: In Defense of One-Vs-All Classification. Journal of Machine Learning Research 5, 101–141 (2004)

24. Kittler, J., Hatef, M., Duin, R., Matas, J.: On Combining Classifiers. IEEE Transactions on Pattern Analysis and Machine Intelligence 20, 226–239 (1998)

25. Jain, A.K., Duin, R.P.W., Mao, J.: Statistical Pattern Recognition: A Review. IEEE Transactions on Pattern Analysis and Machine Intelligence 22, 4–37 (2000)
26. Kuncheva, L.I.: Switching between Selection and Fusion in Combining Classifiers: An Experiment. IEEE Transactions on Systems, Man, and Cybernetics, Part B 32, 146–156 (2002)
27. Kuncheva, L.I., Bezdek, J.C., Duin, R.: Decision Templates for Multiple Classifier Fusion: An Experimental Comparison. Pattern Recognition 34, 299–314 (2001)
28. Nakai, K., Horton, P.: Psort: A Program for Detecting the Sorting Signals of Proteins and Predicting Their Subcellular Localization. Trends Biochem. Sci. 24, 34–36 (1999)

Generalized Schemata Theorem Incorporating Twin Removal for Protein Structure Prediction

Md Tamjidul Hoque, Madhu Chetty, and Laurence S. Dooley

Gippsland School of Information Technology (GSIT)
Monash University, Churchill VIC 3842, Australia
{Tamjidul.Hoque,Madhu.Chetty}@infotech.monash.edu.au,
lsdaussie@ieee.org

Abstract. The schemata theorem, on which the working of Genetic Algorithm (GA) is based in its current form, has a fallacious selection procedure and incomplete crossover operation. In this paper, generalization of the schemata theorem has been provided by correcting and removing these limitations. The analysis shows that similarity growth within GA population is inherent due to its stochastic nature. While the stochastic property helps in GA's convergence. The similarity growth is responsible for stalling and becomes more prevalent for hard optimization problem like *protein structure prediction* (PSP). While it is very essential that GA should explore the vast and complicated search landscape, in reality, it is often stuck in local minima. This paper shows that, removal of members of population having certain percentage of similarity would keep GA perform better, balancing and maintaining convergence property intact as well as avoids stalling.

Keywords: Schemata theorem, twin removal, protein structure prediction, similarity in population, hard optimization problem.

1 Introduction

Protein structure prediction (PSP) using lattice model is regarded as a very hard optimization problem. This is because the prediction using lattice model is proven to be NP-complete [1],[2] and the number of possible valid (i.e., self avoiding walk) conformation is astronomical [3], [4]. We have chosen Genetic Algorithm (GA) as a vehicle for providing solution to the *protein structure prediction* (PSP) problem for its performance in various domains [5], [6], [7], [8], [9], [10], [11], [12], [13], [14]. Crossover, regarded as the key operation of GA, is also being adapted by almost all other promising search approaches [15],[16],[5],[17],[12]. It is considered as the potential operation that can build a promising conformation by cutting and joining the potential sub-parts of more than one conformation. In some cases, the GA population strategy is also being adapted by other approaches. While GA performance is generally very effective it can sometimes stall [18] in a hard optimization problem [19] like PSP with the protein sequences having length above, say 30 [12]. Thus, like other promising approaches, GA too cannot ensure the final generation to contain an

J.C. Rajapakse, B. Schmidt, and G. Volkert (Eds.): PRIB 2007, LNBI 4774, pp. 84–97, 2007.

optimal solution. Even, effective [18],[20] elitism can become ineffectual for PSP problem. This problem is so difficult that unlike other type of problems, a mere application of any of the known approaches will not provide improved results. Therefore, in this paper, we present the generalization of the schemata theorem by incorporating twin removal which is necessary to overcome the limitations of the GA and show the impact of this generalization upon GA operation in order to secure more accurate and efficient PSP solutions. To achieve this, in the initial stage we revisit the idea of identical chromosomes (twins) in the population and relax the concept to embrace similar (strongly-correlated) chromosomes. This helps to generalize the schemata theorem as well as to find the percentage of similarity within the population that can keep in GA optimum search condition.

2 Twins in GA Population

The *schemata theorem* as the basis of a GA, has had its critics as evidenced in [21], [22]. The mathematical derivations in relation to the schemata theorem supports that the schemata with above average fitness values would most likely be sustained as the generations proceed and consequentially the similarity [23], [24], [25],[26],[27] grows within the population. This means that although we can set the crossover rate to a desired value, in many cases, the operation generates no variation due to the similarity. Earlier, it was observed [28] that due to the 'stochastic error' associated with GA's genetic operators, the genetic algorithm tends to converge to a single solution. This can raise two different issues. First, there are certain applications where search interest is not for one but several solutions [29], such as to find Pareto front on a problem using multi-objective optimization application. Second, convergence to a single solution means the search becomes stagnant which can be due to the population losing its diversity. This phenomena is termed as 'genetic drift' [29], [30] due to which, in hard optimization problem such as PSP, the search space is extremely convoluted. It can cause the aforementioned stall effect which could be devastating. The searches can get stuck in local minima without exploring much of the vast space.

The existence of twins and the requirement for their removal in a GA is not new. This matter appears as diversity issue in literature as a result of growth of the twins. The growth of twins was considered [26] in evaluating the cost of duplicate or identical chromosomes which suggested starting each chromosome with different patterns to avoid twins. However, if twin growth is inherent in a GA search, then the effect of initialization using different patterns will quickly decline after starting. In [23], [31], it was advocated that if a population comprised all unique members, tests need to be continually applied to ensure that identical chromosomes did not breed. If chromosome similarity does not grow, then the GA may not converge because the search process becomes random rather than stochastic. While if it does grow, and then finding a non-similar chromosome to mate with, clearly becomes rare because of the inevitable occurrence of more twins, and the increasingly costly exercise of finding dissimilar chromosomes. On the other hand, it was also advocated [28] to allow individuals to reproduce if they are very closely similar. But, we have shown [32] that crossover between identical chromosome is a mutation operation which can turn a stochastic search approach indirectly into a random search, specially for complex problem and therefore the solution of the problem rarely converges [11],[12].

Aforementioned issues related to twin removal provide motivation for the investigations presented in this paper. The need for twin removal was originally highlighted in [25] which emphasized that duplicates chromosomes (*twins*) reduce diversity and ultimately lead to poorer performance. The study was confined solely however, to the detection and removal of identical chromosomes that were unique to each other, with no consideration being given to the removal and impact of similar chromosome or strongly correlated chromosomes. To mitigate the limitations caused by the stall condition, PSP using a GA has principally been confined to developing models based around special operators [33],[34] statistical approaches [5],[33],[35] and special treatment techniques such as *cooling* [11],[12] constraints and hybridization [10],[14],[15],[36] with the consequence that resulting GA-based solutions are both model and sequence dependent but are never generic. Therefore, generic improvement can be coupled for further improvement.

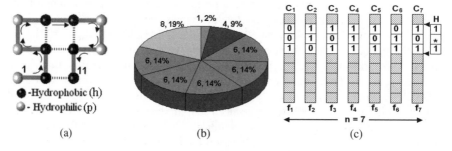

(a)   (b)   (c)

Fig. 1. (a) Conformation of sequence *phhpphhpphh* in 2D HP model [37] is shown by solid line. Any two hydrophobic residues being *topological neighbor* (TN) is indicated by dotted line. Fitness = -(TN Count) = -4, here. Three different arrows indicating *Left* (0), *Right* (1) and *Forward* (2) can be used to for chromosome encoding and it forms 001122110 in this case. **(b)** Pie chart of population having fitness 8, 6, 6, 6, 6, 6, 4 and 1. *Legend*: Fitness, Fitness % (with respect to the sum of the fitness values) **(c)** An example schema, *H* [1∗1] at bits 2 to 4, contained in chromosomes 3, 4, 5 and 7 of population size $Pop_z = 7$ at generation *t*.

A *chromosome correlation factor* (CCF) defines the degree of similarity existing between chromosomes. For similarity measurement between two individuals in the genotype as described in [29], we also measure it by counting the number of bits along each chromosome that are equal in the two individuals being compared. For chromosome presented [32] in the 2D HP (used in this paper) for PSP problem, three bit code 0, 1 and 2 are used for presenting three moves (see Fig. 1 (a) description).

It will be shown that by removing chromosomes having a similarity value greater than or equal to CCF during the search process enables the GA to continue seeking potential PSP solutions and ultimately provide superior results. The improved PSP performance of the algorithm based upon the generalized schemata theorem is analyzed upon accepted benchmark PSP sequences [34],[38]. Randomly-selected single point crossover and mutation operations are used in this paper as well as in the literature [11],[12] for PSP. This is because, as the solution becomes phenotypically compact it can produce more collisions [14],[16], if multi-point crossovers and mutations were involved which would leading to increasing collisions that produce non-self-avoiding walks within the conformation.

3 Preliminaries of Schemata Theorem

While this paper considers only the *Simple GA* (SGA), without any loss of generality, the theoretical framework developed is applicable to all GA variants [39]. Firstly, the initial population is generated, where the ith chromosome C_i is selected based on the fitness f_i with probability $(f_i / \bar{f})$, where $\bar{f}$ is the average fitness of the population. Parents then produce offspring by crossover at a rate p_c for a population of size Pop_z, with the generated offspring chosen with a *selection* probability $(f_i / \bar{f})$ and a mutation rate p_m. Usually, a small percentage of *elite* (high fitness) chromosomes are then copied to the next generation to retain potential solutions, with any remaining chromosomes unaffected by crossover, mutation or elitism moved to the next generation. Assume, an alphabet of cardinality |A| (defined as b_{count} in this paper) is used and hence the cardinality of schema is (|A|+1) including the *don't-care* which is normally applied to cover the unrestricted locus of the schema. The length of the schema $\delta(H)$ is the distance between the position of the first and last non *don't-care* characters, which actually indicates the number of possible crossover positions. For a chromosome length n, there are $((|A|+1)^n - 1)$ possible schema, excluding the combination comprising only *don't cares*, so a population of Pop_z chromosomes evaluates up to $((Pop_z)((|A|+1)^n - 1))$ schema, which provides implicit parallelism within the GA search. The order of schema $o(H)$ is the number of non *don't-care* characters, which governs the impact that any mutation has upon the schema. The number of occurrences of schema H in a population Pop_z at time t (which equals the number of generations) is given by $m(H,t)$. Throughout this paper, *twins* refer to pairs of chromosomes which are, with respect to their conformations, either *i*) identical, so CCF = 1, or *ii*) correlated with CCF $\geq r$, where r is the minimum admissible level of similarity defined for a population. Also, the term *overall similarity* is used to indicate the average of all CCF values of any chromosome with respect to all the other chromosomes in the population.

4 Limitations of the Schemata Theorem

In the following sections, limitations of the working principles of GA, i.e., *schemata theorem* [40], has been explored in the context of twin removal.

Selection: For a chromosome C_k having fitness f_k, the probability of C_k being selected by roulette wheel selection, is given by (1):

$$p_k = f_k \bigg/ \sum_{i=1}^{Pop_z} f_i \qquad (1)$$

The proportionate *selection* probability of the first chromosome (see Fig. 1 (b)) will be $p_1 = (8/43)$, and similarly $p_2 = (6/43)$, ..., $p_8 = (1/43)$. This is fallacious however, as from the pie-chart in Fig. 1 (b), it is clear that assuming chromosomes having the same fitness are identical, the fitness 6 occupies 68% in total, so the probability of a rolling marble randomly selecting a segment having fitness 6 is expressed as $p_{effective_2} = \sum_{i=2}^{6} p_i = 30/43$. The *effective selection* probabilities for C_1 (C_2 or C_3 or C_4 or C_5 or C_6), C_7 and C_8 are thus $8/43$, $30/43$, $4/43$ and $1/43$ respectively. Effectively, any of the fitness 6 occupies 70% of the pie-chart instead of 14%. Now consider an arbitrary schema H [1∗1] from bit position 2 to 4 as shown in Fig. 1 (c). The number of occurrences of such schema at time t is, $m(H, t) = 4$. The expected number of occurrences at time $(t+1)$ is $m(H, t+1)$ which depends on the fitness of the chromosomes containing the schema H such as C_3, C_4, C_5 and C_7. Hence, $\bar{f}(H, t)$ $= (f_3 + f_4 + f_5 + f_7)/4$. The average fitness $\bar{f}$ is now defined as:

$$\bar{f} = \left(\sum_{i=1}^{Pop_z} f_i \bigg/ Pop_z \right) \tag{2}$$

So if $\bar{f}(H, t) > \bar{f}$, then the number of occurrences of schema H in the next generation is likely to increase by $(\bar{f}(H, t)/\bar{f})$. Thus, the expected number of occurrences of schema H at time $(t+1)$ can be expressed as:

$$m(H, t+1) = m(H, t) \frac{\bar{f}(H, t)}{\bar{f}} \tag{3}$$

where, $\bar{f}(H, t)$ is the average fitness of chromosomes containing schema H.

Crossover: The *schemata theorem* computes the probable occurrences of a particular schema H in the next generation, with the proviso that the longer the schema length, the greater probability that the H will be disrupted by either a crossover or mutation operation. For a chromosome of length n there are $(n-1)$ possible crossover positions. Therefore the *disruption* probability is $(\delta(H)/(n-1))$ with the complementary *existence* probability being $(1-(\delta(H)/(n-1)))$, so in general the lower bound of the *existence* probability p_e having a crossover probability p_c is:

$$p_e \geq \left(1 - p_c \frac{\delta(H)}{n-1} \right) \tag{4}$$

The derivation of (4) comes from the fact that if a crossover point lies within the region of schema H, then the schema does not remain intact in the offspring, though this is not always the case. Section **5**, examines all the various possible scenarios:

Mutation: The mutation operation is able to disrupt any schema. With a *mutation* probability p_m, the bit *disruption* probability of a bit or character changing is $(1-p_m)$, so for the schema H having order $o(H)$, the *existence* probability of H is:

$$p_e = (1 - p_m)^{o(H)} \tag{5}$$

For very small values of p_m,

$$p_e \approx (1 - p_m\, o(H)) \tag{6}$$

Schemata Theorem: The number of occurrences of schema H in (3) can be expressed using (4) and (6) as:

$$m(H, t+1) = m(H, t)\frac{\bar{f}(H, t)}{\bar{f}}\left(1 - p_c\frac{\delta(H)}{n-1} - p_m\, o(H)\right) \tag{7}$$

which was the formal mathematical representation of the *schemata theorem*. But, as (4) is incomplete then so also is (7). While it is readily apparent that (7) supports the growth of similarity within a population, it fails to reflect certain anomalies within the original *schemata theorem* that can impact significantly upon GA operations, as the growth in twins and their potential deleterious effect in complex landscape applications such as PSP are considered.

5 Generalization of the Schemata Theorem

To analyse the effect of growing similarity in a population, the following sections directly address the particular limitations highlighted in this Section by firstly generalizing the selection process, resolving the issue of the crossover component

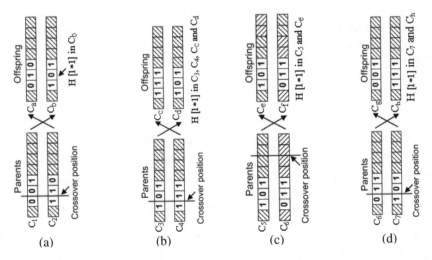

(a) (b) (c) (d)

Fig. 2. (a) Schema H [1*1] produced in offspring even when both parents do not have that schema. **(b)** The offspring always contain schema H [1*1] irrespective of the crossover position when both the parents have the particular schema. **(c)** One of the parents contains schema H [1*1] and the crossover positions lies outside the schema region. **(d)** One of the parents contains schema H [1*1] and the crossover positions is inside the schema region.

contributing in equation (7) and then integrating the solutions into a *generalized schemata theorem* framework.

Selection: The scenario under consideration is that the number of highly fitted chromosome will become larger as they are increasingly selected for crossover and mutation in each generation. The selection procedure will always favor those similar chromosomes that are higher in number in the population, so if w_k is the number of such similar chromosomes having fitness f_k, it will have a lower bound of unity and compared with (1), the effective selection now becomes:

$$p_k = \left(w_k f_k \middle/ \sum_{i=1}^{Pop_z} f_i \right) \tag{8}$$

which is a generalized representation of the original [40] selection procedure with $w_k = 1$. This is fully supported by the comparative examples in **Fig. 1** (b), where the selection process anomaly highlighted in Section **4**, mandates an appropriate twin removal strategy be implemented in order to ensure that as w_k tend to 1, the core *schemata* theory is upheld.

Crossover: The crossover operation however, may in certain cases not be disruptive [24], which can be interpreted as providing an *Accrued Benefit* (*AB*) because the schema of interest H is preserved rather than disrupted, which is not reflected by (4). Three *AB* scenarios are identified:

i) Accrued Benefit$_1$: Neither Parent Contains a Particular Schema
Consider the scenario illustrated in Fig. 2(a) of the crossover between two parents that do not contain schema H, though H may be expected to be created in the offspring. As neither of the parents contain schema H the crossover must occur within the schema region to create such an offspring, so the resulting *AB* can be expressed as in (9).

$$AB_1 = \left[1 - \left(\sum f(H,t) \middle/ \sum_{i=1}^{Pop_z} f_i \right) \right]^2 \left(\frac{\delta(H)}{n-1} \right) \Delta \tag{9}$$

The square parenthesis term is the *selection* probability of those parents that do not contain schema H, with $\sum f(H,t)$ being the aggregated fitness values of those chromosomes containing H, so the *selection* probability using (8) of these particular chromosomes is $\left(\sum f(H,t) \middle/ \sum_{i=1}^{Pop_z} f_i \right)$. The second term in parenthesis represents the probability of the crossover point existing within region H, where it is intuitively reasonable to assume both parents contain some part of schema H close to the crossover point, and this is given by probability Δ. To estimate Δ, assume a single crossover point divides schema H into H_1 and H_2, that is the schema is actually a concatenation of sub-schema so,

$$H = H_1 \bullet H_2 \text{ and } \Delta = \left(b_{count} \right)^{-(o(H_1)+o(H_2))} = \left(b_{count} \right)^{-(o(H))} \tag{10}$$

since $o(H) = o(H_1) + o(H_2)$. In the example in Fig. 2 (a) where the crossover occurs between positions 2 and 3, the schema $H[1*1]$ is divided into $H_1[1]$ and $H_2[*1]$, where $o(H_1) = 1$ and $o(H_2) = 1$. As $A = \{0, 1\}$ then, $b_{count} = 2$ and $\Delta = 2^{-1}.2^{-1} = 0.25$. An important point in (10) is, for a fixed $\delta(H)$, Δ directly depends upon chromosomal encoding and proportional to b_{count}.

ii) Accrued Benefit₂: *Both the Parents Contain a Particular Schema*
If both the parents contain schema H as shown in Fig. 2 (b), then H will never be lost by crossover irrespective of the crossover position. So,

$$AB_2 = \left(\sum f(H, t) \middle/ \sum_{i=1}^{Pop_z} f_i \right)^2 \tag{11}$$

and since the generation continues, this *benefit* increases due to increments in the similarity, which will assist in the growth of twins.

iii) Accrued Benefit₃: *Only One Parent Contains the Schema*
In this case, two options are feasible when one parent contains schema H and the other does not.

(a) Crossover Point is Located Outside the Schema Region
Since the crossover point does not lie within the schema region (**Fig. 2** (c)), then:

$$AB_{3a} = \left(\sum f(H, t) \middle/ \left(\sum_{i=1}^{Pop_z} f_i \right) \right) \left\{ 1 - \left(\sum f(H, t) \middle/ \left(\sum_{i=1}^{Pop_z} f_i \right) \right) \right\} \left(1 - \frac{\delta(H)}{n-1} \right) \tag{12}$$

where the third term in parenthesis indicates the probability that the crossover point is not located within the schema length and region.

(b) Crossover Point Lies Within the Schema Region
The crossover point now lies within the schema region (Fig. 2(d)) and it is further assumed that the crossover point divides the schema H into H_1 and H_2 for single crossover position, so $H = H_1 \bullet H_2$ and:

$$AB_{3b} = \left(\sum f(H, t) \middle/ \left(\sum_{i=1}^{Pop_z} f_i \right) \right) \left\{ 1 - \left(\sum f(H, t) \middle/ \left(\sum_{i=1}^{Pop_z} f_i \right) \right) \right\} \theta \tag{13}$$

$$\text{where, } \theta = \left((\delta(H))/(n-1) \right) \left\{ (b_{count})^{-o(H_1)} \oplus (b_{count})^{-o(H_2)} \right\} \tag{14}$$

where $\oplus$ is the 'Exclusive OR' operation, while θ represents the probability of the formation of schema H from parents. The first bracketed term in (14) is actually the probability of the crossover point occurring within the schema region, while the second term is the probability that part of schema H will come from each parent, so H resides exclusively in one of the offspring. The composite AB_3 for the case where only one parent contains the schema now becomes:

$$AB_3 = AB_{3a} + AB_{3b} \tag{15}$$

Combining the three *Accrued Benefit* from (9), (11) and (15), the *existence* probability p_e of a schema due to crossover occurring at a rate p_c can be expressed as:

$$P_e = p_c (AB_1 + AB_2 + AB_3)$$ (16)

The Generalized Schemata Theorem: The equations delineated in the previous two sections covering chromosome selection and crossover, are now formally embedded into a *generalized schemata theorem* framework. With a *crossover* probability $p_c < 1.0$, those chromosomes unaffected by crossover occur at $(1 - p_c)$. So the original *schemata theorem* in (7) can be rewritten using (10) as in (17):

$$m(H, t+1) = \left[(1 - p_c) \, m(H, t) \, \frac{\bar{f}(H, t)}{\bar{f}} \right] + m(H, t) \, \frac{\bar{f}(H, t)}{\bar{f}} \, p_c (AB_1 + AB_2 + AB_3)$$
$$- m(H, t) \, \frac{\bar{f}(H, t)}{\bar{f}} \, p_m \, o(H)$$ (17)

Now, (17) is a generalized representation of how the GA functions, such that in the case where $p_c = 1.0$, $AB_1 = 0$, $AB_2 = 0$, $AB_{3b} = 0$ and all chromosome selection probabilities are ignored (the first two terms in (17) then it reduces to the classical schemata theorem. Interestingly (17) supports the nonlinear fast growth of the surviving (also referred to as favorable) schema and with the incorporation of the appropriate selection procedure and various crossover scenarios, (17) clearly reveals the obvious expansion of twins in the population. The implications of this growth and the increasing likelihood of converging prematurely into the stall condition are now considered.

Impact of Generalization: The inexorable growth of identical and also progressively more highly-correlated twins as manifest in (17) can lead to the premature convergence or stall [18],[41] in the search process, a situation exacerbated by crossover creating even more twins and the impact of the mutation becoming increasingly ineffectual. These two issues are now respectively considered in the context of the new generalized framework.

(a) *Premature Convergence or Stall Condition*: The *reproduction* probability of twins $(r \le CCF \le 1)$ can be expressed using (8) as:

$$P(C_k, C_k) = (p_k)^2$$ (18)

So, the number of twins that are going to be in the next generation can be written as:

$$Count(C_k, t+1) = P(C_k, C_k)(Pop_z) \, p_c$$ (19)

Now consider the case where the number of similar chromosomes becomes close to the population so $w_k \approx Pop_z$ and $w_k f_k \approx \sum_{i=1}^{Pop_z} f_i$ in (8), so using (18) we get:

$$P(C_k, C_k) \approx 1$$ (20)

which is the stall or premature convergence condition. (19) shows that nearly all offspring generated throughout the population will be similar and go forward to the next generation with the result that there will be no variation in subsequent generations. It is entirely reasonable therefore to surmise that as the *effective crossover rate* $p_c \approx 0$, strategies that facilitate efficient removal of both identical and highly correlated twins will improve the GA performance, a premise that is fully corroborated in the experimental results Section 6.

(b) *Ineffective Mutation*: The growth of correlated twins inevitably weakens the impact of mutation, which despite introducing random variations and thereby different schemas, will quickly disappear in the midst of the common schema of so many correlated chromosomes in the population such that when $w_k \rightarrow Pop_z$, the chromosome selected for mutation ($C_{mutated}$) is very likely to be similar (high CCF value) to the majority of the population. By considering the mutation position, if the conformational change differs with respect to C_k, then two principal scenarios arise: *i*) After mutation, $C_{mutated}$ has a lower fitness ($f_{mutated}$) than average, so it is less likely to be selected, and thus will not be in the next generation. *ii*) After mutation, $C_{mutated}$ has a higher fitness than average, but is not similar to highly populated chromosomes, and so while $f_{mutated} > f_k$, as $w_k \rightarrow Pop_z$ the effect due to (8) becomes $f_{mutated} \ll w_k f_k$, so the chances of $C_{mutated}$ being selected for reproduction in the next generation are lower and it is likely the fitter $C_{mutated}$ will die away, so leading to an effective mutation rate of $p_m \approx 0$. While one possible approach to overcoming these issues is to use *elitism* [23], [42] to preserve a small proportion (5% to 10%) of elite chromosomes through the generations, this can convert the GA into a random rather than stochastic search process, with convergence never guaranteed. A better strategy is to remove both identical and highly correlated chromosomes to not only improve the performance of the GA but also avoid premature convergence.

6 Simulation and Experimental Results

Simulations were undertaken with (*TR-r*) and without (*WT*) the twin removal strategy implemented in the population. For twin removal (*TR-r*), it is performed after the crossover and mutation operations, for a range of CCF settings from $r = 1.0$ (identical chromosomes only) to $r = 0.2$ (the widest chromosome similarity $0.2 \leq CCF \leq 1.0$) in steps of 0.1 (e.g., *TR-60* refers to the removal of all chromosome twins having an admissible similarity value of 0.6 (60%) or above). A knock out system was adopted based on the superior fitness value in a *Correlated Twin Removal* (CTR) algorithm (see Algorithm I), where the chromosome with the lower fitness was removed. CTR uses the minimum admissible correlation value r when comparing chromosome pairs for conformational similarity (Line 4), and if twins are identified, the one with the lower fitness is removed (Lines 5 to7). After the removal, the gap is filled by randomly generated chromosomes, which for simplicity are not crosschecked for further twins. The GA parameters [26], [44] for experiments were set as $Pop_z = 200$, $p_c = 0.8$, $p_m = 0.05$ with elite rate = 0.05. *WT* (without twin removal) runs where same as in [12] but without cooling and *TR-100* is the same approach as in [25]. PSP with complex landscape takes longer time to converge. For this reason a maximum of

Algorithm-I: _Correlated Twin Removal_ (CTR)

Input: Population size= Pop_z , Chromosome (C) length = n , Minimum admissible correlation = r, where, $CCF \geq r$

Output: Population without twins of size $\leq Pop_z$

Assumption: _RetSimilarity_ (i, j) returns % of similarity between C(i) and C(j), where $i \neq j$.

```
1 FOR i = 1 to (Pop_z − 1) DO
2 { IF C(i).MarkDeleted = False THEN
3    FOR j = i+1 to Pop_z DO
4    { IF RetSimilarity(i, j) ≥ r% THEN
5      { IF |C(i).Fitness| < |C(j).Fitness| THEN
6        {Swap (C(i), C(j)) }
7        C(j).MarkDeleted = True } }
8    }}
```

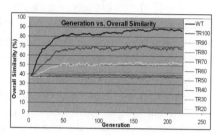

Fig. 1. Generation vs. Overall Similarity (%) plot for PSP of length 50

Table 1. Run results of 5 iterations of PSP for HP sequence length 50; maximum generation = 6000 and minimum fitness = -21. Sequence: H2(PH)3PH4PH(P3H)2P4H(P3H)2PH4P(HP)3H2, [38].

WT	TR-100	TR-90	TR-80	TR-70	TR-60	TR-50	TR-40	TR-30	TR-20
-17	-18	-21 (287)	-21 (1244)	-21 (992)	-21 (4671)	-20	-20	-17	-17
-19	-21 (2776)	-21 (5209)	-21 (2423)	-21 (1721)	-21 (5568)	-20	-19	-17	-16
-18	-20	-20	-21 (488)	-21 (611)	-21 (1668)	-19	-18	-17	-17
-18	-18	-21 (1711)	-21 (928)	-21 (1696)	-20	-20	-17	-18	-16
-19	-20	-20	-21 (345)	-21 (295)	-20	-20	-19	-18	-17

Data format: Maximum |fitness| (Generation number).

Table 2. Average run results of 5 iterations of PDB sequences after conversion into HP sequence; maximum generation = 6000

PDB ID	Length	WT	TR-100	TR-90	TR-80	TR-70	TR-60	TR-50
1PJF	46	-22	-24.6	-24.7	**-25**	-24.5	-24.5	-24
1AAF	55	-13.5	-13.6	**-15**	-14.5	-14.4	-14.3	-14
2PTL	78	-21	-22	-24.6	**-24.9**	24.8	-24.7	24.4
1GH1	90	-22.5	-26	-29	**-29.7**	-29.3	-28.5	-28
2GG1	102	-28.4	-31.8	-35	**35.5**	-34.3	-34.3	-34
2CQO	119	-37.5	-41	-44	**-44.5**	-44	-44	-40

Source: PDB sequences [43].

6000 generations was allocated for these particular series of experiments. PSP sequences [34], [38] shown in Table 1 and Table 2 for the 2D HP model [37]. Unlike Table 2, in Table 1, if during the iterations this optimal value was not reached, the maximum value achieved within generations is displayed. Fig. 3 shows the _Generation_ vs. _Overall similarity_ plot. In Fig. 3, it is shown that for the _WT_ run, the _overall similarity_ reached ≈80% very rapidly (around the 50th generation) from an initial value of ≈35%, before stabilizing at ≈90% *similarity* after the 150th generation. This clearly supports (17) in that without any twin removal policy, the overall population quickly becomes strongly correlated and diversity is lost. The rapid nonlinear growth up to the 50th generation is also supported by AB_1 and AB_2 in (9) and (11) respectively, with AB_2 being the dominant component in the *overall similarity*, because each crossover is more successful in generating twins similar to their parents regardless of the crossover position, and also the biased selection procedure identified in (8) is embedded in AB_2. In the 5 separate iterations (Table 1) *WT* never reached the putative ground value and its maximum fitness stalled, generally before the 250th run, though the simulation ran for the entire 6000 generations. This is a direct consequence of twins with a higher fitness appearing in the population, thereby slowing the convergence over time as the population becomes less diverse. The *overall dissimilarity* or diversity in the chromosomes remained around 10%, which was insufficient to maintain a search capability, and so it became trapped due to premature convergence. It must be emphasized that with such a high number of generations the effect of mutation is negligible even if *elitism* is applied, as highlighted in Section 5. The elite population is clearly not deriving any benefit from the mutation operation. It is also clear that *TR-80* displays the best performance for correlated twins as the population maintains the most favorable balance between the *overall similarity* (chromosome *correlation)* so keeping the search stochastic to aid convergence, and upholding diversity by supporting the growth of dissimilar, but competent chromosomes. The generalized selection procedure delineated in Section 5 supports these newly created chromosomes as well as existing correlated chromosomes by ensuring the entire selection procedure is less biased.

7 Conclusion

The ease of Genetic Algorithm (GA) implementations has made them a popular solution for many optimization problems, with the expectation that they can be effectively and accurately applied to even complex optimization problems such *ab initio* protein structure prediction (PSP). This neglects however, the crucial role of the growth of similarity and chromosome twins has upon the population, which can lead to premature convergence. The twin problem can impair its performance ultimately leading to premature convergence or the stall condition. We have highlighted the fallacies within the selection procedure and shown the 'accrued benefit' from the crossover operation. A generalized schemata theorem has been proposed which highlights the need of twin removal and generalization of the schemata theorem for consistent GA performance. The definition of twins has been relaxed to not only embrace duplicate chromosomes, but also to take cognizance of strongly-correlated chromosomes. It has been observed [27], that while even in relatively simple landscapes, failure to remove twins can lead a GA frequently getting trapped in earlier generations. This problem has been overcome within the generalized framework presented in this paper, with *chromosome correlation factor* (CCF) setting to 0.8, affording the best performance.

References

1. Berger, B., Leighton, T.: Protein Folding in the Hydrophobic-Hydrophilic (HP) Model is NP-Complete. Journal of Computational Biology 5, 27 (1998)
2. Crescenzi, P., Goldman, D., Papadimitriou, C., Piccolboni, A., Yannakakis, M.: On the complexity of protein folding (extended abstract), p. 597. ACM, New York (1998)
3. Chen, M., Lin, K.Y.: Universal amplitude ratios for three-dimensional self-avoiding walks. Journal of Physics A: Mathematical and General 35, 1501 (2002)
4. Guttmann, A.J.: Self-avoiding walks in constrained and random geometries. Elsevier, Amsterdam (2005)
5. Jiang, T., Cui, Q., Shi, G., Ma, S.: Protein folding simulation of the hydrophobic-hydrophilic model by computing tabu search with genetic algorithms. In: ISMB (2003)
6. König, R., Dandekar, T.: Refined Genetic Algorithm Simulation to Model Proteins. Journal of Molecular Modeling 5 (1999)
7. Lamont, G.B., Merkie, L.D.: Toward effective polypeptide chain prediction with parallel fast messy genetic algorithms. In: Fogel, G., Corne, D. (eds.) Evolutionary Computation in Bioinformatics, p. 137 (2004)
8. Pedersen, J.T., Moult, J.: Ab initio protein folding simulations with genetic algorithms: simulations on the complete sequence of small proteins. Proteins 29, 179 (1997)
9. Schulze-Kremer, S.: Genetic Algorithms and Protein Folding, vol. 1996 (2007)
10. Takahashi, O., Kita, H., Kobayashi, S.: Protein Folding by A Hierarchical Genetic Algorithm. In: 4th Int. Symp. AROB (1999)
11. Unger, R., Moult, J.: On the Applicability of Genetic Algorithms to Protein Folding. In: The Twenty-Sixth Hawaii International Conference on System Sciences, p. 715 (1993)
12. Unger, R., Moult, J.: Genetic Algorithms for Protein Folding Simulations. Journal of Molecular Biology 231, 75 (1993)
13. Hoque, M.T., Chetty, M., Dooley, L.S.: Significance of Hybrid Evolutionary Computation for Ab Inito Protein Folding Prediction. Springer, Heidelberg (2006)
14. Hoque, M.T., Chetty, M., Dooley, L.S.: A New Guided Genetic Algorithm for 2D Hydrophobic-Hydrophilic Model to Predict Protein Folding. In: IEEE CEC. IEEE Computer Society Press, Los Alamitos (2005)
15. Bastolla, U., Frauenkron, H., Gerstner, E., Grassberger, P., Nadler, W.: Testing a new Monte Carlo Algorithm for Protein Folding. Nat. Center for Biotech. Info. 32, 52 (1998)
16. Liang, F., Wong, W.H.: Evolutionary Monte Carlo for protein folding simulations. J. Chem. Phys. 115 (2001)
17. Shmygelska, A., Hoos, H.H.: An ant colony optimization algorithm for the 2D and 3D hydrophobic polar protein folding problem. BMC Bioinformatics 6 (2005)
18. Fogel, D.B.: EVOLUTIONARY COMPUTATION Towards a new philosophy of Machine Intelligence. IEEE Press, Los Alamitos (2000)
19. Sareni, B., Krähenbühl, L., Nicolas, A.: Effective Genetic Algorithms for Solving Hard Constrained Optimization Problems. IEEE Transaction on Magetics 36 (2000)
20. Rudolph, G.: Convergence analysis of canonical genetic algorithms. ITNN 5, 96 (1994)
21. Altenberg, L.: The Schema Theorem and Price's Theorem Foundations of Genetic Algorithms 3 (1995)
22. Fogel, D.B., Ghozeil, A.: Schema processing, proportional selection, and the misallocation of trials in genetic algorithms. Information Science 122, 93 (2000)
23. Michalewicz, Z.: Genetic Algorithms + Data Structures = Evolution (1992)
24. Whitley, D.: An Overview of Evolutionary Algorithms. Journal of Information and Software Technology 43, 817 (2001)

25. Ronald, S.: Duplicate Genotypes in a Genetic algorithm. In: IEEE WCCI, p. 793. IEEE Computer Society Press, Los Alamitos (1998)
26. Haupt, R.L., Haupt, S.E.: Practical Genetic Algorithms (2004)
27. Hoque, M.T., Chetty, M., Dooley, L.S.: Critical Analysis of the Schemata Theorem: The Impact of Twins and the Effect in the Prediction of Protein Folding using Lattice Model, GSIT, MONASH University, TR-2005/8 (2005)
28. Deb, K., Goldberg, D.E.: An investigation of niche and species formation in genetic function optimization. In: 3rd Int. Conf. on Genetic Algorithms, p. 42 (1989)
29. Coello, C.A.C.: An Updated Survey of GA-Based Multiobjective Optimization Techniques. ACM Computing Surveys 32, 109 (2000)
30. Rogers, A., Prügle-Bennett, A.: Genetic Drift in Genetic Algorithm Selection Schemes. IEEE Transaction on Evolutionary Computation 3, 298 (1999)
31. Eshelman, L.J., Schaffer, J.D.: Preventing premature convergence in genetic algorithms by preventing incast. In: 4th Int. Conf. on Genetic Algorithms, p. 115 (1991)
32. Hoque, M.T., Chetty, M., Dooley, L.S.: Non-Isomorphic Coding in Lattice Model and its Impact for Protein Folding Prediction Using Genetic Algorithm. In: IEEE Computational Intelligence in Bioinformatics and Computational Biology IEEE CIBCB, Canada (2006)
33. Toma, L., Toma, S.: Contact interactions methods: A new Algorithm for Protein Folding Simulations. Protein Science 5, 147 (1996)
34. Lesh, N., Mitzenmacher, M., Whitesides, S.: A Complete and Effective Move Set for Simplified Protein Folding. In: RECOMB (2003)
35. Bornberg-Bauer, E.: Chain Growth Algorithms for HP-Type Lattice Proteins. In: RECOMB'97 (1997)
36. Backofen, R., Will, S.: A Constraint-Based Approach to Fast and Exact Structure Prediction in Three-Dimensional Protein Models. Kluwer Academic Publishers, Dordrecht (2005)
37. Dill, K.A.: Theory for the Folding and Stability of Globular Proteins. Bio-chemistry 24, 501 (1985)
38. Hart, W.E., Istrail, S.: HP Benchmarks (2005), http://www.cs.sandia.gov/
39. Vose, M.D.: The Simple Genetic Algorithm. MIT Press, Cambridge (1999)
40. Holland, J.H.: Adaptation in Natural And Artificial Systems. MIT Press, Cambridge (2001)
41. Goldberg, D.E.: Genetic Algorithm Search, Optimization, and Machine Learning. Addison-Wesley Publishing Company, Reading (1989)
42. Davis, L.: Handbook of Genetic Algorithm. VNR, New York (1991)
43. PDB, Protein Data Base, vol. 2007 (2007), http://www.rcsb.org/pdb/
44. Digalakis, J.G., Margaritis, K.G.: An experimental Study of Benchmarking Functions for Genetic Algorithms. Intern. J. Computer Math. 79, 403 (2002)

Using Fuzzy Support Vector Machine Network to Predict Low Homology Protein Structural Classes

Tongliang Zhang[1], Rong Wei[3], and Yongsheng Ding[1,2]

[1] College of Information Sciences and Technology
tl.zhang@mail.dhu.edu.cn
[2] Engineering Research Center of Digitized Textile & Fashion Technology,
Ministry of Education
ysding@dhu.edu.cn
Donghua University, Shanghai 201620, P.R. China
[3] College of Sciences, Hebei Polytechnic University,
Hebei Tangshan 063009, P.R. China
wr@heut.edu.cn

Abstract. Prediction of protein structural classes for low homology proteins is a challenging research task in bioinformatics. A dual-layer fuzzy support vector machine (FSVM) network approach is proposed to predict protein structural classes. A protein sample can be represented by nine representation feature vectors: pair couple amino acid (210-D) and eight pseudo amino acid composition vectoers (PseAAC). Eight physicochemical properties of amino acids extracted from AAIndex databank are used to calculate low frequencies of power spectrum density of sequence-order correlation in protein sequence. In the first layer of FSVM network, nine FSVM classifiers are established, which are trained by different protein feature vectors, respectively. The outputs of the first layer are reclassified by FSVM classifier in 2nd layer of the network. The performance of proposed method is validated by low homology (average 25%) dataset covering 1673 proteins. The promising results indicate that the new method may become a useful tool for predicting not only the structural classification of proteins but also their other attributes.

1 Introduction

In structural classification of proteins databank (SCOP) [1-3], proteins are classified into seven structural classes: *all-α, all-β, α+β, α/β*, multi-domain, small protein, and peptide. More than 80% proteins are deposited into the former four classes. Many efforts were focused on the four structural classes, ie., *all-α, all-β, α+β*, and *α/β*.

Numerous prediction methods for protein structural classes have been proposed based on the primary amino acid sequence [4-15], since the work of Klein and Delisi [4]. During the twenty years, the performance of these methods are increasing with the combination of new pattern recognize algorithms and effective protein sequence representation. Perfect accuracy rates (about 95%) have been achieved in some prediction methods. However, these methods were often tested on small datasets, and characterized by different homology of sequences. Kurgan and Homaeian [22] indicated that sequence homology in dataset have a significant impact on the predictive

J.C. Rajapakse, B. Schmidt, and G. Volkert (Eds.): PRIB 2007, LNBI 4774, pp. 98–107, 2007.
© Springer-Verlag Berlin Heidelberg 2007

accuracy. The best achieved prediction accuracy for low homology datasets is about 57%. Wang and Yuan [17] have stated that the prediction method should aim only at proteins with lower 30% homology. So, it is crucial to develop the prediction methods or algorithms for structural classes of protein with lower homology.

Several studies have testified that the performance of ensemble machine learning approaches is superior to individual learning algorithm [7, 18-21]. Recently, the methods of ensembles have been used in this area. Kedarisetti et al. [16] established an ensembles method with heterogeneous classifiers validating on the datasets of varying homology. Chen et al. [7] developed support vector machine fusion network algorithm.

Compared with amino acid composition frequently used in prediction methods of protein structural classes, pseudo amino acid composition (PseAAC) as introduced by Chou [23] can incorporate more information and remarkable enhance prediction performance in various attribute of protein. In this study, a sample of proteins is represented by nine kinds of feature vectors, including pair couple amino acid composition (PcAA) (210-D) and eight PseAAC feature vector. Eight physicochemical properties extracted from AAIndex database [24] are used to calculate sequence-order correlation that introduced by Chou [25]. Low frequencies of power spectrum density of different sequence-order effect are used to construct PseAAC. A dual-layer fuzzy support vector machine (FSVM) network is used as prediction engine. The low homology dataset 25PDB, constructed by Kurgan et al. [22], is applied to verify the new method. Promising results obtained on self-consistent and jackknife cross-validating test methods show that it is effective and practical.

2 Methods

2.1 Protein Sequence Representation

Pair-coupled amino acid composition (PcAA) attempts to extract the information of local order of amino acids in sequence. This concept has been used in protein secondary structure content prediction [7, 26, 27] and other attributes of protein prediction [25]. The PcAA is formulated as follows:

$$\Re_{210} = \left\{ \begin{array}{c} f(AA), f(AC), f(AD),...,f(AY) \\ f(CC), f(CD),...,f(CY) \\ f(DD),...,f(DY) \\ ... \\ f(YY) \end{array} \right\} \tag{1}$$

where, $f(AC)$ is the sum of AC pair occurrence frequency and CA pair occurrence frequency in protein sequence. Thus, the pair-couple AA is a 210-D feature vector.

$$\vec{S}_1 = [x_1, x_2,..., x_{210}] \tag{2}$$

where, $x_1 = f(AA), x_2 = f(AC),..., x_{210} = f(YY)$. The 210-D (dimensional) vector is normalized to meet the condition that the sum is 1.

$$\sum_{i=1}^{210} x_i = 1 \tag{3}$$

Compared with conventional protein composition (20-D), the concept of PseAAC as originally introduced by Chou [23], which is defined in a $(20+\lambda)$-D features space, will contain much more sequence-order information. A protein sample can be represented by $(20+\lambda)$-D vectors, where the λ is the number of additional properties of sequence.

$$x_i = \begin{cases} \dfrac{f_i}{\sum_{j=1}^{20} f_j + w\sum_{j=1}^{\lambda} p_j} & (1 \le i \le 20) \\[4mm] \dfrac{wp_i}{\sum_{j=1}^{20} f_j + w\sum_{j=1}^{\lambda} p_j} & (21 \le i \le 20+\lambda) \end{cases} \tag{4}$$

where, the f_i is the normalized occurrence frequency of the 20 amino acid in the protein, that is amino acid composition (AA), and the P_i is the additional properties of protein sequence. w is weight factor of additional characteristics. In this study, the properties of protein sequence are the low frequencies of power spectrum density of sequence-order effect that introduced by Chou [25]. Protein sequence with N residues can be written as $R = R_1 R_2 \dots R_N$. Protein sequence order effect can be reflected through Eq. (5). It is actually the same as Eq. (2) of Chou [25].

$$\tau_m = \frac{1}{L-m} \sum_{i=1}^{L-m} J_{i,i+m} \tag{5}$$

where, $J_{i,i+m}$ is the correlation factor of residue i and $i+m$. m is the distance of two residues have correlation in sequence. In this study, $J_{i,i+m}$ is defined as the product of physicochemical properties of two residues, see Eq. (6), which is actually the same as Eq. (3) of Chou[25].

$$J_{i,i+\lambda} = h(R_i)h(R_{i+m}) \tag{6}$$

Before substituting the properties into Eq. (6), they are normalized according to the Eq. (7)

$$h(Ri) = \frac{h_0(R_i) - \overline{h_0}}{SD(h_0)} \tag{7}$$

In Eq. (7), $h_0(R_i)$ $(i = 1,2,\dots,20)$ are the original physicochemical properties of 20 amino acids. $\overline{h_0}$ denotes that the average property values of 20 amino acids. $SD(h_0)$ presents the standard deviation.

We use standard function 'PWELCH' in Matlab 7.0 environment to calculate power spectral density of the sequence-order effect. Based on the theory of digital signal processing, the high-frequency components are more noisies, and hence only the low-frequency components are more important. Similarly, low frequencies of Fourier transform of protein sequence-order correlation have been used in prediction of membrane protein types [28, 29]. In our previous work [30], low frequencies of energy spectrum density of protein sequence-order correlation are used to construct PseAAC. This is just like the case of protein internal motions where the low-frequency components are functionally more important [31, 32].

In some prior works, hydrophobicity scale of amino acid was usually used to calculate sequence-order effect [7, 28-30, 33-35]. Except for hydrophobicity, some other physicochemical properties of amino acids are also important in the fold process, such as volume, polarity, average accessible surface area, and so on. Here, eight physicochemical properties extracted from AAIndex database [24] are used to compute the sequence-order correlation of protein sequence through Eq. (6) and (7), respectively. The physicochemical properties are listed in Table 1. Eight PseAAC vectors are obtained and are named as $\vec{S}_i$ $(i = 2,3,...,9)$.

Table 1. The eight physicochemical properties used in this work

No	AA Index	Description
1	PRAM900101	Hydrophobicity
2	COSI940101	Electron-ion interaction potential values
3	RADA880108	Mean polarity
4	PONJ960101	Average volumes of residues
5	KUHL950101	Hydrophilicity scale
6	JANJ790102	Transfer free energy
7	JANJ780101	Average accessible surface area
8	FAUJ880103	Normalized van der Waals volume

2.2 Fuzzy Support Vector Machine

Support vector machine is a typical binary-class classifier based on the statistic learning theory[36]. The task of protein structural classes' prediction is a four classes classification problem. There are many multi-classes SVMs methods to solve the problem, such as one against one, one against others, DAG, etc. However, there are some unclassifiable points still existing in these multi-class SVMs methods. FSVM algorithm as introduced by Abe [37] has capability to solve unclassifiable points effectively.

Compared with conventional SVM algorithm, membership function is defined in FSVM algorithm. When solving k classes classification task, $k(k-1)/2$ SVM classifiers have to be established with one against one method. Toward to the SVM classifier between class i and class j, the decision function of input vector x is

$$D_{ij} = w_{ij}x + b_{ij} \tag{8}$$

where w_{ij} is the m-D vector, b_{ij} is a scalar, and $D_{ij} = -D_{ji}$

For the input vector x, we assemble

$$D_i = \sum_{j \neq i, j=1}^{n} sign(D_{ij}(x)) \tag{9}$$

where,

$$sign(x) = \begin{cases} 1 & x > 0 \\ 0 & x \leq 0 \end{cases} \tag{10}$$

For optimal separating hyperplane $D_{ij} = 0$ $(i \neq j)$, the membership function m_{ij} is defined as below:

$$m_{ij} = \begin{cases} 1 & D_{ij}(x) \geq 1 \\ D_{ij}(x) & otherwise \end{cases} \tag{11}$$

We define the class i membership function of x using the minimum operator:

$$m_i(x) = \min_{j=1,\ldots,n} m_{ij}(x) \tag{12}$$

The shape of the membership function is a truncated polyhedral pyramid. An unknown protein sequence is classified into the class with maximum membership value.

$$\arg \max_{i=1,\ldots,n} m_i(x) \tag{13}$$

In this study, six binary-class SVMs have been developed through the method of one against one for solving protein structural classes prediction (four classes). LS-SVMLab1.5 toolbox [38] in MatLab environment is selected as binary-class SVM classifier which is capable of searching the fittest parameters in SVM automatically. The Radial Basis Function (RBF) kernel is used in SVM. The membership function is calculated according to the Eqs (8)-(12). The output of each FSVM classifier in the first layer is not a rigid class label but a 4-D vector. The vector indicates that the membership values of protein sample belong to four strucutral classes.

2.3 FSVM Network

The SVM fusion uses the FSVM classifier to reclassify the outputs from all sub-classifiers. The protein sample is predicted by FSVM 1 to FSVM 8, and the output

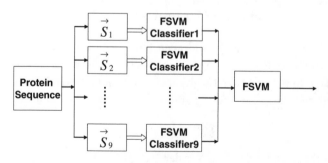

Fig. 1. The work procedure of FSVM network for prediction of protein structural classes

of the $k-th$ FSVM ($1 \leq k \leq 9$) is represented as 4-D vector. Then the input feature vector is defined as $V = [V_1, V_2, ..., V_9]$ to the FSVM fusion classifier for final decision. The process of the new predictive method is illustrated in Fig. 1.

2.4 Dataset and Measure Methods

The dataset constructed by Kurgan and Homaeian [22] is used to validate the performance of the new method, which includes proteins scanned with high resolution and with low on average 25% homology, named as 25PDB. The dataset contains 1673 proteins classified into four structural classes: 443 all-α, 443 all-β, 346 α/β, and 441 α+β.

Three indexes are applied to evaluate the prediction accuracy, that is, sensitivity (Sn), specificity (Sp), and Mathew's correlation coefficient (MCC).

$$S_n = \frac{TP}{TP + FN} \quad (14)$$

$$S_p = \frac{TP}{TP + FP} \quad (15)$$

$$MCC = \frac{TP \times TN - FP \times FN}{[(TP+FN)(TP+FP)(TN+FN)(TN+FP)]^{1/2}} \quad (16)$$

where, TP (true positives) is the protein number of right prediction in a structure class, FN (false negatives) is the protein number of wrong prediction in a structure class, and FP (false positives) is the number of the proteins in other classes to be predicted in this class. TN (true negatives) is the number of proteins observed in other classes that are not predicted in this class. S_n represents the accuracy, and S_p represents the reliability in procedure of prediction. The MCC is a single parameter characterizing the matching extent between the observed and predicted structural classes. Self-consistency and jackknife test methods are used to test the performance of the new approach. The jackknife test is thought the most rigorous and objective one [see [39] for a comprehensive review in this regard], and hence has been used by more and more investigators in examining the power of various prediction methods.

3   Results and Discussion

In the first layer of the FSVM network, nine FSVM classifiers trained by different representation methods of protein are established. The three indexes values of each FSVM classier are tested under self-consistency and jackknife cross-validating test methods. The number of addition characteristics λ and factor parameter w in Eq. (4) are important in determining the performance of PseAAC. Different parameters $\{\lambda, w\}$ in Eq. (4) are tested, and the PseAAC feature vector is input into FSVM classifier. The parameters $\{\lambda, w\}$ are determined when the accuracy of jackknife test method is the highest. The jackknife test results of nine FSVM classifier trained by different PseAAC feature vectors are listed in Table 2. The highest accuracy is 55.9%, and the lowest is

Table 2. Performance of each FSVM classifier

Representation	$\vec{S}_1$	$\vec{S}_2$	$\vec{S}_3$	$\vec{S}_4$	$\vec{S}_5$	$\vec{S}_6$	$\vec{S}_7$	$\vec{S}_8$	$\vec{S}_9$
λ	-	11	11	7	9	15	11	20	3
w	-	0.15	0.05	0.02	0.1	0.1	0.1	0.15	0.2
Accuracy (%)	32.7	53.2	55.6	53.4	47.1	55.9	50.6	43.9	53.4

32.7%. The outputs of all FSVM classifiers are combined into a vector, and used as input of the FSVM classifier in the second layer of the network.

The prediction results of four structural classes on self-consistency and jackknife test methods are listed in Table 3. The three indexes are calculated respectively. The overall accuracy of self-consistency is 91.87%, and that of jackknife test is 62.9%. Class all-β achieves the highest accuracy (79.9%) in four structural classes, and that of class all-α is more than 70%. However, accuracy of class α/β is 36.7%, and that of class $\alpha+\beta$ is 56%. The Sp value of class all-α is 53.9%. It demonstrate that some proteins in other structural class are incorrect classified into class all-α. According to the concept of Levitt and Chothia [30], it is true that the class $\alpha+\beta$ and α/β are more complex than class all-α and all-β. It might be the reason to explain the phenomenon that accuracies of class all-α and class all-β are higher than that of class $\alpha+\beta$ and class α/β.

Table 3. Results on self-consistency and jackknife test methods

Structural classes	Self-consistency test			Jackknife test		
	S_n	S_p	MCC	S_n	S_p	MCC
all-α	95.7%	85.3%	0.86	73.1%	53.9%	0.59
all-β	87.8%	95.3%	0.88	79.9%	75.5%	0.53
α/β	87.6%	94.1%	0.88	36.7%	52.0%	0.32
$\alpha+\beta$	95.4%	94.4%	0.93	56.0%	68.8%	0.41
Total	1537/1673=91.87%			1052/1673=62.9%		

The results of proposed method are compared with that of others using the same dataset, 25PDB. The accuracy rates of self-consistency and jackknife test methods together with the protein representation are deposited in Table 4. In same measure method and validation dataset, proposed method obtains 4.2% and 3% improvement on jackknife test when compared with the results of Logistic regression method [22] and StackingC ensemble [16], respectively. Meanwhile, the highest accuracy is achieved on self-consistency test method compared with other two methods. StackingC ensemble algorithm uses four heterogeneous classifiers. Protein sequence is initially represented using comprehensive set of 122 features which is reduced to 34 features through application of several feature selection algorithms [16]. The proposed method use different protein representation and same classifier. FSVM classifier has capability to resolve unclassifiable region effectively. FSVM is a strong classifier with excellent classification performance. It is widely accepted that ensemble method can enhance classification performance than single classifier [31]. Multiple FSVM classifiers are combined into FSVM network where various physicochemical properties of amino

Table 4. Comparison with other prediction methods on same dataset

Classification algorithm	Representation	Accuracy rates (%)	
		Self-consistency	Jackknife
Logistic regression[22]	66 feature	62.2	57.1
StackingC ensemble[16]	34 feature	87.6	59.9
This paper	Multi-feature	91.87	62.9

acid are taken into account. The promising prediction results illuminate that the FSVM network incorporating various physicochemical properties of amino acid is effective and practical. It might become potential tool for prediction of protein structural classes and other attributes of protein.

4 Conclusions

A dual-layer FSVM network is established to predict protein structural classes. Protein sample is represented by nine kinds of feature vectors including pair-couple amino acid composition (210-D) and eight PseAAC vectors. Eight physicochemical properties extracted from AAIndex databank are used to calculate sequence-order correlation in protein sequence. Low-frequencies of power spectrum density of different sequence-order correlations are used to construct PseAAC vectors. In the first layer of FSVM network, nine FSVM classifiers are obtained which are trained by different feature representation methods, respectively. The low homology dataset (average about 25% homology) is applied to verify the proposed method. The results of jackknife test in different structural classes illuminate that accuracies of class all-α and class all-β are higher than class $\alpha+\beta$ and class α/β. The phenomenon meets the truth of more complexity existing in class $\alpha+\beta$ and class α/β. Compared with other two methods tested on same dataset, the proposed methods obtain the highest accuracy. Promising results show the new method is effective and practical. It might become potential tool for protein structural class and other attribute of proteins.

Acknowledgments

This work was supported in part by Program for New Century Excellent Talents in University from Ministry of Education of China (No. NCET-04-415), the Cultivation Fund of the Key Scientific and Technical Innovation Project from Ministry of Education of China (No. 706024), International Science Cooperation Foundation of Shanghai (No. 061307041), and Specialized Research Fund for the Doctoral Program of Higher Education from Ministry of Education of China (No. 20060255006).

References

1. Murzin, A.G., Brenner, S.E., Hubbard, T., Chothia, C.: J. Mol. Biol. 247, 536–540 (1995)
2. Lo Conte, L., Brenner, S.E., Hubbard, T.J.P., Chothia, C., Murzin, A.: Nucl. Acid Res. 30(1), 264–267 (2002)

3. Andreeva, A., Howorth, D., Brenner, S.E., Hubbard, T.J.P., Chothia, C., Murzin, A.: Nucl. Acid Res. 32, D226–D229 (2004)

4. Klein, P., Delisi, C.: Prediction of Protein Structural Class from the Amino Acid Sequence. Biopolymers 25, 1659–1672 (1986)

5. Cai, Y.D., Liu, X.J., Xu, X., Zhou, G.P.: Support Vector Machines for Predicting Protein Structural Class. BMC Bioinformatics 2, 3–7 (2001)

6. Cao, Y., Liu, S., Zhang, L., Qin, J., Wang, J., Tang, K.: Prediction of Protein |Structural Class with Rough Sets. BMC Bioinformatics 7, 20 (2006)

7. Chen, C., Zhou, X., Tian, Y., Zou, X., Cai, P.: Predicting Protein Structural Class with Pseudo Amino Acid Composition and Support Vector Machine Fusion Network. Anal. Biochem. 357, 116–121 (2006)

8. Chou, K.C., Cai, Y.D.: Predicting Protein Structural Class by Functional Domain Composition. Biochemical and Biophysical Research Communications (Corrigendum: ibid., 2005, Vol.329, 1362) 321, 1007–1009 (2004)

9. Du, Q.S., Jiang, Z.Q., He, W.Z., Li, D.P., Chou, K.C.: Amino Acid Principal Component Analysis (AAPCA) and Its Applications in Protein Structural Class Prediction. Journal of Biomolecular Structure and Dynamics 23, 635–640 (2006)

10. Feng, K.Y., Cai, Y.D., Chou, K.C.: Boosting Classifier for Predicting Protein Domain Structural Class. Biochemical and Biophysical Research Communications 334, 213–217 (2005)

11. Luo, R.Y., Feng, Z.P., Liu, J.K.: Prediction of Protein Structural Class by Amino Acid and Polypeptide Composition. Eur. J. Biochem. 269, 4219–4225 (2002)

12. Niu, B., Cai, Y.D., Lu, W.C., Zheng, G.Y., Chou, K.C.: Predicting Protein Structural Class with AdaBoost learner. Protein & Peptide Letters 13, 489–492 (2006)

13. Shen, H.B., Yang, J., Liu, X.J., Chou, K.C.: Using Supervised Fuzzy Clustering to Predict Protein Structural Classes. Biochemical and Biophysical Research Communications 334, 577–581 (2005)

14. Sun, X.D., Huang, R.B.: Prediction of Protein Structural Classes Using Support Vector Machines. Amino Acids 30, 469–475 (2006)

15. Xiao, X., Shao, S.H., Huang, Z.D., Chou, K.C.: Using Pseudo Amino Acid Composition to Predict Protein Structural Classes: Approached with Complexity Measure Factor. Journal of Computational Chemistry 27, 478–482 (2006)

16. Kedarisetti, K.D., Kurgan, L., Dick, S.: Classifier Ensemble s for Protein Structural Class Prediction with Varying Homology. Biochemical and Biophysical Research Communications 348, 981–988 (2006)

17. Wang, Z.X., Yuan, Z.: How Good is the Prediction of Protein Structural Class by the Component-coupled Method? Proteins 38, 165–175 (2000)

18. Chou, K.C., Shen, H.B.: Hum-PLoc: A Novel Ensemble Classifier for Predicting Human Protein Subcellular Localization. Biochem. Biophys. Res. Commun. 347, 150–157 (2006)

19. Nanni, L., Lumini, A.: MppS: An Ensemble of Support Vector Machine Based on Multiple Physicochemical Properties of Amino Acids. Eurocomputing 69, 1688–1690 (2006)

20. Nanni, L., Lumini, A.: Ensemblator: An Ensemble of Classifiers for Reliable Classification of Biological Data. Pattern Recognition Letters 28, 622–630 (2007)

21. Peng, Y.H.: A Novel Ensemble Machine Learning for Robust Microarray Data Classification. Computers in Biology and Medicine 36, 553–573 (2006)

22. Kurgan, L., Homaeian, L.: Prediction of Structural Classes for Protein Sequences and Domain: Impact of Prediction algorithms, Sequence Representation and Homology, and Test Procedures on Accuracy. Pattern Recognition 39, 2323–2343 (2006)

23. Chou, K.C.: Prediction of Protein Structural Classes and Subcellular Locations. Curr. Protein Peptide Sci. 1, 171–208 (2000)
24. Kawashima, S., Ogata, H., Kanehisa, M.: AAindex: Amino Acid Index Database. Nucleic Acids Res. 27, 368–369 (1999)
25. Chou, K.C.: Prediction of Protein Cellular Attributes Using Pseudo Amino Acid Composition. PROTEINS: Structure, Function, and Genetics (Erratum: ibid., 2001, Vol.44, 60) 43, 246–255 (2001)
26. Cai, Y.D., Liu, X.J., Xu, X.B., Chou, K.C.: Artificial Neural Network Method for Predicting Protein Secondary Structure Content. Computers and Chemistry 26, 347–350 (2002)
27. Chou, K.C.: Using Pair-coupled Amino Acid composition to Predict Protein Secondary Structure Content. J. Protein Chem. 18, 473–480 (1999)
28. Liu, H., Wang, M., Chou, K.C.: Low-frequency Fourier Spectrum for Predicting Membrane Protein Types. Biochem. Biophys. Res. Commun. 336, 737–739 (2005)
29. Liu, H., Yang, J., Wang, M., Xue, L., Chou, K.C.: Using Fourier Spectrum Analysis and Pseudo Amino Acid Composition for Prediction of Membrane Protein Types. The Protein Journal 24, 385–389 (2005)
30. Zhang, T.L., Ding, Y.S.: Using Pseudo Amino Acid Composition and Binary-tree Support Vector Machines to Predict Protein Structural Classes. Amino Acids (2007) 10.1007/s00726-007-0496-1
31. Chou, K.C.: Review: Low-frequency Collective Motion in Biomacromolecules and Its Biological functions. Biophysical Chemistry 30, 3–48 (1988)
32. Chou, K.C.: Low-frequency Resonance and Cooperativity of Hemoglobin. Trends in Biochemical Sciences 14, 212–213 (1989)
33. Shen, H.B., Chou, K.C.: Ensemble Classifier for Protein fold pattern recognition. Bioinformatics 22, 1717–1722 (2006a)
34. Shen, H.B., Chou, K.C.: Using Ensemble Classifier to Identify Membrane Protein Types. Amino Acids (2006) 10.1007/s00726-006-0439-2
35. Shen, H.B., Yang, J., Chou, K.C.: Fuzzy KNN for Predicting Membrane Protein Types from Pseudo Amino Acid Composition. Journal of Theoretical Biology 240, 9–13 (2006)
36. Vapnik, V.N.: The Nature of Statistical Learning Theory. Springer, Heidelberg (1995)
37. Abe, S.: Fuzzy LP-SVM for multiClass problems. In: ESANN 2004 proceedings- European symposium on artificial neural networks Bruges (Belgium), 28-30 April 2004 d-side public, pp. 429–434 (2004) ISBN 2-930307-04-8
38. Suykens, J.A.K., Van Gestel, T., De Brabanter, J., De Moor, B., Vandewalle, J.: Least Squares Support Vector Machines. World Scientific, Singapore (2002)
39. Chou, K.C., Zhang, C.T.: Review: Prediction of Protein Structural Classes. Critical Reviews in Biochemistry and Molecular Biology 30, 275–349 (1995)

SVM-BetaPred: Prediction of Right-Handed ß-Helix Fold from Protein Sequence Using SVM

Siddharth Singh[1], Krishnan Hajela[2], and Ashwini Kumar Ramani[1]

[1] School of Computer Science
[2] School of Life Science, Devi Ahilya University,
Khandwa Road, Indore-452001, India
{sidssingh,hajelak}@gmail.com, ramani.iips@dauniv.ac.in

Abstract. The right-handed single-stranded ß-helix proteins are characterized as virulence factors, allergens, toxins that are threat to human health. Identification of these proteins from amino acid sequence is of great importance as these proteins are potential targets for anti-bacterial and fungal agents. In this paper, support vector machine (SVM) has been used to predict the presence of ß-helix fold in protein sequences using dipeptide composition. An input vector of 400 dimensions for dipeptide compositions is used to search for the presence of putative rungs or coils, the conserved secondary structure, found in ß-helix proteins. An average accuracy of 89.2% with Matthew's correlation coefficient of 0.75 is obtained in a 5-fold cross-validation technique. In addition, a PSSM was also used to score the query sequence of proteins identified as ß-helices by SVM. The method recognizes right-handed ß-helices with 100% sensitivity and 99.6% specificity on test set of known protein structures.

Keywords: ß-helix fold, ß-sheet stacking, pectate lyase, fold recognition, secondary structure, Support Vector Machines, SVM, PSSM.

1 Introduction

The right-handed single-stranded parallel ß-helices [RßH] are perfect examples of protein fold that display high three-dimensional structural similarity in spite of low sequence homologies across the families [1]. These proteins display diverse functions ranging from the enzymatic activities to specific structural roles [2,3]. The characteristic feature of these proteins is the presence of a coiled polypeptide backbone having almost same circular arrangement of β-sheet structure elements in space [1,4,5]. The backbone alternates between ß-strands and turns with the ß-sheets running perpendicular to the axis of protein molecule as shown in Figure 1. As a result of this progressive coiling of backbone, the buried hydrophobic cores of ß-helix domains is extended, rather than globular, and is characterized by the distinctive stacking of side chains that occur at equivalent position in successive coils called rungs or turns. A comparative analysis of different known ß-helices has revealed strong positional preference for specific amino acid residues towards the interior of parallel ß-helices [6,7]. The ß-helix fold lacks an obvious sequence repeats, although presence of some sequence patterns has been reported across the super-families [3,8].

J.C. Rajapakse, B. Schmidt, and G. Volkert (Eds.): PRIB 2007, LNBI 4774, pp. 108–119, 2007.
© Springer-Verlag Berlin Heidelberg 2007

In recent past, a shift from classical fold recognition approach to machine learning has been observed. For example, Leslie et al [9], have developed a sequence similarity based kernel for protein fold classification. Cheng et al [10], present a good overview of these fold recognition methods. A number of attempts have been made to predict the ß-helix fold across the family of known structure. Heffron et al [3] developed a sequence profile based on pectate lyase family under SCOP classification [11], however, the method could only predict few proteins from pectate lyase and pectin lyase families as RßH proteins. Other general fold prediction methods like PSI-BLAST [12] and HMMer [13], fail to identify sequences across the RßH fold [1,14]. BetaWrap, which depends mainly on ß-strand interactions, shows some success in predicting correct fold across the families in the RßH fold. However, the method when tested on Swiss-Prot & TrEmBL [15] predicts a number of LRR, LβH and other structurally related proteins as right-handed ß-helix proteins. Immunoglobulin-like ß-sandwiches, TIM beta/alpha barrels, the FAD-NAD(P)-binding domain, acid proteases, the subtilisin-like fold and various beta-propeller motifs also appear as false positives [1]. Although these protein classes bear less structural homology with ß-helix proteins, many of these contain amphiphatic ß-sheets that are highly favored by BetaWrap algorithm. Recently, threading has emerged as a very successful method for the prediction of protein folds and structure across the parallel ß-helix fold [16].

In this paper, a systematic attempt has been made to achieve high prediction accuracy for ß-helix fold from proteins sequences. A SVM module, SVM-BetaPred, is developed based on dipeptide compositions (e.g. ala–ala, ala–leu, val–ser) of known ß-helix proteins to correctly identify rungs or coils in protein sequence. These rungs are the characteristic feature of ß-helix proteins, each displaying a conserved B2-T2-B3 pattern. The initial screening indicates that the prediction accuracy of the dipeptide composition based SVM module is superior to the amino acid

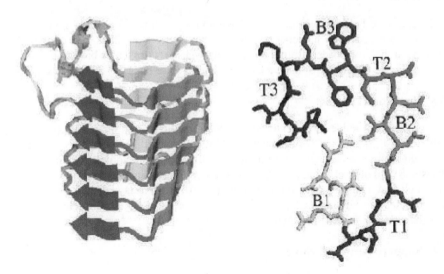

Fig. 1. Ribbon diagram of right-handed single-stranded ß-helix protein (1plu) and schematic representation of a single rung or coil [1]

composition and tripeptide composition based modules. In addition, a position specific scoring matrix (PSSM) is constructed using multiple sequence alignments of known ß-helix proteins in SCOP database. This PSSM is used to calculate a log odds score for each putative rung identified by the dipeptide SVM module. The predicted rungs are wrapped into a ß-helix fold using a number of rules learned from known ß-helices. Finally, a summed score of all the putative rungs identified in a protein is used as a threshold cut off to predict the ß-helix proteins. The method performs at par with some of the state-of-art methods for ß-helix fold prediction.

2   Materials and Methods

2.1   Training and Testing Data Set

All positive and negative datasets for training and testing SVM modules are derived from ß-helix database that consists of right-handed single-stranded proteins in SCOP 1.69 database. The RßH fold consists of seven superfamilies, however, insect anti-freeze proteins (AFPs) are not considered in this study, as their sequence and structure patterns do not represent an ideal ß-helix fold. The method, however, successfully predicts AFPs as ß-helix proteins. The ß-helix database is represented by 28 unique sequences. The SCOP structures are used to annotate the rungs in all proteins. The positive data set consists of 253 rungs of length nine amino acids that are derived from the known ß-helices. The negative data set consists of 261 loop regions of length nine derived from the known ß-helix database. The SVM module is trained with various window size that range from 9 to 15.

2.2   Evaluation Data Set

The negative ß-helix protein dataset, PDB-minus, is composed of 5049 non-β-helical proteins sequences with no sequences having homology more than 33% is used. This dataset can be downloaded from the following website: http://cubic.bioc.columbia.edu/eva/res/unique_list.html. All known ß-helix proteins were removed for this dataset.

2.3   Prediction Data Set

Potential new ß-helices are identified by SVM-BetaPred from the UniProtKB/Swiss-Prot Release 51.2 of 28-Nov-2006 with 243975 sequence entries. This database is filtered to eliminate sequence with more than 40% sequence homology. The resulting dataset, SW40 dataset, contains 67,879 sequences. Incremental redundancy filtering is accomplished using the CD-HIT program [17]. All sequences of length less than 100 amino acids are eliminated from SW40 dataset, as ideal ß-helix proteins cannot be less than 100 amino acids in length.

2.4   Support Vector Machine

Support Vector Machines [SVM] are universal approximators based on statistical and optimization theory [18,19]. A Support Vector Machine (SVM) performs classification by constructing a N-dimensional hyperplane that optimally separates the data into two categories. SVMs have emerged as a high performance tool to solve classification problems. A number of problems in computational biology are classification problems and SVM can be applied to these problems effectively due to its ability to handle noise, large dataset and large input spaces [18,19]. Vapnik et al [20] have described SVM in details. In the current studies, SVMlight is used to predict putative rungs in a protein sequence. This software is freely downloadable from http://svmlight.joachims.org. SVMlight is an implementation of Vapnik's Support Vector Machine [21] for the problem of pattern recognition, regression and learning a ranking function. The software enables the users to define a number of parameters and also allows a choice of inbuilt kernel function, including linear, RBF and Polynomial.

We have developed a dipeptide composition based binary classifier to handle the rung prediction problem. The problem of right-handed β-helix fold prediction can be considered as a classification problem. The main aim is to predict the presence of rungs in a given protein sequence. In this classification problem a rung can be considered as an object $\vec{x}$, which is a part of protein amino acids sequence. A class +1 can be assigned to this object if the center of $\vec{x}$ lies inside a rung or −1 otherwise.

2.5 SVM Features

Amino acid, dipeptide and tripeptide frequency are used to train RBF kernels at default parameters, the accuracies are 72.1%, 89.3% and 78.9% respectively. These frequencies are obtained from the positive rungs and negative loop regions discussed above. The frequencies are obtained using the following equation:

$$Frequency\ of\ k-mer = \frac{Total\ number\ of\ k-mer}{Total\ number\ of\ all\ possible\ k-mers} \tag{1}$$

where k =1, 2 and 3 which give a fixed length pattern of 20, 20x20 and 20x20x20 respectively. SVM accuracy was determined to be highest for dipeptide frequency thus all further training and predictions were restricted to dipeptide frequency. Some of the important dipeptides reported are NV, SG, VT, TI, IT, NI, TV, GA, GG, DV, NS, VI.

2.6 Kernel and Parameters

The selection of the kernel function parameters is an important step for SVM training and testing. All types of kernels i.e. RBF, Polynomial, Sigmoid and Linear were tested, and RBF kernel was identified to have the best performance for rung prediction. A number of parameters need to be determined to generate optimal results, the most important of these are the regularization parameter C and γ for RBF kernel. The prediction accuracy of various parameter values is shown in Table 1. The dipeptide based SVM modules with RBF kernel, C = 18 and γ = 900 is used for training.

Table 1. Prediction accuracy comparison for different parameters for RBF kernel function

SVM Model	Parameters	Accuracy (%)
1	C = default, γ = default	89.93
2	C = 0.1, γ = 1000	96.51
3	C = 1, γ = 1000	97.83
4	C = 10, γ = 1000	98.49
5	C = 18, γ = 900	98.82
6	C = 24, γ = 900	98.13

2.7 SVM Input

Increasing the window size can provide more local sequence information. The window size w is defined as the residue numbers involved in the local sequence windows centered on fifth residue of a rung of length 9, i.e. w = 9, 10, 11,12, 13,14 and 15 in this study. Different local window sizes are used to build the SVM models in order to find one, which could lead to the best performance. The prediction accuracy is shown in Table 2. As expected, the overall prediction accuracy Q2 [22] increases with the enlarging window size and attain its peak at 13. As window size increases so does the background noise while smaller window size results in less useful information. Accordingly, the optimal window size is fixed to 13 for further analysis in this study. The rungs are extended both upstream and downstream to obtain different window size.

Table 2. Predictive performance of SVM based on single sequence inputs of different local window sizes

Window Size	Prediction accuracy (%)			
	Sensitivity	Specificity	Q_2	MCC
9	52.8	94.8	81.7	0.54
10	52.8	95.6	82.2	0.56
11	54.7	96.5	83.4	0.60
12	62.2	98.3	86.9	0.69
13	64.2	100.0	88.7	0.74
14	60.3	99.1	86.9	0.70
15	60.3	97.4	85.8	0.66

2.8 SVM Output

The output from SVM light is in form of real number, a positive value denotes a rung while a negative value denotes a loop region.

2.9 Performance Evaluation

To evaluate the prediction performance of the SVM module, a 5-fold cross-validation method is used. The dataset were randomly divided into five groups, with each group containing roughly equal numbers of proteins. Each group was singled out in turns as the testing dataset, while the remaining proteins in other groups were used as the training dataset. Four different measurements have been used to measure the prediction performance of SVM module [22], these are:

$$sensitivity = \frac{TP}{TP + FN} \tag{2}$$

where, TP is the number of true positives and FN is the number of false negatives or under-predictions.

$$specificity = \frac{TN}{TN + FP} \tag{3}$$

where, TN is the number of true negatives, and FP is the number of false positives or over-predictions.

The overall prediction accuracy Q_2 is given by

$$Q_2 = \frac{TP + TN}{TP + TN + FP + FN} \tag{4}$$

The Matthews Correlation Coefficient (MCC) [23] is given by

$$MCC = \frac{TP \times TN - FP \times FN}{\sqrt{(TP + FP)(TP + FN)(TN + FP)(TN + FN)}} \tag{5}$$

The value of MCC is 0 for a random assignment and 1.0 for a perfect prediction.

2.10 Position Specific Scoring Matrix

All rungs of size 13 were extracted from the β-helices proteins available in SCOP database. These rungs were used to generate a position specific weight matrix. The main aim is to score the putative rungs identified by the dipeptide frequency based SVM module. The number of occurrences of each amino acid at a given position is compiled. Prophecy module in EMBOSS package [24] is used to generate the frequency matrix for amino acid occurrence at a specific position in the rungs. These counts are converted to frequencies that are used to calculate log odds scores. The odds score [25] is the frequency observed divided by the theoretical frequency expected (the background frequency of the amino acid, usually averaged over the protein ~0.05/aminoacid). For example, if the amino acid frequency is 0.79 and the estimated background frequency is 0.25, the odds score would be 0.79/0.25 = 3.16. Finally, odds scores were converted to log odds scores by taking the logarithm base 2.

$$W_{i.j} = \log_2(\frac{F_{i,j}}{P_i}) \tag{6}$$

where $W_{i,j}$ is the scoring matrix value of amino acid i in position j. $F_{i,j}$ is the frequency of base i in position j and P_i is the background frequency of base i.

As the logarithm of zero is infinity, a zero occurrence of a particular base in the matrix creates a problem. In this case, a very low frequency value of 0.0001 is assigned to amino acid count at such a position in a scoring matrix. PSSM is available as supplementary material at http://www.scs.dauniv.ac.in/research.php.

3   The Algorithm

3.1   Rung Prediction in Protein Sequence

The dipeptide composition based SVM module as described above is used to predict rungs in protein sequences. A window of size 13 is used to predict rungs across a protein sequence. All predictions with score of 1 or above are selected as valid rungs. This cutoff was derived from known β-helix proteins. These putative rungs are wrapped into β-helix using a minimum distance threshold of 17 and a maximum distance threshold of 70 amino acids between adjacent rungs. In case, if two adjacent rungs are closer than the minimum distance threshold, then the rung with higher score is selected. Similarly, if the distance between two adjacent rungs is more than the maximum allowed distance and number of valid rungs is less than 5, then valid rungs counter is reset to zero and the existing rungs are rejected. The process is repeated for the rest of the protein sequence till at least a single wrap of 5 valid rungs is obtained. Proteins with a wrap of less than 5 rungs are rejected. Once the wraps are generated they are filtered for the presence of charged residues in rungs. This filter assigns a heavy demerit to rungs for presence of highly charged residues at inward pointing positions in rungs.

3.2   Log Odds Score for Wrap

A log odds score is determined for every predicted wrap using the PSSM. All adjacent rungs are aligned to generate a pairwise log odds score. The PSSM is used to refer the frequencies of occurrence of amino acid X at a position while calculating the score of aligned rung pair. A substitution score is calculated for each position in the two adjacent rungs, for example, log odds score is calculated for occurrence of amino acid X at position 1 in rung A as to the occurrence of amino acid Y at the position 1 in rung B. The log ratio for all the thirteen position of rungs is summed to give a pairwise log odds score. A random score of −1000 is assigned to alignment of a rung with itself. Once a pairwise log odds score is calculated for all the adjacent rung pairs, a log odds score for a wrap is calculated by taking the sum of pairwise log odds scores. All proteins with a wrap log odds score of -35 or more are further processed. The log odds score derived using PSSM helps in filtering a number of proteins that appear as false positives in SVM predictions, for example, LRR and amphiphatic proteins that have amino acids composition similar to β-proteins.

3.3 Completing the Parse

Now that the position of rungs have been determined using SVM module, the position of B1 strands are determined using PSIPRED [26] predictions. Small β-sheets of average length four amino acids cannot be effectively predicted by SVM module, thus the secondary structure predictions from PSIPRED are used to place B1 strands between adjacent B2-T2-B3 rungs. A rung that lacks a predicted B1 strand is heavily penalized.

3.4 α-Helix Filter

The secondary structure predictions output from PSIPRED is used to screen for α-helices in the β-helical wrap. If the α-helical content for a given wrap is more than 20% then the wrap is disqualified and the protein is rejected. The threshold value of 20% is determined from known β-helix proteins. The known β-helix proteins show an average α-helical content of 8-14% [27]. All proteins that pass α-helix filter and have a log odds score more than -35 are predicted as right-handed β-helix proteins. Similarly, if a putative rung shows presence of 4 or more amino acids to have a α-helix secondary structure in PSIPRED results, then the rung is rejected.

4   Results

SVM-BetaPred performs at par in comparison to the state-of-art right-handed β-helix prediction methods. It recognizes the RβH-helix fold with 100% sensitivity and 99.6% specificity as compared to BetaWrapPro [14] with 94.1% sensitivity and 99.4% specificity for known β-helix and non-β-helix proteins data set discussed in the materials and methods section. The method also succeeds in predicting the accurate rung position across the β-helix fold. It achieves 63.0% sensitivity and 68.1% specificity as compared to 58.0% sensitivity and 55.0% specificity for BetaWrapPro and 65.0% sensitivity and 46.3% specificity for BetaWrap on same set of known RβH proteins. Homology search tools like PSI-BLAST and HMMER fail to recognize RβH fold across superfamilies [1,14], thus results for SVM-BetaPred method are compared only with most accurate publicly available. A comparison of SVM-BetaPred results is shown in Table 3. It should be noted that both BetaWrapPro and BetaWrap use HMMER to filter Left-handed β-helix and Leucine Rich Repeat proteins that appear as false positive predictions [1,14]. No such filter is used to the current SVM method.

The performance of dipeptide composition based SVM module is shown in Table 4. The performance of SVM modules was evaluated through 5-fold cross-validation. The dipeptide composition based SVM module (kernel = RBF, $\gamma = 18$ and C = 900) was able to predict rung with an average Q2 of 89.2% and MCC of 0.75.

SVM-BetaPred predicts 498 proteins as RßH proteins. The pectate lyases and galacturonases are well represented among the predicted proteins. However, few of the high-scoring putative proteins that include ribosomal proteins, polymerase and other DNA replication enzymes; repeat proteins like WD repeat proteins and

Table 3. Comparison of SVM-BetaPred, BetaWrapPro and BetaWrap results. Table denotes the percentage sensitivity (Sn) and specificity (Sp) obtained for known β-helices and rungs, dataset used for prediction (DB), number of proteins in dataset, number of predicted β-helix proteins (TP), false prediction in PDB minus dataset (FPmin) and PDB database (FP PDB) and incorrectly predicted folds (Folds). * denotes information in McDonnell et al [14].

Tool	Sn	Sp	Rung Sn	Rung Sp	DB	# of proteins	TP	FP min	FP PDB	Folds
SVM - Beta Pred	100	99.6	63.0	68.1	Swiss-Prot 40 (Release 51.2)	67,835	498	18	42	β– san dwich/ barrel, jelly rolls, repeat protein
Beta Wrap Pro	94.1	99.4	58.0	55.0	Swiss-Prot 40 (Release 44.0)	48,269	774	23	57	β–prop eller/sa ndwich /barrel, α helix, α+β pr otein
Beta Wrap	93.8	98.5	65.0	46.3	Swiss-Prot (39.6) and TrEMB L (14.11)	595,890	--	>300 *	-	β–prop eller/sa ndwich /barrel, α helix, α+β pr otein

Table 4. The predictive performance of dipeptide composition based SVM to predict rungs in protein sequences. The results were obtained by 5-fold cross-validation.

Data set	Prediction accuracy (%)			
	Sensitivity	Specificity	Q_2	MCC
1	64.2	100	88.7	0.74
2	68.0	100	90.1	0.77
3	63.3	100	88.8	0.73
4	65.3	100	89.4	0.75
5	63.3	100	88.8	0.73

transmembrane proteins are likely false positives predictions from independent evidences and high homology to known structures. Table 5 available as supplementary material at http://www.scs.dauniv.ac.in/research.php lists all predicted right-handed ß-helix proteins, each ranked by p-value, Z-score, score, accession number, ID, source organism, description and the wrap position.

4.1 Recognition of Unknown Sequences

SVM-BetaPred identifies a number of probable right-handed ß-helices in the SW40 data set. Some of these include Pectate lyase precursor from *Pseudomonas fluorescens* (P0C1A7), *Bacillus subtilis* (P39116); Pectin lyase from *Saccharomyces cerevisiae* (P47180), *Agrobacterium tumefaciens* (P27644), *Gibberella fujikuroi* (Q07181), *Actinidia chinensis* (P35336) and *Pseudomonas sp.* (P58598); Chondroitinase from *Pedobacter heparinus* (Q46079); Carrageenase from *Alteromonas carrageenovora* (P43478) and Dextranase from *Penicillium minioluteum* (P48845). As reported earlier by Jenkins et al [5], there is a clear bias for the occurrence of RßH fold across the major group of organisms. Only ~20% of the proteins predicted to contain RßH fold belong to eukaryotes, furthermore only a few archeal and viral proteins show the presence of this fold. We found that proteins with p-value < 0.65 have a strong likelihood to display right-handed β-helix fold.

SVM-BetaPred successfully identifies newly solved β-helical protein hemoglobin protease (1wxr) from *Escherichia coli* with a p-value of 0.225 and β-roll subunit from C5 epimerase (2agm) from *Azotobacter vinelandii* with a p-value of 0.522 as RßH protein. Interestingly both BetaWrapPro and BetaWrap fail to identify 2agm as right-handed β-helix.

5 Discussions

Machine learning methods like SVMs and neural networks are highly successful for residue state prediction where fixed window/pattern length is used [28]. In order to make optimal use of these techniques for protein structure prediction a fixed-length pattern must be generated. Amino acids composition that gives a fixed pattern length of 20 is commonly used by AI techniques for the classification of proteins. More information can be supplied by using by dipeptide composition. It gives a fixed pattern length of 400. Dipeptide composition has been widely used for the development of fold prediction methods [29] to achieve higher accuracies than that of amino acid composition based methods. A further step would be the use of tripeptide frequencies, however, AI techniques are unable to handle the noise due to the large number of input units and number of missing tripeptides in a protein. Thus, in this paper, we have constructed a SVM module on the basis of the dipeptide composition of β-helix proteins. This module is able to predict the rungs in a protein with overall accuracy of 90.1%, as shown in Table 4.

To further improve prediction accuracy, a PSSM is developed to encapsulate more comprehensive information of β-helix proteins protein. This position weight matrix is used to score the rungs identified by SVM module and thus further enhance the β-helix fold prediction accuracy. The results confirmed that our approach is capable of capturing more information about super secondary structures like rungs that are vital to β-helix fold prediction. The method would complement the existing prediction tools for β-helix prediction.

References

1. Bradley, P., Cowen, L., Menke, M., King, J., Berger, B.: BETAWRAP: Successful prediction of parallel beta helices from primary sequence reveals an association with many microbial pathogens. Proc. Natl. Acad. Sci. 98, 14819–14824 (2001)
2. Yoder, M.D., Jurnak, F.: The parallel β helix and other coiled folds. FASEB J. 9(5), 335–342 (1999)
3. Heffron, S., Moe, G., Sieber, V., Mengaud, J., Cossart, P., Vitali, J., Jurnak, F.: Sequence profile of the parallel beta helix in the pectate lyase superfamily. J. Struct. Biol. 122, 223–235 (1998)
4. Yonder, M., Keen, N., Jurnak, F.: New domain motif: The structure of pectate lyase C, a secreted plant virulence factor. Science 260(5113), 1503–1507 (1993)
5. Jenkins, J., Shevchik, V.E., Hugouvieux-Cotte-Pattat, N., Pickersgill, R.W.: The crystal structure of Pectate Lyase Pel9A from Erwinia chrysabthemi. J. Biol. Chem. 279(10), 9139–9145 (2004)
6. Iengar, P., Joshi, N.V., Padmanabhan, B.: Conformational and Sequence Signatures in β Helix Proteins. Structure 14(3), 529–542 (2006)
7. Kreisberg, J.F., Betts, S.D., King, J.: βeta-helix core packing within the triple-stranded oligomerization domain of P22 tailspike. Protein Sci. 9(12), 2338–2343 (2000)
8. Jenkins, J., Mayans, O., Pickersgill, R.: Structure and evolution of parallel helix proteins. Journal of Struct. Biol. 122, 236–246 (1998)
9. Leslie, C., Eskin, E., Noble, W.S.: The spectrum kernel: a string kernel for SVM protein classification. Pac. Symp. Biocomput., 564–575 (2002)
10. Cheng, J., Baldi, P.: A machine learning information retrieval approach to protein fold recognition. Bioinformatics 22(12), 1456–1463 (2006)
11. Murzin, A., Brenner, S., Hubbard, T., Chothia, C.: SCOP: a structural classification of proteins database for investigation of sequences and structures. J. Mol. Bio. 297, 536–540 (1995)
12. Altschul, S., Madden, T., Schaffer, A., Zhang, J., Zhang, Z., Miller, W., Lipman, L.: Gapped BLAST and PSI-BLAST: a new generation of protein database search programs. Nucleic Acids Res. 25, 3389–3402 (1997)
13. Eddy, S., Mitchison, G., Durbin, R.: Maximum discrimination hidden Markov models of sequence consensus. J. Comput. Biol. 2, 9–23 (1995)
14. McDonnell, A.V., Menke, M., Palmer, N., King, J., Cowen, L., Berger, B.: Prediction and comparative modeling of sequences directing beta-sheet proteins by profile wrapping. Proteins: Structure, Function, and Bioinformatics 63, 976–985 (2006)
15. Bairoch, A., Apweiler: The SWISS-PROT protein daabse and its supplement TrEMBL in 2000. Nucleic Acids Res. 28, 45–48 (2000)
16. Govaerts, C., Wille, H., Prusiner, S.B., Cohen, F.E.: Evidence for assembly of prions with left-handed beta-helices into trimers. Proc. Natl. Acad. Sci. USA 101(22), 8342–8347 (2004)
17. Li, W., Jaroszewski, L., Godzik, A.: Sequence clustering strategies improve remote homology recognitions while reducing search times. Protein Eng. 15(8), 643–649 (2002)
18. Zavaljevski, N., Stevens, F.J., Reifman, J.: Support vector machines with selective kernel scaling for protein classification and identification of key amino acid positions. Bioinformatics 18, 689–696 (2002)
19. Bhasin, M., Raghava, G.P.S.: ESLpred: SVM-based method for subcellular localization of eukaryotic proteins using dipeptide composition and PSI-BLAST. Nucleic Acids Research 32, W414–W419 (2004)

20. Vapnik, V.N.: The Nature of Statistical Learning Theory. Springer, Heidelberg (1995)
21. Joachims, T.: Making large-scale SVM learning practical. In: Scholkopf, B., Burges, C., Smola, A. (eds.) Advances in Kernel Methods—Support Vector Learning. MIT Press, Cambridge, MA, London, England (1999)
22. Song, J., Burrage, K., Yuan, Z., Huber, T.: Prediction of cis/trans isomerization in proteins using PSI-BLAST profiles and secondary structure information. BMC Bioinformatics 7, 124 (2006)
23. Matthews, B.W.: Comparison of predicted and observed secondary structure of T4 phage lysozyme. Biochim. Biophys. Acta. 405, 442–451 (1975)
24. Rice, P., Longden, I., Bleasby, A.: EMBOSS: The European Molecular Biology Open Software Suite. Trends in Genetics 16(6), 276–277 (2000)
25. Qiu, P., Cai, X.Y., Wang, L., Greene, J., Malcolm, B.: Hepatitis C virus whole genome position weight matrix and robust primer design. BMC Microbiology 2, 29 (2002)
26. Bryson, K., McGuffin, L.J., Marsden, R.L., Sodhi, J.S., Jones, D.T.: Protein structure prediction servers at University College London. Nucleic Acids Res. 1, 33 (2005)
27. Freiberg, A., Morona, R., Bosch, L., Baxa, U.: The Tailspike Protein of Shigella Phage Sf6. J. Biol. Chem. 278(3), 1542–1548 (2003)
28. Krogh, A., Riis, S.K.: Prediction of b sheets in protein. In: Touretzky, D.S., Mozer, M.C., Hasaselmo, M.E. (eds.) Advances in Neural Information Processing System 8, pp. 917–923. MIT Press, Cambridge, MA (1996)
29. Reczko, M., Bohr, H.: The DEF database of sequence based protein fold class prediction. Nucleic Acid Res. 22, 3616–3619 (1995)

Protein Fold Recognition
Based Upon the Amino Acid Occurrence

Y.-h. Taguchi[1] and M. Michael Gromiha[2]

[1] Department of Physics, Chuo University, 1-13-27 Kasuga, Bunkyo-ku, Tokyo
112-8551, Japan
tag@granular.com

[2] Computational Biology Research Center (CBRC), National Institute of Advanced
Industrial Science and Technology (AIST), AIST Tokyo Waterfront Bio-IT Research
Building, 2-42 Aomi, Koto-ku, Tokyo 135-0064, Japan
michael-gromiha@aist.go.jp

Abstract. We have investigated the relative performance of amino acid
occurrence and other features, such as predicted secondary structure, hydrophobicity, normalized van der Waals volume, polarity, polarizability,
and real/predicted contact information of residues, for recognizing protein folds. We observed that the improvement over other features is only
marginal compared with amino acid occurrence. This is because amino
acid occurrence, indirectly, can consider varieties of physical properties
which are useful to discriminate protein folds. If we consider only proteins which are well aligned structurally with each other, the accuracy of
discrimination is drastically improved. In order to discriminate protein
folds more accurately, we need to consider anything other than structure
alignment.

1   Introduction

Deciphering the native conformation of a protein from its amino acid sequence
known as protein folding problem is a challenging task. The recognition of proteins of similar folds is a key intermediate step for protein structure prediction.
Alignment profiles are widely used for recognizing protein folds [1,2]. Recently,
Cheng and Baldi [3] proposed a machine learning algorithm using secondary
structure, solvent accessibility, contact map and β-strand pairing for fold recognition, which showed the pair wise sensitivity of 27%. On the other hand, it
has been reported that the amino acid properties are the key determinants of
protein folding and are used for discriminating membrane proteins [4], identification of membrane spanning regions [5], prediction of protein structural classes
[6], protein folding rates [7], protein stability [8] etc. Towards this direction, Ding
and Dubchak [9] proposed a method based on neural networks and support vector machines for fold recognition using amino acid composition and five other
properties, and reported a cross-validated sensitivity of 45 %.

Recently [10], we have used the amino acid occurrence (not composition)
of proteins belonging to 30 major folds for recognizing protein folds. We have

J.C. Rajapakse, B. Schmidt, and G. Volkert (Eds.): PRIB 2007, LNBI 4774, pp. 120–131, 2007.
© Springer-Verlag Berlin Heidelberg 2007

developed a method based on linear discriminant analysis (LDA), which showed an accuracy of 37% for recognizing 1612 proteins from 30 different folds, which is comparable with other methods in the literature, in spite of the simplicity of our method and the large number of proteins considered.

In this paper, we have compared the performance of other features with that of amino acid occurrence. We have found that amino acid occurrence outperform other features to discriminate protein folds. Even if other features are considered together with amino acid occurrence, the ability to discriminate protein folds is hardly improved. On the hand, if we exclude pairs of proteins with poor 3D structural alignment, we have found that discrimination by amino acid occurrence is drastically improved. In conclusion, amino acid occurrence turns out to be the best feature to discriminate protein folds.

2   Materials and Methods

We have used three data sets to test the performance of our method. The first data set is that used by Ding and Dubchak [9]. It is available from their web site. The second data set is that used in the previous study [10]. It consists of 1612 amino acid sequence among which there are less than 25 % sequence identity. These amino acid sequence is taken from SCOP [11] and belong to one of major 27 folds. It is available from our prediction server [12]. We also used several feature (contact) vectors corresponding to these data set. The third one is taken from CATH [13]. It consists of 4146 amino acid sequences with less than 40 % mutual sequence identity. These belong to one of major 39 topologies. The selection of major 39 topologies is based upon Gubbi *et al* [14].

Since the method is described in our previous report [10], we have briefly outlined our methods. First, we have counted the number of amino acid residues in each amino acid sequence. Thus, we have 20 dimensional integer vector for each protein. Then, LDA is applied to this set of vectors. LDA we used is lda module in R [15]. Although there are many ways to weight the discrimination [10], in this paper we weight each fold(topology) equally. In other words, prior probability of each fold (topology) is assumed to be equal. As a measure of performance, we employ accuracy Q,

$$Q = \frac{\sum_i TP_i}{N}, \tag{1}$$

where TP_i is the number of proteins correctly discriminated in ith fold (topology) and N is total number of proteins considered. In the following, Q by leave one out cross validation wil be reported.

3   Results

3.1   Accuracy of Discrimination Using Amino Acid Occurrence and Other Features

Ding and Dubchak [9] has discriminated 27 folds for 311 proteins. They have used support vector machine (SVM) and/or nerual networks (NN) with voting

Table 1. Accuracy Q as a function of used features

Features	Q [%]	References
Composition	35	
Composition + length	38	
Composition + five features	39	
Composition + five features	45	Ding and Dubchak [9]
Occurrence	42	Our previous report [10]
Occurrence + five features	44	

system. Features they used differ from amino acid occurrence. They have reported that their method achieved $Q = 45\%$ as 10-fold cross validation results. In the previous report [10], we have shown that our method can achieve $Q = 42\%$ in spite of simplicity of our method. Since ours are leave one out cross validation, our Q value does not have any statistical errors. However, Ding's value is 10-fold cross validation. If we consider this, our value $Q = 42\%$ is comparative with Ding's value $Q = 45\%$. Since our method uses solely amino acid occurrence while Ding and Dubchak used many other features than amino acid occurrence together, it is natural to expect that considering other features together with amino acid occurrence can improve accuracy Q. In Table 1, we have summarized Q as a function of used features. When we use more than two features to discriminate folds, we simply apply LDA to merged feature vectors. This means, if there are two features vectors $\boldsymbol{f}_n$ with n components and $\boldsymbol{f}_m$ with m features,

$$\boldsymbol{f}_n = (f_{n1}, f_{n2}, \ldots, f_{nn}) \tag{2}$$
$$\boldsymbol{f}_m = (f_{m1}, f_{m2}, \ldots, f_{mm}), \tag{3}$$

then we merge these two and apply LDA to

$$\boldsymbol{f}_{m+n} = (f_{n1}, f_{n2}, \ldots, f_{nn}, f_{m1}, f_{m2}, \ldots, f_{mm}). \tag{4}$$

Additional five features, i.e., predicted secondary structure, hydrophobicity, normalized van der Waals volume, polarity, polarizability, are those Ding and Dubchak [9] used. Since their method is sophisticated, and it utilized all features (i.e., composition + five features), their Q is better than us by 6 %. In spite of that, our simple method employing amino acid occurrence and five features has almost fulfilled this gap ($Q = 44\%$). If we consider the simplicity of our method, our method is even better than Ding and Dubchak's method.

If we see Table 1 more detail, we can find many interesting things. For example, if we consider only composition, Q is only 35 %, which is 7 % smaller than $Q = 42\%$ when we consider only occurrence. On the other hand, if we consider composition and length, i.e., the first 20 components of feature vectors consist of composition and the 21th component is amino acid length, Q raises from 35% to 38 %. In spite of that, if we consider composition and five features, Q becomes 39 %, which is as the same as $Q = 38\%$ when composition and length are considered. Thus, solely considering length is comparative with considering all of five features. This definitely demonstrate the importance of considering amino acid length. This is the reason why considering occurrence instead of composition can improve accuracy by 7%.

3.2 SCOP

Since our method is simple, we can deal with larger data set. In our previous report [10] we have applied our method to 1612 proteins belonging to 30 major folds in SCOP. Q which we have achieved was 33 %. In this subsection, we have compared Q when considering other features than amino acid occurrence with Q when only amino acid occurrence is considered. In Table 2, we have listed Q obtained using other features than amino acid occurrence. Other features we used is average contacts in different sequence intervals, for example, 3-4, >4, 5-10, 11-20, 21-30, 31-40, 41-50 and >50. The contacts are predicted by several different contact prediction servers [16,17,18] and are taken from real structure. For some predicted contacts, number of amino acid sequence considered is less than 1612. Clearly, the real contact information outperformed in discrimination. On the other hand, the performance of amino acid occurrence is better than that with predicted contacts.

We also consider dipeptide occurrence. Since there are 20 amino acids, dipeptide occurrence n_{ij} are 400 kinds, where n_{ij} is the dipeptide occurrence for ith and jth amino acid ($1 \leq i, j \leq 20$). In Table 3, we have shown Q for considering dipeptide occurrence. Consideration of dipeptide occurrence does not improve Q at all. Even if we consider dipeptide occurrence together with amino acid occurrence, Q is not improved.

It may be assumed that the consideration of dipeptide occurrence would improve the Q value. However, we observed that the Q value is less than that with

Table 2. Accuracy Q as a function of used feature

Feature	number of sequences	Q [%]	References
Occurrence	1612	33	Our previous report [10]
Composition	1612	26	Our previous report [10]
cornet	1530	22	Ref. [16]
nick	1555	13	Ref. [17]
gpcpred	1612	15	Ref. [18]
real structure	1612	50	

Table 3. Accuracy Q for discrmination by dipeptide

Feature	$Q[\%]$	References
Occurrence	33	Our previous report [10]
Dipeptide	29	
Dipeptide+Occurrence	31	

amino acid occurrence. It might be due to the fact that dipeptide occurrence is not an independent information of amino acid occurrence, because

$$n_i = \sum_j n_{ij}. \tag{5}$$

Further, Table 3 shows that dipeptide occurrence cannot have more information than that solely amino acid occurrence can provide.

3.3   CATH

In order to see if our method can discriminate other fold classification than SCOP, we have considered topologies in CATH. In Table 4, we have shown the

Table 4. Accuracy Q [%] for SCOP and CATH with various sorts of definition of Q and weighting. Bold numbers are the same as those in other tables.

	with re-weighting		without re-weighting	
	over all fold	average	over all fold	average
CATH	**26**	34	43	24
SCOP	**33**	32	37	28

comparison between CATH and SCOP. It is clear that the performance depends upon the definition of methods/accuracy which data base can be discriminated better by our method. Especially, CATH is very sensitive to the variety of definition of methods/accuracy used in these databases and our method couldwell discriminate the folds. The lowest Q is 26 % for CATH while the highest one is 43 %. Thus, the later is larger than the former by more than 50 %. This results show how difficult to decide what the *best* discrimination is. When we consider the definition of Q used in the present research, CATH ($Q = 26\%$) is harder to discriminate than SCOP ($Q = 33\%$). Although it is generally true that our method can discriminate folds no matter how they are defined, the performance is strictly dependent how we measure the goodness. Although we do not consider CATH in more detail here, one has to be careful how we can measure the goodness of discrimination.

4   Discussion

4.1   Why Does Occurrence Work So Well?

In the previous section, we have shown that our method (amino acid occurrence + LDA) can discriminate protein folds up to 30 to 40 % for up to thousands proteins and up to 40 folds (topology). We have also shown that considering other feature than amino acid occurrence generally can hardly improve accuracy Q. In this subsection, we would like to discuss why amino acid occurrence works well.

First of all, it is natural that occurrence is better than composition in contrast to the first impression. Suppose we have some protein belonging to one fold. Then, try to duplicate its amino acid sequence. Clearly, there will be very few possibility that duplicated protein belong to same fold. This discussion definitely show that composition cannot detect this effect at all, because duplication cannot change composition. The importance of protein length can be seen in Table 1. Consideration of protein length in addition to composition can improve Q by 3 %, which is as large as half of difference between composition and occurrence. In conclusion, we had better to consider occurrence than composition to discriminate protein folds.

Second, one may think it is strange that consideration other feature than occurrence cannot improve accuracy Q so much. However, *any* physical feature can be more or less expressed by amino acid occurrence. Thus, linear combination of amino acid occurrence can express more or less many of physical properties of proteins. In order to see this, we have computed the correlation coefficients between 49 physical, chemical energetic ans conformational properties of each amino acid [19,20,21] and the first discriminate function for the second data set case (i.e., 1612 proteins belonging to 30 major folds in SCOP). Each property consists of 20 dimensional vector, like

$$\boldsymbol{P}^k = (P_1^k, P_2^k, \ldots, P_i^k, \ldots, P_{20}^k), \tag{6}$$

where P_i^k describe kth physical properties of ith amino acid. Since discriminant function is also 20 dimensional vector each component of which describe contribution from each amino acid, we can take correlation coefficient between them.

As can be seen in Table 5, 23 out of 49 properties have correlation coefficients with less than 5 % q-values (i.e., FDR corrected p-values). We can find 24 out of 49 properties have less than 5% q-value if we apply the same procedure to the third data set(CATH), although the number of commonly selected properties is as large as those by chance. (Not shown here). Thus, it is clear that linear discriminant function can express many of physical properties, at least, partly. Thus, even if we do not consider physical properties directly, amino acid occurrence can express them if some of physical properties are important for the discrimination of folds. This is the reason why the consideration of amino acid occurrence can discriminate folds (topologies for CATH) well.   As another example of how well amino acid occurrence can express other physical properties, we consider contact information in Table 2. In order to check if amino acid occurrence can express contact information taken from real structure (the last row in Table 2), which achieved 50 % accuracy Q, we have applied multiple linear regression,

$$f^\ell = \sum_{i'} C_{i'} n_{i'}^\ell + C_0^\ell, \tag{7}$$

where f^ℓ is the contact information of lth protein, n_i^ℓ is ith amino acid occurrence for ℓth protein. In Table 6, we have shown squared partial correlation coefficients based upon (7). Although these values are not so high, they are too large to

Table 5. Brief descriptions of 49 selected physico-chemical, energetic and conformational properties, their correlation coefficient with the first discriminate function for SCOP, and q-value. Asterisks in the last column shows q-value is less than 5 %.

No. Description	Corr. Coef.	q-value [%]	$q \leq 5\%$
1. Compressibility	0.03	44.5	
2. Thermodynamic transfer hydrophobicity	0.44	5.4	
3. Surrounding hydrophobicity	0.72	0.3	*
4. Polarity	0.47	4.5	*
5. Isoelectric point	0.19	25.8	
6. Equilibrium constant with reference to the ionization property	0.07	41.4	
7. Molecular weight	0.05	42.8	
8. Bulkiness	0.36	10.4	
9. Chromatographic index	0.65	0.4	*
10. Refractive index	0.19	25.8	
11. Normalized consensus hydrophobicity	0.42	6.1	
12. Short and medium range non-bonded energy	0.17	28.9	
13. Long-range non-bonded energy	0.75	0.3	*
14. Total non-bonded energy	0.68	0.3	*
15. Alpha-helical tendency	0.08	39.4	
16. Beta-helical tendency	0.68	0.3	*
17. Turn tendency	0.54	2.3	*
18. Coil tendency	0.45	5.0	*
19. Helical contact area	0.24	21.8	
20. Mean rms fluctuational displacement	0.66	0.3	*
21. Buriedness	0.68	0.3	*
22. Solvent accessible reduction ratio	0.68	0.3	*
23. Average number of surrounding residues	0.67	0.3	*
24. Power to be at the N-terminal of alpha helix	0.45	5.0	*
25. Power to be at the C-terminal of alpha helix	0.58	1.2	*
26. Power to be at the middle of alpha helix	0.04	44.5	
27. Partial-specific volume	0.28	16.7	
28. Average medium-range contacts	0.02	46.0	
29. Average long-range contacts	0.71	0.3	*
30. Combined surrounding hydrophobicity (globular and membrane)	0.72	0.3	*
31. Solvent accessible surface area for denatured protein	0.13	33.7	
32. Solvent accessible surface area for native protein	0.56	1.8	*
33. Solvent accessible surface area for protein unfolding	0.51	3.0	*
34. Gibbs free energy change of hydration for unfolding	0.31	14.4	
35. Gibbs free energy change of hydration for denatured protein	0.42	6.1	
36. Gibbs free energy change of hydration for native protein	0.49	3.7	*
37. Unfolding enthalpy change of hydration	0.01	46.2	
38. Unfolding entropy change of hydration	0.51	3.0	*
39. Unfolding hydration heat capacity change	0.69	0.3	*
40. Unfolding Gibbs free energy change of chain	0.16	29.9	
41. Unfolding enthalpy change of chain	0.26	18.7	
42. Unfolding entropy change of chain	0.50	3.1	*
43. Unfolding Gibbs free energy change	0.37	9.7	
44. Unfolding enthalpy change	0.35	10.7	
45. Unfolding entropy change	0.33	11.9	
46. Volume (number of non-hydrogen side chain atoms)	0.10	37.0	
47. Shape (position of branch point in a side-chain)	0.20	25.6	
48. Flexibility (number of side-chain dihedral angles)	0.22	23.4	
49. Backbone dihedral probability	0.33	11.9	

neglect. Actually speaking, p-values for these is less than 1×10^{-14}. If we consider higher order,

$$f^{\ell} = \sum_{i'} \left[C_{i'} n_{i'}^{\ell} + C_{i'}^{2} \left(n_{i'}^{\ell} \right)^{2} \right] + C_{0}^{\ell}, \tag{8}$$

Table 6. Squared Partial Correlation Coefficients for (7) and (8)

Contact range	Squared Partial Correlation Coefficients	
	1st order (7)	2nd order (8)
3-4	0.29	0.37
>4	0.25	0.32
5-10	0.20	0.25
11-20	0.23	0.28
21-30	0.12	0.17
31-40	0.08	0.12
41-50	0.07	0.10
>50	0.19	0.25

squared partial correlation coefficients has increased (Here we have confirmed that Akaike Information Criterion (AIC) has decreased by considering higher order in order to avoid over fittings). This again demonstrates that amino acid occurrence can express physical properties which are useful for discrimination of protein folds.

In conclusion, in contrast to the intuition, amino acid occurrence can express, at least partially, variety of physical properties with which protein folds can be discriminated.

4.2 Folds vs Structural Alignments

Although we have considered many other features than amino acid occurrence, accuracy Q cannot be improved so much. This is because the amino acid occurrence can have ability to express other physical features as discussed in the previous subsection. In this subsection, we try to estimate the relationship between the goodness of structural alignment and the goodness of fold recognition. If structural alignment between proteins belonging to the same fold is poor, it is natural that fold recognition is not successful.

In order to check this point, we have employed the third data set (4146 proteins belonging to 39 major folds). We have randomly picked up 100 pairs of proteins from each of intra/inter topology pairs. For example, when we consider inter topology pairs from topology I and J, pairs of proteins are taken such that one of pair belongs to topology I while another of pair belongs to topology J. On the other hand, when considering intra topology pairs, both of proteins are taken from the same topologies.

Then for selected pairs, structural alignment has been done using Matalign [22] which can get structural alignment even for chopped sequence which frequently appears in CATH. We employ $Nscore$ as a measure of goodness of structural alignment,

$$Nscore \equiv \frac{3N_a}{1 + \Delta} \frac{1}{\min(\text{length1}, \text{length2})},\qquad(9)$$

where N_a is number of aligned residues, Δ is root mean squared deviation (RMSD) and length1 and length2 are number of residues of two aligned proteins. Larger *Nscore* means better structural alignment. Then, we have found many intra topology pairs have less than or equal to *Nscores* of inter topology pairs (Fig. 1(a)). This means, solely the goodness of structural alignment cannot decide if a pair of proteins belong to the same topology or not.

Here, we have considered the pair of topologies (3 10 129) and (3 30 360) for which our method get the least accuracy *Q*. Then by applying structural alignment to all pairs among these two topologies, we have excluded intra protein pairs which have poorer structural alignment. Although *Nscores* within (3 10 129) are always larger than those between two topologies (Fig. 1 (b)), those

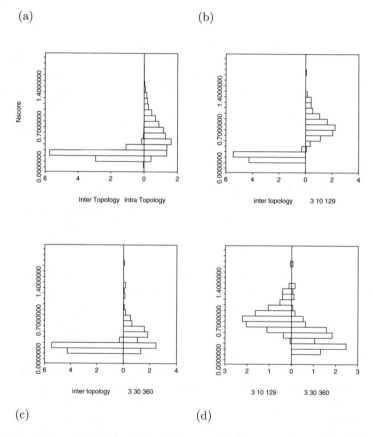

Fig. 1. Comparisons of histogram of *Nscore*. (a) Inter topology (left half) vs intra topology (right half) for all of 39 topologies considered. Each histrgam is normalized such that total area is unity no matter how many pairs are considered. Alignment has been done randomly sampled 100 pairs of proteins for each of pairs of topologies. (b) *Nscore* between topologies (3 10 129) and (3 30 360) (left half) vs that within (3 10 129) (right half) (c) *Nscore* between topologies (3 10 129) and (3 30 360) (left half) vs that within (3 30 360) (right half) (d) Nscore within (3 10 129) (right half) vs that within (3 30 360) (right half).

Table 7. Sensitivity and accuracy for pair wise discrimination between topologies (3 10 129) and (3 30 360). Before : before exclusion of poorly structural aligned pairs of proteins. After : after exclusion of poorly structural aligned pairs of proteins.

	Sensitivity		
	(3 10 129)	(3 30 360)	Q
Before			
Number of Proteins	19	20	39
	0.58	0.40	0.48
After			
Number of Proteins	19	6	25
	0.76	0.83	0.76

within (3 30 360) are not (Fig. 1 (c)). Clearly, those within (3 30 360) are less than those within (3 10 129) (Fig. 1 (d)). Then we found that our method can have better Q by exclusion of pairs of proteins that have poor suructurally alignment. In Table 7, we have shown the comparison of sensitivity and accuracy Q before and after exclusion of badly aligned pairs of proteins. Although no proteins are removed from topology (3 10 129), 14 out of 20 proteins are excluded from (3 30 360). Then sensitivity and accuracy drastically increases. Especially, it is remarkable that sensitivity for (3 10 129) also increases although there are no proteins removed. Thus, it is clear that our method can discriminate topologies if they are well structurally aligned. In other words, we have to consider something other than structural alignment to discriminate topologies in CATH. It is very important to find what we should consider.

5   Conclusion

In this paper, we have investigated the relative performance of amino acid occurrence and other features to recognize protein folds. We found that consideration of other features than amino acid occurrence cannot improve accuracy Q so much. The reason is because amino acid occurrence can have ability to consider variety of physical properties which are useful to discriminate protein folds. It is conformed that our method can better discriminate topologies if proteins within each topology have good structural alignment. In order to improve accuracy Q, we have to consider something other than structural alignment.

Acknowledgement

This work has been supported by the Grant-in-Aid for Creative Scientific Research No.19500254 of the Ministry of Education, Culture, Sports, Science and Technology (MEXT) from 2007 to 2008. We are grateful for their support.

References

1. Shi, J., Blundell, T.L., Mizuguchi, K.: FUGUE: sequence-structure homology recognition using environment-specific substitution tables and structure-dependent gap penalties. J. Mol. Biol. 310, 243–257 (2001)
2. Zhou, H., Zhou, Y.: Fold recognition by combining sequence profiles derived from evolution and from depth-dependent structural alignment of fragments. Proteins 58, 321–328 (2005)
3. Cheng, J., Baldi, P.: A machine learning information retrieval approach to protein fold recognition. Bioinformatics 22, 1456–1463 (2006)
4. Gromiha, M.M., Suwa, M.: A Simple statistical method for discriminating outer membrane proteins with better accuracy. Bioinformatics 21, 961–968 (2005)
5. Hirokawa, T., Boon-Chieng, S., Mitaku, S.: SOSUI: classification and secondary structure prediction system for membrane proteins. Bioinformatics 14, 378–379 (1998)
6. Chou, K.C.: Prediction of protein structural classes and subcellular locations Curr. Protein Pept. Sci. 1, 171–208 (2000)
7. Gromiha, M.M., Selvaraj, S., Thangakani, A.M.: A Statistical method for predicting protein unfolding rates from amino acid sequence. J. Chem. Inf. Model 46, 1503–1508 (2006)
8. Gromiha, M.M., Oobatake, M., Kono, H., Uedaira, H., Sarai, A.: Relationship between amino acid properties and protein stability: Buried Mutations. J. Protein Chem. 18, 565–578 (1999)
9. Ding, H.Q.D., Dubchak, I.: Multi-class protein fold recognition using support vector machines and neural networks. Bioinformatics 17, 349–358 (2001)
10. Taguchi, Y.-h., Gromiha, M.M.: Comparison of amino acid occurrence and composition for predicting protein folds. IPSJ SIG Technical Report 2007-BIO-008, pp. 9–16 (2007)
11. Murzin, A.G., Brenner, S.E., Hubbard, T., Chothia, C.: SCOP: a structural classification of proteins database for the investigation of sequences and structures. J. Mol. Biol. 247, 536–540 (1995)
12. PROLDA: http://www.granular.com/PROLDA/
13. Pearl, F.M., Bennett, C.F., Bray, J.E., Harrison, A.P., Martin, N., Shepherd, A., Sillitoe, I., Thornton, J., Orengo, C.A.: The CATH database: an extended protein family resource for structural and functional genomics. Nucleic Acids Research 31, 452–455 (2003)
14. Gubbi, J., Shilton, A., Parker, M., Palaniswami, M.: Protein Topology Classification Using Two-Stage Support Vector Machines. Genome Informatics 17, 259–269 (2006)
15. R: http://www.R-project.org/
16. Olmea, O., Valencia, A.: Improving contact predictions by the combination of correlated mutations and other sources of sequence information. Folding Design 2, S25–S32 (1997). Fariselli, P., Casadio, R.: A neural network based predictor of residue contacts in proteins. Protein Eng. 12, 15–21(1999)
17. Nick and Thomas' Protein Contact Prediction Server, http://foo.acmc.uq.edu.au/nick/Protein/contact.html
18. MacCallum, R.M.: Striped sheets and protein contact prediction. Bioinformatics 20(suppl. 1), I224–I231 (2004)
19. Gromiha, M.M., Oobatake, M., Sarai, A.: Important amino acid properties for enhanced thermostability from mesophilic to thermophilic proteins. Biophysical Chemistry 82, 51–67 (1999)

20. Gromiha, M.M., Oobatake, M., Kono, H., Uedaira, H., Sarai, A.: Importance of Mutant Position in Ramachandran Plot for Predicting Protein Stability of Surface Mutations. Biopolymers 64, 210–220 (2002)
21. Grmiha, M.M.: Importance of Native-state Topology for Determining the Folding Rate of Two-state Proteins. J. Chem. Inf. Comp. Sci. 43, 1481–1485 (2003)
22. Zeyar, A., Kian-Lee, T.: MatAlign: Precise Protein Structure Comparison by Matrix Alignment. J. Bioinform. Comp. Biol. 6, 1197–1216 (2006)

Using Efficient RBF Network to Identify Interface Residues Based on PSSM Profiles and Biochemical Properties

Yu-Yen Ou[1], Shu-An Chen[1], Chung-Lu Shao[2], and Hao-Geng Hung[2]

[1] Department of Computer Science and Engineering,
Graduate School of Biotechnology and Bioinformatics,
Yuan-Ze University, Chung-Li, Taiwan
[2] Department of Computer Science and Information Engineering,
National Taiwan University, Taipei, Taiwan

Abstract. Protein-protein interactions play a very important role in many biological processes, for example, information transfer along signaling pathways, and enzyme catalysis. Recently, scientists tried to predict the protein-protein interaction interface from sequences. Since the number of protein 3D structure still increase slowly comparing to the number of protein sequences, it may be a good idea to predict the protein-protein interface from sequences directly.

In this paper, the compositions and conserved functions of the amino acids in the protein interface are studied, and the information of secondary structures is added. In addition, we used radio basis function network to predict the protein interface with adding some useful biochemical features.

1 Introduction

Protein-protein interactions play a very important role in many biological processes, for example, information transfer along signaling pathways, and enzyme catalysis. Recently, scientists tried to predict the protein-protein interaction interface from sequences[1,2]. Since the number of protein 3D structure still increase slowly comparing to the number of protein sequences, it may be a good idea to predict the protein-protein interface from sequences directly.

In [2], authors have developed a two-stage support vector machine (SVM) based method using amino acid sequence information to discriminate interface residues and non-interface residues from surface residues, and showed good results. In this paper, we try to use an efficient Radial Basis Function Network (RBFN) classifier and PSSM profiles to enhance the prediction results. In addition, we adopt secondary structure information and some biochemical properties to improve the prediction accuracy. The experimental results showed that the additional information are useful for prediction.

The radial basis function network (RBFN) is a special type of neural networks with several distinctive features [3,4,5,6]. Since its first proposal, the RBFN has

J.C. Rajapakse, B. Schmidt, and G. Volkert (Eds.): PRIB 2007, LNBI 4774, pp. 132–141, 2007.

attracted a high degree of interest in research communities. An RBFN consists of three layers, namely the input layer, the hidden layer, and the output layer. The input layer broadcasts the coordinates of the input vector to each of the nodes in the hidden layer. Each node in the hidden layer then produces an activation based on the associated radial basis function. Finally, each node in the output layer computes a linear combination of the activations of the hidden nodes. How an RBFN reacts to a given input stimulus is completely determined by the activation functions associated with the hidden nodes and the weights associated with the links between the hidden layer and the output layer. The general mathematical form of the output nodes in an RBFN is as follows:

$$c_j(x) = \sum_{i=1}^{k} w_{ji}\phi(||x - \mu_i||\,;\sigma_i), \tag{1}$$

where $c_j(x)$ is the function corresponding to the j-th output unit (class-j) and is a linear combination of k radial basis functions $\phi()$ with center μ_i and bandwidth σ_i. Also, w_j is the weight vector of class-j and w_{ji} is the weight corresponding to the j-th class and i-th center. The general architecture of RBFN is shown in Fig 1.

In this paper, we select the spherical Gaussian function as our basis function of RBFN, so the Eq.1 becomes:

$$c_j(x) = \sum_{i=1}^{k} w_{ji} \exp\left(-\frac{||x - \mu_i||^2}{2\sigma_i^2}\right). \tag{2}$$

From Eq.2, we can see that constructing an RBFN involves determining the number of centers, k, the center locations, μ_i, the bandwidth of each center, σ_i, and the weights, w_{ji}. That is, training an RBFN involves determining the values of three sets of parameters: the centers (μ_i), the bandwidths (σ_i), and the weights (w_{ji}), in order to minimize a suitable cost function.

Nevertheless, the essential task in constructing a RBFN classifier is to optimize the weights associated with the radial basis functions. In this paper, we proposes an efficient algorithm for determining the weights associated with the RBFN by exploiting the regularization theory [7] and the Cholesky decomposition [8]. The general observation is that the RBFN constructed is capable of delivering the same level of prediction accuracy as the SVM, while enjoying significant execution efficiency during the phase to construct the classifier.

2   Constructing the Radial Basis Function Network

In this paper, we focus on the calculation of the weights, so we conduct the simplest method to determine the centers and bandwidths. We have adopted all training instances as centers in the our experiments. Also, we employ the simplest method which is use the fixed bandwidth of each kernel function, and set the bandwidth as 5 for each kernel function.

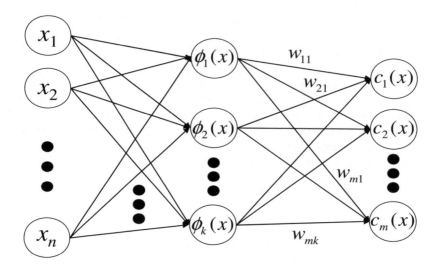

Input Layer   Hidden Layer   Output Layer

Fig. 1. General Architecture of Radial Basis Function Networks

After the centers and bandwidths of the kernel functions in hidden layer have been determined, the transformation between the inputs and the corresponding outputs of the hidden units is now fixed. The network can thus be viewed as an equivalent single-layer network with linear output units. Then, we use the lease mean square error (LMSE) method to determine the weights associated with the links between the hidden layer and the output layer.

In the following section, we will show how the LMSE method have been used in data classification field, and then propose a method which has a better theoretical foundation and practical use.

Assume h is the output of the hidden layer.

$$h = \begin{bmatrix} \phi_1(x) & \phi_2(x) & \dots & \phi_k(x) \end{bmatrix}^T, \tag{3}$$

where k is the number of centers, $\phi_1(x)$ is the output value of first kernel function with input x. Then, the discriminant function $c_j(x)$ of class-j can be expressed by the following:

$$c_j(x) = w_j^T h, j = 1, 2, \dots, m \tag{4}$$

where m is the number of class, and w_j is the weight vector of class-j. We can show w_j as:

$$w_j = \begin{bmatrix} w_{j1} & w_{j2} & \dots & w_{jk} \end{bmatrix}^T. \tag{5}$$

After calculating the discriminant function value of each class, we choose the class with the biggest discriminant function value as the classification result. We will discuss how to get the weight vectors by using least mean square error method in the following subsections.

2.1   Traditional Least Mean Square Error Method

The traditional LMSE method was proposed by Broomhead and Lowe [9]. This method is originally proposed for function approximation, and is the most popular supervised learning method of constructing the weights of RBFN [4,5,10,11]. In this method, the objective function of class-j can be shown as:

$$\min \sum_{i=1}^{n} [c_j(x_i) - v_j(x_i)]^2 , \tag{6}$$

where

$$v_j(x_i) = \begin{cases} 1 & \text{if } x \in class\text{-}j, \\ 0 & \text{otherwise.} \end{cases} \tag{7}$$

This system is overconstrained, being composed of n equations with k unknown weights, then the optimal solution of w_j can be written as

$$w_j^* = \Phi^+ y_j, \tag{8}$$

where $y_j = [\, v_j(x_1)\, v_j(x_2) \ldots v_j(x_n)\,]^T$, $\Phi_{li} = \phi_i(x_l)$ and Φ^+ is the pseudoinverse of Φ. The matrix Φ is rectangular $(n \times k)$ and its pseudoinverse can be computed as

$$\Phi^+ = (\Phi^T \Phi)^{-1} \Phi^T ,$$

provided that $(\Phi^T \Phi)^{-1}$ exists. The matrix $(\Phi^T \Phi)$ is square and its dimensionality is k, so that it can be inverted in time proportional to k^3.

The cost of computing Φ^+ is very high. Especially, we need to store Φ of size $(n \times k)$ in the memory. The value of n in some classification problems is very large, such that it may be impractical to have such large amounts of memory space for storage. Therefore, this method may not be suitable for the use of classification problem.

2.2   Least Mean Square Error Method with Statistics Techniques

The improved LMSE method for data classification was proposed by Devijver et. al.[12]. The idea of this method is basically the same with the traditional method, but [12] used the statistics techniques to analyze the whole problem. From this aspect, we can get the following results, and the major advantage from this approach is that we don't need to store Φ in the memory.

For a classification problem with m classes, let V_i designate the i-th column vector of an $m \times m$ identity matrix and W be an $k \times m$ matrix of weights:

$$W = [\, w_1\ w_2 \ldots w_m \,] . \tag{9}$$

Then the objective function to be minimized is

$$J(W) = \sum_{j=1}^{m} P_j E_j \left\{ \|W^T h - V_j\|^2 \right\} , \tag{10}$$

where P_j and $E_j\{\}$ are the a priori probability and the expected value of class-j, respectively.

To find the optimal W that minimizes J, we set the gradient of $J(W)$ to be zero:

$$\nabla_W J(W) = 2 \sum_{j=1}^{m} P_j E_j \left\{ hh^T \right\} W - 2 \sum_{j=1}^{m} P_j E_j \left\{ h \right\} V_j^T$$

$$= [0], \tag{11}$$

where $[0]$ is a $k \times m$ null matrix.

Let K_i denote the class-conditional matrix of the second-order moments of h, i.e.

$$K_i = E_i \left\{ hh^T \right\}. \tag{12}$$

If K denotes the matrix of the second-order moments under the mixture distribution, we have

$$K = \sum_{j=1}^{m} P_j K_j. \tag{13}$$

Then Eq. 11 becomes

$$KW = M, \tag{14}$$

where

$$M = \sum_{j=1}^{m} P_j E_j \left\{ h \right\} V_j^T. \tag{15}$$

If K is nonsingular, the optimal W can be calculated by

$$W^* = K^{-1}M. \tag{16}$$

When compared to the traditional method, the size of K, $k \times k$, is much smaller than the Φ matrix of size $(n \times k)$ described in the previous subsection. Therefore, this method requires less memory space for storing the matrix.

However, there is a critical drawback of this method. That is, K may be singular and this will crash the whole procedure. By observing the matrix hh^T, we are aware of that the matrix hh^T is symmetric positive semi-definite (PSD) matrix with rank $= 1$. Since K is the summation of hh^T for each training instance, K is also a PSD matrix with rank $\leq n$. However, PSD matrix may be a singular matrix, so we should add the regularization term to make sure the matrix will be invertible.

In the regularization theory [7], it consists in replacing the objective function as follows:

$$J(W) = \sum_{j=1}^{m} P_j E_j \left\{ \| W^T h - V_j \|^2 \right\} + \lambda \sum_{j=1}^{m} w_j^T w_j, \tag{17}$$

where λ is the regularization parameter. Then the Eq. 14 becomes

$$(K + \lambda I)W = M. \tag{18}$$

If we set $\lambda > 0$, $(K + \lambda I)$ will be a positive definite (PD) matrix and therefore is nonsingular. The optimal W^* can be calculated by

$$W^* = (K + \lambda I)^{-1} M. \tag{19}$$

However, the PD matrix has many good properties, and one of them is a special and efficient triangular decomposition, Cholesky decomposition. By using Cholesky decomposition, we can decompose the $(K + \lambda I)$ matrix as follows,

$$(K + \lambda I) = LL^T, \tag{20}$$

where L is a lower triangular matrix. Then, the Eq. 18 becomes

$$(LL^T)W = M. \tag{21}$$

Actually, we can solve the linear system efficiently by using backsubstitution twice. In our experiments, Cholesky decomposition is about 10-20 times faster than alternative methods for matrix inversion. For example, in our experiments, we only used 25.36 seconds for inverting a 2558×2558 matrix in letter data set, while the traditional method used 754.93 seconds. It's about 30 times faster than the traditional method.

Finally, we can get the optimal w_j^* for class-j from W^*, and then the optimal discriminant function $c_j(x)$ for class-j is derived. By using the regularization theory, the optimal weights can be obtained analytically and efficiently.

3   Experimental Results of Interface Residues Prediction

3.1   Datasets

We adapt the same dataset from Yan et al [2]. The dataset is originally from Chakrabarti et al. [13]. Yan et al. selected 77 protein chains from 70 protein-protein complexes.

3.2   PSSM Profiles

Recently, scientists try to use the Position Specific Scoring Matrix (PSSM) profiles as features in residues level function or structure prediction. [14,15] In this paper, we also adapt the PSSM profiles as our primal feature set instead of sequence residue type only. We obtain the PSSM profiles by using PSI-BLAST and non-redundant (NR) protein database. Also, every element has been scaled by $\frac{1}{1+e^{-x}}$.

3.3   Biochemical Properties

We tried 7 biochemical properties, which are hydrophobic, polar, small, aliphatic, aromatic, positive, and negative. In addition, we consider the properties toward interface-based or surface-based. That is, if the amino acid appears in interface more frequent than in surface, we think this amino acid is interface-based. Otherwise, we think the amino acid is surface-based. We list the properties with interface-based/surface-based in Table 1.

Table 1. Amino Acid interface-based(I)/surface-based(S) Properties

Property	I	L	V	C	A	G	M	F	Y	W	H	K	R	E	Q	D	N	S	T	P
Hydrophobic_S	O	O	O	*	*	*	O	O	O	O	*	O	O	O	O	O	O	*	*	O
Hydrophobic_I	*	*	*	O	O	O	*	*	*	*	*	O	O	O	O	O	O	O	O	O
Polar_S	O	O	O	O	O	O	O	O	O	O	*	O	*	*	*	*	*	*	*	O
Polar_I	O	O	O	O	O	O	O	*	*	*	O	*	O	O	O	O	O	O	O	O
Small_S	O	O	O	*	O	*	O	O	O	O	O	O	O	O	O	*	*	*	*	*
Small_I	O	O	*	O	*	O	O	O	O	O	O	O	O	O	O	O	O	O	O	O
Aliphatic_I	*	*	*	O	O	O	O	O	O	O	O	O	O	O	O	O	O	O	O	O
Aromatic_I	O	O	O	O	O	O	O	*	*	*	*	O	O	O	O	O	O	O	O	O
Positive_S	O	O	O	O	O	O	O	O	O	O	O	*	O	O	O	O	O	O	O	O
Positive_I	O	O	O	O	O	O	O	O	O	O	*	O	*	O	O	O	O	O	O	O
Negative_S	O	O	O	O	O	O	O	O	O	O	O	O	O	*	O	*	O	O	O	O

3.4 Secondary Structure Information

To further improve the prediction performance, we combined the PSSM profiles, biochemical properties and the predicted secondary structure from PSIPRED [14]. Secondary structure play a very important role on protein folding and 3D structure, and generally believe that the function of protein is basically determined by its structure. We think the secondary structure information may useful on interface residues prediction. The experimental results show that the secondary structure information did improve the prediction performance.

4   Results

We used recall, precision, f-score, MCC (Matthew's correlation coefficient), ACC (accuracy) to measure the prediction performance. TP, FP, TN, FN are true positive number, false positive number, true negative number, and false negative number, respectively.

$$\text{Recall} = \frac{\text{TP}}{\text{TP} + \text{FN}} \tag{22}$$

$$\text{Precision} = \frac{\text{TP}}{\text{TP} + \text{FP}} \tag{23}$$

$$\text{F-score} = \frac{2 \times \text{Recall} \times \text{Precision}}{\text{Recall} + \text{Precision}} \tag{24}$$

$$\text{ACC} = \frac{\text{TP} + \text{TN}}{\text{TP} + \text{FP} + \text{TN} + \text{FN}} \tag{25}$$

$$\text{MCC} = \frac{\text{TP} \times \text{TN} - \text{FP} \times \text{FN}}{\sqrt{(\text{TP} + \text{FN})(\text{TP} + \text{FP})(\text{TN} + \text{FP})(\text{TN} + \text{FN})}} \tag{26}$$

First of all, we compared our proposed RBF network classifier with Yan's results on [2]. Mr. Yan provided us one of datasets of his experiment, and told us that the dataset may have some improper information within the training

Table 2. Comparison results of Proposed Method and Yan Method

	Proposed		Yan's Method [2]	
	1st stage	2nd stage	1st stage	2nd stage
Precision	0.42	0.77	0.44	0.58
Sensitivity	0.52	0.79	0.43	0.39
Accuracy	0.63	0.86	0.66	0.72
MCC	0.19	0.68	0.19	0.30

Table 3. Comparison results of adding different biochemical properties

	Sensitivity	Precision	Accuracy	MCC	F-score
PSSM	0.587	0.361	0.601	0.174	0.447
+Aliphatic_I	0.594	0.374	0.616	0.197	0.459
+Aromatic_I	0.597	0.379	0.621	0.204	0.463
+Positive_I	**0.608**	**0.375**	**0.614**	**0.201**	**0.464**
+Small	0.602	0.375	0.615	0.199	0.462
+hydrophobic	0.599	0.378	0.619	0.203	0.463
+Negative_S	0.591	0.378	0.620	0.201	0.461
+Polar_I	0.599	0.377	0.618	0.202	0.462

data. As Table 2 shows, our proposed classifier performs significantly better than results on [2], especially on second stage results. However, this dataset seems not fair on training and testing data, and than easily overfit with second stage process. We analyzed the problem, and concluded that the problem may be caused by residues based 5-fold cross validation, so we divided the 77 protein chains into 5 groups, and than used the new divided dataset as the comparison standard.

In Table 3, we listed the different results with the new divided dataset. The "PSSM" row listed the results by using PSSM profiles as features. Also, the following rows are the results by adding different biochemical properties as features. We can see the best results are from PSSM profile with 3 additional biochemical features, Aliphatic_I, Aromatic_I, and Positive_I.

In Table 4, we can see the secondary structure information can enhance the prediction accuracy. The final results show that the method proposed in this paper can achieve the 0.471 of F-score and 0.214 of MCC.

Table 4. Comparison results of different additional features

	Sensitivity	Precision	Accuracy	MCC	F-score
PSSM	0.587	0.361	0.601	0.174	0.447
PSSM+Biochemical	0.608	0.375	0.614	0.201	0.464
PSSM+Biochemical+SSE	0.614	0.381	0.621	0.214	0.471

5   Conclusion

In this paper, we proposed an efficient method to construct an RBFN classifier by using the improved LMSE method for constructing an RBFN optimized for data classification and bioinformatics applications. The method proposed by [12] is more efficient than the traditional one, but it may suffer the singular matrix problem and fails to build the classifier in such case. We solved the singular matrix problem by using the regularization theory, and used the Cholesky decomposition to speedup the matrix inversion process. This provides a good framework for constructing an RBFN in classification problems, and the proposed method can obtain the optimal weights analytically and efficiently.

We have applied our proposed approach to the prediction of interface residues. The interface residues prediction is one of the most important problems in computational biology and bioinformatics. Experimental results showed that combining proposed classifier and additional biochemical properties and secondary structure information can significant improve the prediction accuracy.

References

1. Zhou, H., Shan, Y.: Prediction of protein interaction sites from sequence profile and residue neighbor list. Proteins Structure Function and Genetics 44, 336–343 (2001)
2. Yan, C., Dobbs, D., Honavar, V.: A two-stage classifier for identification of protein-protein interface residues. Bioinformatics 20, i371–i378 (2004)
3. Park, J., Sandberg, I.W.: Universal approximation using radial-basis-function networks. Neural Computation 3, 246–257 (1991)
4. Poggio, T., Girosi, F.: A theory of networks for approximation and learning. Technical Report A.I. Memo 1140, Massachusetts Institute of Technology, Artificial Intelligence Laboratory and Center for Biological Information Processing, Whitaker College (1989)
5. Ghosh, J., Nag, A.: An overview of radial basis function networks. In: Howlerr, R.J., Jain, L.C. (eds.) Radial Basis Function Neural Network Theory and Applications (2000)
6. Mitchell, T.M.: Machine Learning. McGraw-Hill, New York (1997)
7. Tikhonov, A.N., Arsenin, V.Y.: Solutions of Ill-Posed Problems. V.H. Winston & Sons, John Wiley & Sons, Washington D.C (1977)
8. Press, W.H.: Numerical Recipes in C, 2nd edn. Cambridge University Press, Cambridge (1992)
9. Broomhead, D.S., Lowe, D.: Multivariable functional interpolation and adaptive networks. Complex Systems 2, 321–355 (1988)
10. Orr, M.J.L.: Introduction to radial basis function networks. Technical report, Center for Cognitive Science, University of Edinburgh, UK (1996)
11. Tarassenko, I., Roberts, S.: Supervised and unsupervised learning in radial basis function classifiers. In: IEE Proceedings-Vision, Image and Signal Processing, vol. 141, pp. 210–216 (1994)
12. Devijver, P.A., Kittler, J.: Pattern recognition: a statistical approach. Prentice-Hall, Englewood Cliffs (1982)

13. Chakrabarti, P., Janin, J.: Dissecting protein-protein recognition sites. Proteins Structure Function and Genetics 47, 334–343 (2002)
14. Jones, D.T.: Protein secondary structure prediction based on position-specific scoring matrices. J. Mol. Biol. 292, 195–202 (1999)
15. Guo, J., Chen, H., Sun, Z., Lin, Y.: A novel method for protein secondary structure prediction using dual-layer svm and profiles. Proteins 54, 738–743 (2004)

Dynamic Outlier Exclusion Training Algorithm for Sequence Based Predictions in Proteins Using Neural Network

Shandar Ahmad

National Institute of Biomedical Innovation,
Saito Asagi, Ibaraki-shi, Osaka, Japan
shandar@netasa.org

Abstract. Many structural and functional properties of proteins can be described as a one-dimensional one-to-one mapping between residues of protein sequence and target structure or function. These residue level properties (RLPs) have been frequently predicted using neural networks and other machine learning algorithms. Here we present an algorithm to dynamically exclude from the neural network training, examples which are most difficult to separate. This algorithm automatically filters out statistical outliers causing noise and makes training faster without losing network ability to generalize. Different methods of sampling data for neural network training have been tried and their impact on learning has been analyzed.

Keywords: Binding sites, Neural networks, Sequence information, Outliers.

1 Introduction

Sequence-structure-function relationship of proteins has been historically one of the most important issues in bioinformatics for a very long time [1-3]. However, despite an intense effort to predict protein structure from the amino acid sequence, the task has remained difficult and far from complete. Compared to that ambitious goal of predicting everything from sequence or structure, it seems much more plausible to predict the so-called one-dimensional properties of protein structure such as secondary structure, solvent accessibility and coordination number on the one hand and biological functions such as binding with specific ligands or DNA bases on the other. Both one-dimensional structural features of proteins and probability of binding of an amino acid with other molecules have been predicted from the information of amino acid sequence with good success [4-9] and have in many ways led the way for an eventual ab initio structure and function prediction without homology or structure models. One of the most widely used method for mapping sequence information on to functional and structural target properties has been neural network. Neural networks provide a very efficient tool to model almost any non-linear relationship between sequence data and their target properties. These models have been successful in predicting secondary structure, solvent accessibility and binding sites. As larger data sets of binding sites and structural properties become available, their processing with

J.C. Rajapakse, B. Schmidt, and G. Volkert (Eds.): PRIB 2007, LNBI 4774, pp. 142–147, 2007.

neural networks will become slower albeit more powerful. Faster algorithms and efficient analysis of feature vectors and their relationshop with target properties are needed to address these problems. One of the problems is poor predictability of some of the patterns even when most of the samples are well predicted. We have developed an algorithm to dynamically select training examples for neural network and flag them as prediction outliers. In this algorithm a neural network is not trained on the entire data set, but the error scores are computed for each data example and then the examples contributing the most to the error score are eliminated from the training process. We report the resulting learning curves, amount of excluded data and their impact on the ability of the neural network to generalize prediction.

2 Methods

2.1 Definition of an Outlier

A statistical outlier is generally known to be a pattern with too high or too small value of its corresponding attribute. In the context of feature-based predictions of target properties, we define a statistical outlier to be a pattern in which the relationship between its feature vector and its target property does not follow the same relationship as done by the overall data set. Formally, a pattern will be classified as an outlier if the prediction error (ε_i) in that sample is much more than the overall variance in the data i.e.

$$\varepsilon_i > \varepsilon_{av} + \alpha. \, \sigma \, (\varepsilon) \tag{1}$$

Where ε_{av} is the average absolute error in the overall data, $\sigma \, (\varepsilon)$ is the standard deviation in the pattern-wise absolute error and ε_i is the error in the ith sample, to be tested for being an outlier or not and α is an adjustable parameter to determine the strictness of the flagging criterion.

2.2 Treatment of Outliers

Once the training examples have been flagged as outliers, there are at least two methods of treating them. First, instead of assigning them high error values returned by the predictor, their predicted values may be reassigned such that their contribution to error does not exceed the criterion set by (1). Alternatively, the outliers may be totally removed from the data set and they do not contribute at all to the performance scores. Later leaves behind a smaller data and and the calculation of the error gradient becomes faster in the process. We have used the both these criterion to analyze learning behavior but report the results obtained from the second one.

2.3 Dynamic Identification of Outliers

Using the outlier identification criterion given by (2), the identification of outliers has to be done for every epoch as data points move from normal to outlier categories and vice versa as the training progresses. In particular, the random initialization of weights produces large variance in error and therefore very few outliers according to the above definition. As the training progresses, both mean error and their variance decrease with a clearer picture of outliers emerging. A typical variation in the number of patterns identified as outliers with training (epoch number) has been shown in Figure 2 (see results section).

2.4   Data Sets and RLP Types

Three types of predictions are attempted viz. Solvent accessibility (class-type predictions and real value predictions) [5-6], DNA-binding site predictions [7-8] and Carbohydrate-binding site predictions [9]. Data sets used for these predictions have been explained in the corresponding previous publications. In this work, we have used 512 proteins for analyzing ASA prediction, 40 proteins for analyzing sugar-binding sites and 62 proteins for assessing DNA-binding sites. Similar results have been obtained for these data sample, but the results discussed in this paper are based on solvent accessibility data because its values are distributed in a range from 0 to 1, instead of binary values in the case of binding sites and hence analyzing performance in solvent accessibility prediction is easier.

2.5   Neural Networks

In all our prediction experiments, a layered neural network with single hidden layer containing two units was used. The input layer consists of 60 units representing a tripeptide with a target residue at the center and one sequence neighbors on either side included as context information. Output layer is a single neuron with real valued outputs, transformed into binary values with a simple threshold function. Activation function for the hidden layer is *arctan*, and for the output layer it is a *sigmoidal* function. Neural network is trained using generalized delta-rule and weights are updated in the direction of maximum gradient after presenting all patterns at the end of each epoch according to the following learning rule:

$$\Delta W_{ijk} = \eta \; \partial E / \partial W_{ijk} \qquad (2)$$

Learning rate was maintained at 1.0 for all these calculations.

3   Results and Discussion

3.1   Outlier Exclusion Does Not Affect Generalization

Figure 1 shows learning curves of a neural network trained for 200 epochs using generalized delta rule, using different criterion of data exclusion. Mean absolute error of prediction in the test data, used for determining the stopping point for training was used as a measure of generalization. This particular graph shows the learning curves for solvent accessibility and similar curves were also obtained with binding data of DNA and carbohydrates. We observe that the neural network training carried out at $\alpha=3$ has almost the same prediction error as the one carried out on full data sets. Training performed with a more strict criterion of data inclusion ($\alpha=1$) suffered from poor training performance as it excluded too many data points. An interesting

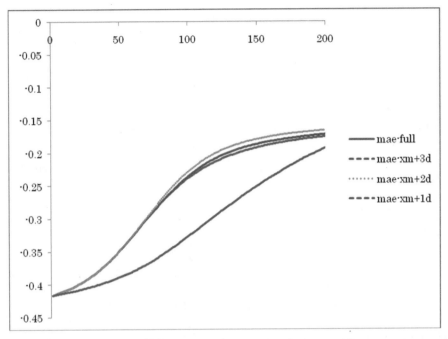

Fig. 1. Learning history of mean absolute error in the test data (generalization). Abbreviation (mae-full: Mean Absolute Error in test data without error exclusion, mae-xm+α.d: MAE in test data when training points with $\varepsilon_i > \varepsilon_{av} + \alpha. \sigma (\varepsilon)$ were excluded). MAE values are marked as negative to contrast them from correlation and other accuracy scores to indicate that a smaller MAE means better prediction.

observation was made for $\alpha=2$. There was a small improvement in prediction performance of the neural network at this value, suggesting that a suitably selected value of α may actually improve the generalizing ability of neural network. However, DNA and Carbohydrate-binding sites data did not show a similar improvement, probably because the amount of data available in these categories was not large enough to take advantage of this situation.

3.2 Error Distribution and History of Outlier Frequency

In Figure 2, we show the outlier frequency variations in different stages of neural network training. In the early stages of neural network training errors are randomly distributed leading to a large value of variance and hence no outliers can be identified in the early training. As the neural network learns the variance in prediction error decreases and outliers can be identified. With a strict criterion of outliers (large α), very few outliers are detected and at small values of α, too many patterns are excluded from training. A large number of rejected data for $\alpha=1$, is clearly responsible for poor generalization of prediction (Figure 1). A value of $\alpha=2$ is suitable for generalization and also excluding sufficient number of data points to speed up the process of learning.

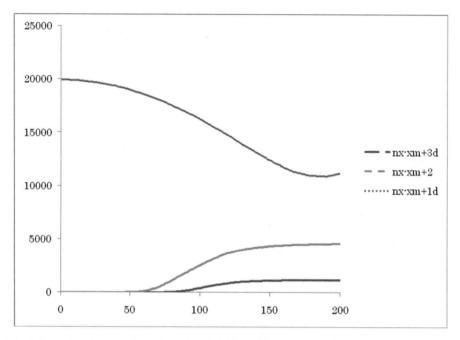

Fig. 2. Learning history and number of excludable outliers. Abbreviations: nx-xm+α.d (number of outliers with) $\varepsilon_i > \varepsilon_{av} + \alpha. \sigma(\varepsilon)$.

3.3 Role of Data Sets

Solvent accessibility and binding sites employ similar neural networks and hence similar results were obtained by using outlier exclusion criterion. However, target vectors in binding site problem are binary valued, whereas ASA is a real-valued function. Mean absolute error in case of binding sites does not carry much physical meaning like ASA which can take continuous values. Thus the neural network for these problems was also trained to maximize coefficient of correlation between predicted and observed values (data not shown). Variance in the prediction error for these two binary class predictions was found to be smaller than ASA data and no outliers could be detected at $\alpha=3$. However a value of $\alpha=2$, was found to be optimum at which significant number of outliers could be removed.

3.4 Biological Basis of Prediction Outliers

Machine learning relies on pattern recognition and a neural network tries to recognize patterns which it has seen during training. Thus if a pattern has not been seen before, the neural network fails to recognize it. Conversely, a pattern which is present in the training data but has no similar patterns in the validation data does not contribute to the performance. A poorly predicted pattern within the training data is just a noise which might tend to over-train the neural network without leading to generalization, thus increasing the unnecessary computational overhead. Furthermore, the nature of relationship between selected features and their target property for some patterns may not follow a general trend for which neural networks are trained. From a point of view of protein structure, this may be caused by some unusual bonds (e.g. disulfide bond), presence of some ligand in the neighboring region or some features of the biochemical or thermodynamic environment, which is not usually seen by proteins or which cannot be determined from local sequence and evolutionary information.

4 Conclusion

A new algorithm for filtering noisy sequence data from neural network training has been developed which shows promise for applications in RLP predictions. Outlier removal can speed up neural network training without loss of generalization. Different definitions of prediction outliers have been employed and structural basis of the same has been discussed.

References

1. Rost, B., Liu, J., Nair, R., Wrzeszczynski, K.O., Ofran, Y.: Automatic prediction of protein function. Cell Mol. Life Sci. 60(12), 2637–2650 (2003)
2. Moult, J.: A decade of CASP: progress, bottlenecks and prognosis in protein structure prediction. Curr. Opin. Struct. Biol. 15(3), 285–289 (2005)
3. Wolfson, H.J., Shatsky, M., Schneidman-Duhovny, D., Dror, O., Shulman-Peleg, A., Ma, B., Nussinov, R.: From structure to function: methods and applications. Curr. Protein Pept. Sci. 6(2), 171–183 (2005)
4. Schlessinger, A., Rost, B.: Protein flexibility and rigidity predicted from sequence. Proteins 61(1), 115–126 (2005)
5. Nguyen, M.N., Rajapakse, J.C.: Prediction of protein relative solvent accessibility with a two-stage SVM approach. Proteins 59(1), 30–37 (2005)
6. Ahmad, S., Gromiha, M.M., Sarai, A.: A Real value prediction of solvent accessibility from amino acid sequence. Proteins 50(4), 629–635 (2003)
7. Ahmad, S., Sarai, A.: PSSM-based prediction of DNA binding sites in proteins. BMC Bioinformatic 6, 33–35 (2005)
8. Ahmad, S., Gromiha, M., Sarai, A.: Analysis and Prediction of DNA-binding proteins and their binding residues based on Composition, Sequence and Structural Information. Bioinformatics 20, 477–486 (2004)
9. Malik, A., Ahmad, S.: Sequence and structural features of carbohydrate binding in proteins and assessment of predictability using a neural network. BMC Structural B

Bioinformatics on β-Barrel Membrane Proteins: Sequence and Structural Analysis, Discrimination and Prediction

M. Michael Gromiha

Computational Biology Research Center (CBRC)
National Institute of Advanced Industrial Science and Technology (AIST)
AIST Tokyo Waterfront Bio-IT Research Building
2-42 Aomi, Koto-ku, Tokyo 135-0064, Japan
michael-gromiha@aist.go.jp

Abstract. The analysis on the amino acid sequences of transmembrane beta barrel proteins (TMBs) provides deep insights about their structure and function. We found that the occurrence of Ser, Asn and Gln is significantly higher in TMBs than globular proteins, which might be due to their importance in the formation of β-barrel structures in the membrane, stability of binding pockets and the function of TMBs. Utilizing this information, we have devised statistical methods and machine learning techniques to discriminate TMBs from other folding types of globular and membrane proteins and we obtained the maximum accuracy of 96%. Further, we have devised protocols for identifying the membrane spanning β-strand segments and detecting TMBs in genomic sequences.

Keywords: β-barrel membrane protein, amino acid composition, sequence analysis, discrimination, prediction, genome.

1 Introduction

The β-barrel membrane proteins (TMBs) perform a variety of functions, such as mediating non-specific, passive transport of ions and small molecules, selectively passing the molecules like maltose and sucrose and are involved in voltage dependent anion channels. These proteins contain β-strands as their membrane spanning segments and are found in the outer membranes of bacteria, mitochondria and chloroplast. The assembly of TMBs is somewhat more complex when compared to the assembly of transmembrane helical proteins having α-helices as transmembrane parts. This is probably due to the difference of amino acid sequences in the transmembrane part strands and helices; transmembrane helical proteins (TMH) contain a stretch of hydrophobic amino acid residues whereas transmembrane strand proteins are intervened by several charged and polar residues. Because of this feature, most predictive schemes, which are successful in predicting transmembrane helical segments, fail to predict the transmembrane strand segments and discriminating β-barrel membrane proteins.

J.C. Rajapakse, B. Schmidt, and G. Volkert (Eds.): PRIB 2007, LNBI 4774, pp. 148–157, 2007.
© Springer-Verlag Berlin Heidelberg 2007

We have systematically analyzed the amino acid compositions of TMBs, TMH and globular proteins, and observed that the residues Ser, Asn and Gln are predominant in TMBs. Utilizing amino acid and dipeptide compositions, we have devised statistical methods and machine learning techniques for discriminating TMBs. Further, we have developed a rule based approach and neural networks based method for identifying the membrane spanning segments. A novel protocol has been proposed for detecting TMBs in genomic sequences and a database has been set up for the annotation of TMBs in genomes.

2 Materials and Methods

2.1 Dataset

We have constructed several sets of data for the analysis, discrimination and prediction: (i) a dataset of 377 well annotated TMB sequences obtained from PSORT database (1) and a subset of 208 non-redundant TMB sequences with less than 40% sequence identity obtained with CD-HIT algorithm (2), (ii) non-redundant dataset of 19 known TMB structures with the sequence identity of less than 25%, (iii) 674 globular proteins belonging to different structural classes (155 all-α, 156 all-β, 184 α+β and 179 α/β proteins), (iv) non-redundant data set of 1602 globular proteins belonging to 30 different folds obtained from Protein Data Bank (3) (v) a dataset of 268 well-annotated TMH sequences and a subset of 206 non-redundant TMH sequences obtained from PSORT and (vi) the amino acid sequences of 275 completed genomes from NCBI database (http://www.ncbi.nih.gov/). This includes 23 genomes from archaea, 237 from bacteria and 15 from eukaryote. The total number of proteins in these three kingdoms of life is 52241, 686562 and 165186, respectively with the total of 903,989 sequences.

2.2 Computation of Amino Acid and Dipeptide Compositions

The amino acid composition for a residue type (e.g. Ala) in a protein is the number of amino acids of specific type normalized with the total number of residues. It is defined as:

$$Comp(i) = \Sigma\, n_i/N \qquad (1)$$

where, i stands for the 20 amino acid residues. n_i is the number of residues of each type and N is the total number of residues.

The composition of dipeptides is a measure to quantify the preference of amino acid residue pairs in a sequence. This has been computed using the following expression:

$$Dipep(i,j) = \Sigma N_{ij}*100/\,(\Sigma N_i + \Sigma N_j) \qquad (2)$$

where i,j stands for the distribution of 20 amino acid residues at positions i and i+1. $N_{i,j}$ is the number of residues of type i followed by the residue j. ΣN_i and ΣN_j are the total number of residues of type i and j, respectively.

The concept of motifs provides the information about the preference of residue pairs with a gap (any residue between the pair of residues). This has been computed using the same expression that we used for dipeptide composition (Eqn. 2). The main difference is that the residues i and j are the distribution of 20 amino acid residues at positions i and i+1 for dipeptides, and i and i+2, i and i+3 etc. for motifs.

2.3 Discrimination Methods

We have used statistical methods and machine learning algorithms for discriminating TMBs. In these methods, we have used the compositions of amino acids and dipeptides as attributes. The protocol used to discriminate TMBs using amino acid composition is given below: The amino acid composition has been computed for standard datasets of both globular ($Comp_{glob}$) and TMBs ($Comp_{TMB}$). For a new protein, X, firstly, we have calculated the amino acid composition using Eqn. 1. Then we have calculated the total absolute difference of amino acid composition between protein X and the amino acid composition of globular proteins, and that between protein X and TMBs. The protein X is predicted to be a TMB if the deviation is lowest with $Comp_{TMB}$ and vice versa (4).

We have followed the below mentioned steps to discriminate TMBs using residue pair preference/motif: (i) calculated the dipeptide composition for both globular ($Dipep_{glob}$) and TMBs ($Dipep_{TMB}$) and the difference between them ($\sigma_{TMB-glob}$); (ii) for a new protein, X, they have calculated the dipeptide composition using Eqn. 2 and given weights to the dipeptide composition with $\sigma_{TMB-glob}$; (iii) calculated the sum of weighted dipeptide composition and (iv) the protein X is predicted to be an TMB if the total weighted dipeptide composition is positive and globular protein otherwise.

Further, machine learning techniques including Bayes functions, Neural networks, Logistic functions, Support vector machines, Regression analysis, Nearest neighbor methods, Meta learning, Decision trees and Rules have been used to discriminate the TMBs.

2.4 Assessment of the Validity of the Method

We have performed a 5-fold cross-validation test for assessing the validity of the present work. In this method, the data set is divided into five groups, four of them are used for training and the rest is used for testing the method. The same procedure is repeated for five times and the average is computed for obtaining the accuracy of the method.

We have used different measures to assess the accuracy of discriminating TMBs, non-TMBs and combination of the two. The term, sensitivity shows the correct prediction of TMBs, specificity about the non-TMBs and accuracy indicates the overall assessment. These terms are defined as follows:

Sensitivity = TP/(TP+FN)
Specificity = TN/(TN+FP)
Accuracy = (TP+TN)/(TP+TN+FP+FN),

where, TP, FP, TN and FN refer to the number of true positives (TMBs identified as TMBs), false positives (non-TMBs identified as TMBs), true negatives (non-TMBs identified as non-TMBs), and false negatives (TMBs identified as non-TMBs), respectively.

3 Results and Discussion

3.1 Amino Acid Composition

We have computed the amino acid compositions for the 20 amino acid residues in 377 TMBs and the results are displayed in Fig. 1. In this figure, we have also included the data for 674 globular proteins for comparison. We observed that the composition of Glu, His, Ile and Cys are higher in globular proteins than TMBs and an opposite trend is observed for Ser, Asn and Gln (4). The formation of disulfide bonds between Cys residues requires an oxidative environment and such disulfide bridges are not usually found in intracellular proteins (5). Further, the analysis on the three-dimensional structures of 15 β-barrel TMBs showed the presence of just eight (0.1%) Cys residues and none of them are in membrane part. Hence, the occurrence of Cys is significantly higher in globular proteins than in TMBs. Glu is a strong helix former and this tendency influences the higher occurrence of it in globular proteins than TMBs. The comparative analysis on the occurrence of Ile in the β-strand segments of globular and TMBs revealed that the preference of Ile in TMBs is less than that in globular proteins, which may increase the occurrence of it in globular proteins.

The structural analysis of several TMBs shows that the residues, Ser, Asn and Gln play an important role to the stability and function of TMBs. In OmpA, the interior of β-strands contain an extended hydrogen bonding network of charged and polar residues and especially, the side chains of the residues, Ser22, Gln228 and Asn258 in OmpT, located above the membrane form hydrogen bonds to main chain atoms in the β-barrel. Interestingly, none of the residues, which have high composition in globular proteins (Glu, His, Ile and Cys), are involved in such pattern (6,7). In FecA, Yue et al. (8) showed that the binding pockets for diferric dicitrate involve the hydrogen bonds from the three residues, Gln176, Gln570 and Asn721 as shown in Figure 2. Similar

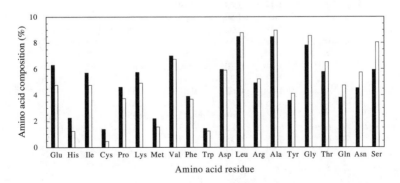

Fig. 1. Amino acid composition of the 20 amino acid residues in globular (filled bars) and TMBs (open bars)

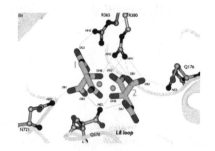

Fig. 2. The binding pockets for diferric dicitrate in FecA. The hydrogen bonds are shown as dotted lines.

trend is observed in other TMBs and this analysis revealed that the high occurrence of Ser, Asn and Gln in TMBs is required in the formation of β-barrel structures in the membrane, stability of binding pockets and the function of TMBs.

3.2 Discrimination of β-Barrel Membrane Proteins

We have used the compositions of amino acids, dipeptides and motifs for discriminating TMBs using statistical methods. An example to discriminate TMBs using amino acid composition is shown in Table 1.

For 1ADT (adenovirus DNA binding protein), the deviation of amino acid composition from globular protein (34.18) is less than that of TMB (39.89) and hence this protein is predicted as a non-TMB. On the other hand, for OutD protein, the deviation from TMB (16.09) is less than that from globular protein (23.70) and hence it is identified as an TMB.The amino acid composition based method could correctly identify 89% of the TMBs (334/377) and exclude 79% of globular proteins (531/674).

Table 1. Steps to discriminate globular and outer membrane proteins in two typical proteins

Residue	N	Comp	σ_{glob}	σ_{TMB}	N	Comp	σ_{glob}	σ_{TMB}
Adenovirus DNA-Binding Protein (1ADT)					**OutD protein**			
Ala	10	11.11	2.64	2.16	54	8.31	0.16	0.64
Asp	4	4.44	1.53	1.47	40	6.15	0.18	0.24
Cys	1	1.11	0.28	0.64	1	0.15	1.24	0.32
Glu	7	7.78	1.46	3.00	31	4.77	1.55	0.01
Phe	5	5.56	1.65	1.88	21	3.23	0.68	0.45
Gly	3	3.33	4.49	5.21	46	7.08	0.74	1.46
His	4	4.44	2.18	3.19	3	0.46	1.80	0.79
Ile	1	1.11	4.60	3.66	35	5.38	0.33	0.61
Lys	7	7.78	2.02	2.85	28	4.31	1.45	0.62
Leu	10	11.11	2.63	2.33	53	8.15	0.33	0.63
Met	4	4.44	2.23	2.88	19	2.92	0.71	1.36
Asn	4	4.44	0.10	1.30	43	6.62	2.08	0.88
Pro	3	3.33	1.30	0.41	21	3.23	1.40	0.51
Gln	4	4.44	0.62	0.31	33	5.08	1.26	0.33
Arg	3	3.33	1.60	1.91	37	5.69	0.76	0.45
Ser	3	3.33	2.61	4.72	55	8.46	2.52	0.41
Thr	6	6.67	0.88	0.13	47	7.23	1.44	0.69
Val	6	6.67	0.35	0.09	64	9.85	2.83	3.09
Trp	2	2.22	0.78	0.98	6	0.92	0.52	0.32
Tyr	3	3.33	0.25	0.80	12	1.85	1.73	2.28
Total			**34.18**	39.89			23.70	**16.09**
Discrimination		*Globular protein*				*β-barrel membrane protein*		

N: number of residues; σ_{glob} = |comp - comp(glob)|; σ_{TMB} = |comp − comp(TMB)|.

The dipeptides have more information than just amino acid composition and we observed an increase in accuracy. This information could correctly identify 95% of the TMBs and exclude 79% of the globular proteins (9). The performance of motifs is better to exclude globular proteins and the accuracies of identifying 377 TMBs and excluding 674 globular proteins are 95.8% and 82.2%, respectively (10).

Further, we have analyzed different machine learning techniques for discriminating TMBs. These methods could discriminate a set of 1088 TMBs and globular proteins with the accuracy in the range of 89-92% using amino acid composition (11). We have also used a set of 49 amino acid properties for discrimination, which improved the accuracy up to 94% for the same set of proteins (12). Interestingly, this will also have the ability of correctly excluding 1612 proteins belonging to 30 major folds of globular proteins with the accuracy of 99% as seen in Figure 3. The inclusion of PSSM profiles enhanced the accuracy of discriminating TMBs (in a dataset of 206 TMBs and 1045 non-TMBs obtained with less than 40% sequence identity) up to 96.4% (13).

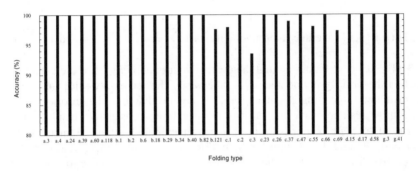

Fig. 3. The accuracy of excluding 30 major folding types of globular proteins. The SCOP classification (14) is used to denote the folding types.

Table 2. Predictive performance of different methods for discriminating TMBs

Method	Accuracy (%)	Reference
Sequence alignment profile	80	[15]
Amino acid composition (β-strand segments)	84	[16]
Amino acid composition (TMB and globular proteins)	87	[5]
Hidden Markov model and alignment profies	88	[17]
Hidden Markov model	88	[18]
Neural networks (amino acid composition)	91	[11]
Support vector machines (amino acid composition)	92	[19]
k-nearest neighbor	92	[20]
Support vector machines (amino acid and dipeptide compositions)	94	[19]
Neural networks (amino acid properties)	94	[12]
RBF network and PSSM profiles	96	[13]

The comparison of statistical and machine learning techniques for discriminating TMBs is presented in Table 2. We noticed that the accuracy is remarkably higher with machine learning techniques than with statistical methods. Further, the statistical methods could correctly identify the TMBs whereas the performance of machine learning techniques is better for excluding globular and TMH proteins than identifying TMBs.

3.3 Prediction of Membrane Spanning Segments

We have developed a "rule based approach" for predicting transmembrane β-strands using three features, (i) preference of amino acid residues in membrane spanning β-strands (conformational parameters), (ii) hydrophobic character and (iii) amphipathicity (21). A set of five primary rules have been designed to assign the priority of each residue to be in transmembrane β-strand and four secondary rules to pick up the membrane spanning segments. The primary rules for assigning the priority

of each residue, i, are: $\beta(i) > 1.0$ (average conformational parameter), $1/6 \sum_{i=1}^{6} \beta(i) >$

1.0, $H_p(i) > 13.34$ (average hydrophobicity), $1/6 \sum_{i=1}^{6} H_p(i) > 13.34$ and $1/2 \sum_{i=1}^{2} H_p(i)$

$= 13.34 \pm 0.5$ (oscillating around the average hydrophobicity). If these conditions are satisfied the priority is one and zero, otherwise. The secondary rules for picking up the membrane spanning segments are: if any residue has the priority of 5, two consecutive residues have the priority of 4 or three consecutive residues have the priority of ≥ 3 there is a possibility of a transmembrane β-strand segment around the residue(s). Extend the length in both directions so that there may not be two consecutive low priority residues (less than 3) or a residue of zero priority. If the segment is longer than 20 residues cut into two smaller segments at the residue of highest hydrophobicity. This method is mainly applicable to bacterial porins and it could predict the membrane spanning segments with the accuracy of 82%.

Further, we have set up a method using neural networks for predicting membrane spanning regions in TMBs. In this method, a three-layered neural network with one hidden layer has been used for predictions. Input layer reads the input information about a residue and its sequence neighbors from the neural network through a running window. Each residue is represented by a 21-bit vector (20 units for the amino acids and one unit for describing the terminal position of the protein). This input information is then fed forward through linear activation function, and the final signal received at the single unit of the output layer is transformed via a sigmoidal function to yield a value between 0 and 1. Our method could predict the membrane spanning regions of 13 TMBs with the accuracy of 73% using only the sequence information (22). In addition, our method would provide the probability of each residue to be in the transmembrane segment.

3.4 Annotation of β-Barrel Membrane Proteins in Genomic Sequences

We have developed a novel method for detecting TMBs in genomic sequences. We have followed the below mentioned steps for detecting TMBs as depicted in Figure 4: (i) identification of TMBs using the preference of residue pairs in globular, TMH and

TMBs, (ii) exclusion of TMH proteins using SOSUI, a prediction system for TMH proteins, (iii) elimination of globular/TMH proteins that show the sequence identity of more than 70% for the coverage of 80% residues with known structures in PDB and (iv) elimination of globular/TMH proteins that have the sequence identity of more than 60% with known sequences in SWISS-PROT. This method showed good agreement with experimental observations. An example is shown below for E. coli.

The complete genome of E. coli has 4237 proteins and the comparison of residue pair preferences identified 1036 proteins as TMBs (step i). Further, globular and TMH proteins were eliminated with steps (ii-iv) and finally we obtained 87 sequences as TMBs. Interestingly, all the 11 TMBs of known structures from E. coli have been identified by our method. Further, our approach could detect representative sequences in all the 15 families of TMBs deposited in Transport Classification Database (23).

TMB finding pipeline

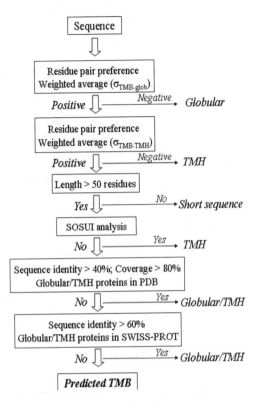

Fig. 4. Pipeline for detecting TMB proteins in genomic sequences

We have developed a database, TMBETA-GENOME , for annotated TMBs in 275 genomic sequences and it is available at http://tmbeta-genome.cbrc.jp/ annotation/. TMBETA-GENOME includes several features, such as, the service for detecting TMBs in genomic sequences using various methods, related references, statistics for the detected TMBs by different methods for each genome, details about all algorithms used to detect TMBs, relative links to other databases and a help page (24). An example is shown in Figure 5. In this figure the results are shown for Escherichia coli. K12 genome. The method, "New approach" has been selected for obtaining the annotated TMBs. This search picked up 87 entries and the TMBs identified by the new approach are shown with the identification number. In addition, the results obtained with other methods are also given for comparison. This database is a valuable resource for finding annotated TMBs in genomic sequences.

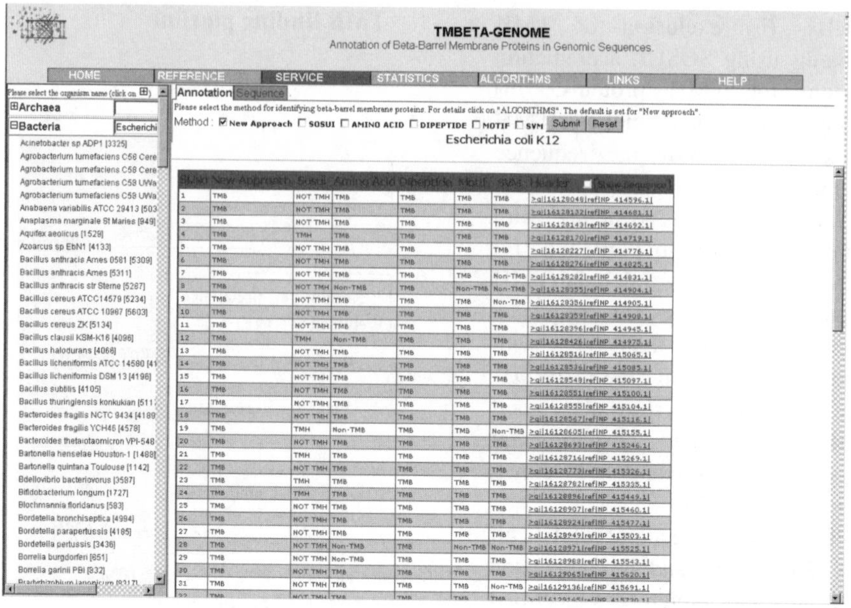

Fig. 5. Annotated TMBs in E. coli using our approach and deposited in TMBETA-GENOME database

4 Conclusions

We have systematically analyzed the characteristic features of amino acid residues in the sequences of TMBs and globular proteins and revealed the differences between them. Utilizing this information, we have developed statistical and machine learning techniques for discriminating TMBs from other folding types of globular and membrane proteins. Further, rule based and neural networks methods have been proposed for identifying membrane spanning segments. A new approach has been developed to detect TMBs in genomic sequences and a data base has been set up for the annotated TMBs in genomes.

Acknowledgements. The author wishes to thank Dr. Makiko Suwa, Dr. Shandar Ahmad and Dr. Yukimitsu Yabuki for helpful discussions. The travel support received from Japan Society for the Promotion of Sciences (JSPS) is gratefully acknowledged.

References

1. Gardy, J.L., Spencer, C., Wang, K., Ester, M., Tusnady, G.E., Simon, I., Hua, S., de Fays, K., Lambert, C., Nakai, K., Brinkman, F.S.: PSORT-B: Improving protein subcellular localization prediction for Gram-negative bacteria. Nucleic Acids Res. 31, 3613–3617 (2003)
2. Li, W., Jaroszewski, L., Godzik, A.: Clustering of highly homologous sequences to reduce the size of large protein databases. Bioinformatics 17, 282–283 (2001)
3. Berman, H.M., Westbrook, J., Feng, Z., Gilliland, G., Bhat, T.N., Weissig, H., Shindyalov, I.N., Bourne, P.E.: The Protein Data Bank. Nucleic Acids Res. 28, 235–242 (2000)

4. Branden, C., Tooze, C.: Introduction to protein structure. Garland Publishing Inc., New York (1999)
5. Gromiha, M.M., Suwa, M.A.: Simple statistical method for discriminating outer membrane proteins with better accuracy. Bioinformatics 21, 961–968 (2005)
6. Pautsch, A., Schulz, G.E.: High-resolution structure of the OmpA membrane domain. J. Mol. Biol. 298, 273–282 (2000)
7. Vandeputte-Rutten, L., Kramer, R.A., Kroon, J., Dekker, N., Egmond, M.R., Gros, P.: Crystal structure of the outer membrane protease OmpT from Escherichia coli suggests a novel catalytic site. EMBO J. 20, 5033–5039 (2001)
8. Yue, W.W., Grizot, S., Buchanan, S.K.: Structural evidence for iron-free citrate and ferric citrate binding to the TonB-dependent outer membrane transporter FecA. J. Mol. Biol. 332, 353–368 (2003)
9. Gromiha, M.M., Ahmad, S., Suwa, M.: Application of residue distribution along the sequence for discriminating outer membrane proteins. Comput. Biol. Chem. 29, 135–142 (2005)
10. Gromiha, M.M.: Motifs in outer membrane protein sequences: Applications for discrimination. Biophys. Chem. 117, 65–71 (2005)
11. Gromiha, M.M., Suwa, M.: Discrimination of outer membrane proteins using machine learning algorithms. Proteins: Struct. Funct. Bioinf. 63, 1031–1037 (2006)
12. Gromiha, M.M., Suwa, M.: Influence of amino acid properties for discriminating outer membrane proteins at better accuracy. Biochim. Biophys. Acta 1764, 1493–1497 (2006)
13. Ou, Y.-Y., Gromiha, M.M., Chen, S.-A., Suwa, M.: Discrimination of beta barrel membrane proteins using RBF networks and PSSM profiles. Proteins: Struct. Funct. Bioinf. (in press)
14. Murzin, A.G., Brenner, S.E., Hubbard, T., Chothia, C.: SCOP: a structural classification of proteins database for the investigation of sequences and structures. J. Mol. Biol. 247, 536–540 (1995)
15. Gnanasekaran, T.V., Peri, S., Arockiasamy, A., Krishnaswamy, S.: Profiles from structure based sequence alignment of porins can identify beta stranded integral membrane proteins. Bioinformatics 16, 839–842 (2000)
16. Liu, Q., Zhu, Y., Wang, B., Li, Y.: Identification of beta-barrel membrane proteins based on amino acid composition properties and predicted secondary structure. Comput. Biol. Chem. 27, 355–361 (2003)
17. Martelli, P.L., Fariselli, P., Krogh, A., Casadio, R.: A sequence-profile-based HMM for predicting and discriminating beta barrel membrane proteins. Bioinformatics 18, S46–S53 (2002)
18. Bagos, P.G., Liakopoulos, T.D., Spyropoulos, I.C., Hamodrakas, S.J.: A Hidden Markov Model method, capable of predicting and discriminating beta-barrel outer membrane proteins. BMC Bioinformatics 5, 29 (2004)
19. Park, K.J., Gromiha, M.M., Horton, P., Suwa, M.: Discrimination of outer membrane proteins using support vector machines. Bioinformatics 21, 4223–4229 (2005)
20. Garrow, A.G., Agnew, A., Westhead, D.R.: TMB-Hunt: a web server to screen sequence sets for transmembrane beta-barrel proteins. Nucleic Acids Res. 33, W188–W192 (2005)
21. Gromiha, M.M., Majumdar, R., Ponnuswamy, P.K.: Identification of membrane spanning beta strands in bacterial porins. Protein Eng. 10, 497–500 (1997)
22. Gromiha, M.M., Ahmad, S., Suwa, M.: TMBETA-NET: Discrimination and prediction of membrane spanning?-strands in outer membrane proteins. Nucleic Acids Res. 33, W164–W167 (2005)
23. Busch, W., Saier Jr., M.H.: The transporter classification (TC) system, 2002. Crit. Rev. Biochem. Mol. Biol. 37, 287–337 (2002)
24. Gromiha, M.M., Yabuki, Y., Kundu, S., Suharnan, S., Suwa, M.: TMBETA-GENOME: database for annotated beta-barrel membrane proteins in genomic sequences. Nucleic Acids Res. 35, D314–D316 (2007)

Estimation of Evolutionary Average Hydrophobicity Profile from a Family of Protein Sequences

Said H. Ahmed and Tor Flå

Dept of Mathematics and Statistics, University of Tromsø, 9037 Tromsø- Norway
{said.hassan.ahmed,tor.fla}@matnat.uit.no
http://www.uit.no

Abstract. Hydrophobicity has long been considered as one of the primary driving forces in the folding of proteins. We discuss here the evolutionary average of the hydrophobicity profile in an aligned family of proteins and found a patchy mean hydrophobicity profile. This is in contrast to Bastolla et al (2005b) results for the large superfamily of globular proteins. The idea is to use singular value decomposition and cavity filtering in order to remove the eigensequences burried in the evolutionary noise

1 Introduction

It is well known that hydrophobicity is a major determinant of protein stability and evolution. With respect to sequence-structure correlation, the evolutionary average of hydrophobicity profiles of sequences with the same fold correlates with principal eigenvector of fold's contact matrix (PE) much strongly than the hydrophobicity profile (HP) of its single sequence [1]. In the Structurally Constrained Neutral (SCN) model of protein evolution [2,3,4] the correlation is perfect (almost one), and yields

$$h^s_{evol} \equiv \sum_{a=1}^{20} \pi^s_a h_a = \sqrt{\frac{\langle h^2_{evol}\rangle - \langle h_{evol}\rangle^2}{(\langle c^2\rangle - \langle c\rangle^2)}} \left(c_s - \langle c\rangle\right) + \langle h_{evol}\rangle \ , \qquad (1)$$

where h_{evol} is the position specific evolutionary average of the HP, π^s_a is the position specific amino acid distribution at site s resulting from the evolutionary process (a indicates one of the 20 amino acid types) and c_s is the PE component of the contact matrix of the family. Assuming this equation is the only relevant condition, the amino acid distribution at site s is predicted to be the distribution of maximal entropy [11] with mean given above , i.e.

$$\pi^s_a = \frac{\exp[-\beta_s h_a]}{\sum_{a'=1}^{20} \exp[-\beta_s h_{a'}]} \ . \qquad (2)$$

The site specific Boltzmann parameters ('inverse temperature') β_s determine the width of the amino acid distribution. The width parameter varies from site

J.C. Rajapakse, B. Schmidt, and G. Volkert (Eds.): PRIB 2007, LNBI 4774, pp. 158–165, 2007.

to site and measures the tolerance of site s to accept mutations over very long evolutionary time. In principle it can catch external parameter dependence of the distribution due to say temperature, regulatory effects, e.t.c.

In this paper we estimate the evolutionary average hydrophobicity sequence from a set of aligned protein sequences from elastase family. The idea is to use eigensequences related to the inter-species hydrophobicity sequence correlation matrix to remove the evolutionary noise from the sequences and hence avoid inspection of large database to compute the mean hydrophobicity. For example, Bastolla et al. (2005a) used thousands of globular sequences from the PFAM, the FSSP, and the SCN databases in order to compute the evolutionary mean hydrophobicity profile. Since the aligned sequences are represented through hydrophobic profiles by quantifying each of the amino acids in the sequences using for example Kyte and Doolittle hydropathy scale it can be viewed as multidimensional heterogenous hydrophobicity sequences. We then use Singular Value Decomposition (SVD) and cavity filtering in order to decorrelate and remove the eigensequences buried in evolutionary noise. The average hydrophobicity profile is then computed from the first few useful eigensequences corresponding to the largest eigenvalues of the cross species hydrophobicity covariance matrix.

2 Dataset and Methods

2.1 Dataset

The dataset consists of $L = 32$ aligned sequences of length $N = 247$ (including gaps) from elastase family. The sequences were located from a search in the NCBI and SWISSPROT protein data banks. Elastase is a member of the large family of serine proteinases which includes trypsin and chemotrypsin, and is synthesized initially in the pancreas as an inactive precursor. The 3D structure of these molecules has also abeen modeled at the department of chemistry, university of Tromsø. The dataset can be obtained on request from us.

We represented the sequences through hydrophobic profiles by quantifying each of the amino acids in the sequences using Kyte and Doolittle hydropathy scale [7]. That is the hydrophobicity of residue a at position s in a sequence is given by

$$H_{a(s)} = \mathbf{Y}_{a(s)}^T \mathbf{f} \qquad (3)$$

where $\mathbf{Y}_{a(s)} = (0, 0, \ldots, 1_{a(s)}, 0, \ldots, 0) \in \mathbb{R}^{21}$ is a count vector for residue $a^1 = \{1, 2, \ldots, 21\}$ at site s and $\mathbf{f}$ is the hydrophobic index in Kyte and Doolittle. For all the consecutive amino acids in sequence l we have

$$H_l = \mathbf{Y}_l^T \mathbf{f} \qquad (4)$$

where $\mathbf{Y}_l = \{\mathbf{Y}_{l,a(s)}\}_{s=1}^N \in \mathbb{R}^{21 \times N}$ is a dummy matrix that consists of N unit count vectors. Hence $\mathbf{H}$ is $L \times N$ elastase sequences represented through hydrophobic profiles. Plot of hydrophobicity level of ela-pig (PDB:1qnj) is shown in Figure 1.

---

[1] Gaps were treated as if they were a 21st amino acid type.

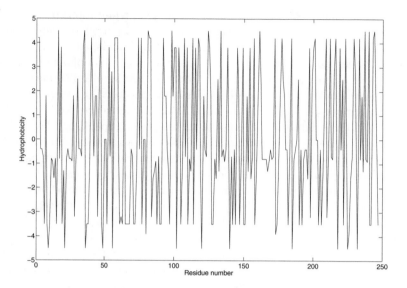

Fig. 1. Hydrophobicity profile of one of the elastase sequences,1QNJ, generated by quantifying each of the amino acids in the sequence using Kyte and Doolittle hydropathy scale

2.2 Estimating Average Hydrophobicity Profile (HP)

The problem of computing the average HPs from $\mathbf{H}$ can be considered as extracting mean hdrophbocity sequence from a noisy one[2]. We assume two types of noise contributions in our data - one along the sequence chain (due to for example the stochasticity of the folded protein chain) and the other across the sequences (due to evolutionary noise). In order to decorrelate and remove the eigensequences burried in evolutionary noise we eigen decompose (SVD - Singular Value Decomposition) the estimate of the noise covariance matrix $\hat{\boldsymbol{\Sigma}}$,

$$\hat{\boldsymbol{\Sigma}} = \mathbf{U}\boldsymbol{\Lambda}\mathbf{U}^{\mathbf{T}} \tag{5}$$

where $\mathbf{U} \in \mathbf{R}^{L \times L}$ is an orthogonal matrix (i.e., $\mathbf{U}^{\mathbf{T}}\mathbf{U} = \mathbf{I}$), the columns of $\mathbf{U}$ form an orthonormal basis for the HPs of the sequences and $\boldsymbol{\Lambda} = diag(\lambda_1, \lambda_2, \ldots, \lambda_L)$ is a diagonal matrix with entries λ_l, eigenvalues in decreasing order. The noise sequences are approximated by subtracting a smoothed (denoised) mean HP from each of the observed HPs. We choose a deterministic gaussian 'cavity filtering' procedure [10] due to local amino acid interactions along the protein sequence. It has also the non-enhancement property of local extrema: values of local maxima cannot increase and respective values of local minima cannot decrease [8]. The

---

[2] We are presently developing a Boltzmann lattice approximation for discrete evolutionary sequence noise and protein observables in an aligned phylogenetic protein family.

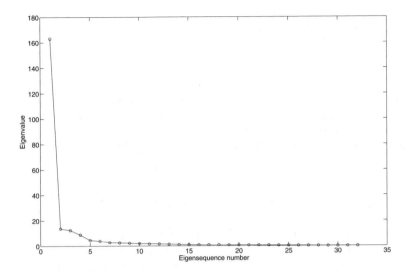

Fig. 2. Scree plot: A plot of eigenvalues λ_l, in decreasing order. The plot is used to decide the number of eigensequences that are useful (eigensequences to the left of the elbow or bend).

hydrophobicity profiles of sequences, $\mathbf{H}$ are then projected into new coordinates to obtain the eigensequences

$$\mathbf{Q} = \mathbf{H}^T \mathbf{U} \ . \tag{6}$$

The eigensequences due to evolutionary noise are then filtered out by using the first $K = 3$ eigensequences. K is determined by the point at which the remaining eigenvalues are relatively small and all about the same size. One way to determine K, the number of eigensequences $\mathbf{Q} = [\mathbf{q_1 q_2} \ldots \mathbf{q_K}]$ to retain is by use of a scree plot [6], a plot of λ_l (the eigenvalues in deccreasing order) versus l. A scree plot for the HPs of elastases, $\mathbf{H}$ is shown in Figure 2. To determine K, we look for an 'elbow' (bend) in the scree plot. The eigensequences whose eigenvalues plot to the right of such 'elbow' are ignored since they are defined here to be due to evolutionary noise. Thus the information in the scree plot indicates that we extract the first three eigensequences.

The denoised eigensequences $\hat{\mathbf{Q}}$ are inverse projected to obtain a denoised version $\hat{\mathbf{H}}$ of $\mathbf{H}$:

$$\hat{\mathbf{H}}^T = \hat{\mathbf{Q}}^T \mathbf{U}^T \ . \tag{7}$$

The site specific average hydrophobicity profile of the aligned elastases is calculated by taking the mean of the denoised HPs of the sequences:

$$\overline{h}_s = \frac{1}{L} \sum_{l=1}^{L} \hat{H}_{l,s} = (\overline{h}_1, \overline{h}_2, \ldots, \overline{h}_N) \ . \tag{8}$$

Finally we perform cavity field on the average hydrophobicity profile. The cavity fields is defined as [10]

$$\overline{h}_s = \sum_{t \neq s} J_{st}\overline{h}_t \tag{9}$$

where the couplings J_{st} are taken to be translational invariant gaussian. We choose a deterministic cavity field since our J_{st} parameters are assumed to have small variance compared to their mean. The cavity field describes the local internal filed which the amino acid 'sees'.

The algorithm to estimate the site specific average hydrophobicity profile can then be divided into seven main steps:

1. Estimate the noise hydrophobicity sequences by subtracting a cavity filtered cross species mean HP from all the HPs of the sequences.
2. Compute the estimated noise covariance matrix $\hat{\boldsymbol{\Sigma}}$.
3. Diagonalize $\hat{\boldsymbol{\Sigma}} = \mathbf{U}\boldsymbol{\Lambda}\mathbf{U}^{\mathbf{T}}$, where $\mathbf{U} \in \mathbf{R}^{L \times L}$ is an orthogonal matrix (singular vectors), $\boldsymbol{\Lambda} = diag(\lambda_1, \lambda_2, \ldots, \lambda_L)$ are the eigenvalues. Decorrelate the HPs of the sequences by projecting them into new coordinates to obtain the eigensequences, i.e., $\mathbf{H}^T\mathbf{U}$

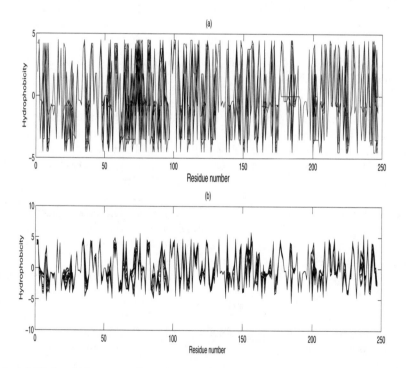

Fig. 3. (a) Hydrophobicity profiles of all the aligned elastase sequences. The hydrophobicity profiles were generated by assigning a hydrophobicity value to each of the amino acids in each sequence using Kyte and Doolittle hydropathy scale. (b) SVD denoised version of HPs of the sequences. The HPs were reconstructed using only the first three useful eigensequences that account 82.4% of the total variance.

4. Remove the eigensequences due to evolutionary noise by choosing the first K useful eigensequences (use for example a scree plot to decide the number of eigensequences to retain).
5. Reconstruct the hydrophobicity sequences, $\hat{\mathbf{H}}$ from the denoised eigensequences (multiply by $\mathbf{U}^T$).
6. Calculate the site specific average hydrophobicity profile from the denoised HPs of the sequences using (8).
7. Perform cavity filtering on the average hydrophobicity profile using (9).

3   Results and Discussion

We demonstrated our method using the aligned protein sequences from elastase family represented through their HPs (see Materials and Methods). Figure 3 shows plot of HPs of all the aligned elastase sequences and their SVD denoised version. From the figure we see a lot of variations (evolutionary noise) in the original sequences while in the second plot much of the evolutionary noise is removed. Only the first three eigensequences that account 82.4% of the total variace were used in the reconstruction. The site specific average hydrophobicity profile was then estimated from the reconstructed denoised eigensequences (first

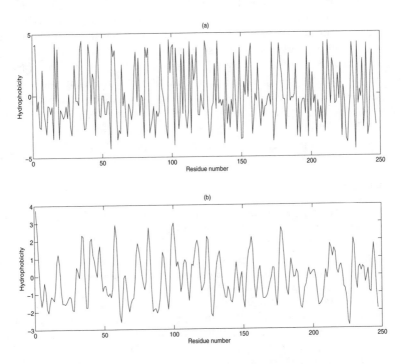

Fig. 4. (a) Site specific average HP estimated from the reconstructed denoised eigensequences. (b) Result of cavity 'filtering', short range interaction - three amino acid local interactions along the mean HP sequence.

three eigensequences) using equation 3. Finally cavity filtering was applied on the average HP. Short range amino acid interactions (three local amino acid interactions) along the sequence profile was used. Figure 4 shows plot of average HP and its cavity filtered version. From Figure 4(a) we see that the estimated average hydrophobicity profile is still patchy. This might be due to variation along the sequences. We therefore used cavity filtering to smooth this variation (see Figure 4(b)).

We have analyzed (not yet published) the correlation between this estimated average hydrophobicity and average surface exposure of our proteins and found that the correlation is stronger than when the average hydrophobicity is computed by just averaging the HPs of the sequences or estimated using wavelet based smoothing methods.

4 Conclusions and Further Work

In this paper, we developed a method to estimate average hydrophobicity sequence from a set of aligned sequences from one protein family. We tested this method on aligned sequences from elastase family. The method has removed the evolutionary noise effectively. We are still working further to test how effective the method is by analyzing the correlation between mean hydrophobicity and surface-exposure or principal eigenvector of fold's contact matrix for varioius families. This mean profile can be improved if we for example use more physico-chemical properties like charge, electrostatic interactions, e.t.c.

So far we have computed an estimate of a mean HP profile but our future aim is to estimate the site specific Boltzmann paramters, β_s from this mean HP. This width parameter in principle can catch external parameter dependence of the distribution due to for example temperature. So we think that this parameters can be used to classify proteins within a family, for example identify significant diferences between mesophilic and psychrophilic populations [13].

References

1. Bastolla, U., Porto, M., Roman, H.E., Vendruscolo, M.: The principal eigenvector of contact matrices and hydrophobicity profiles in proteins. Proteins 58, 22–30 (2005a)
2. Bastolla, U., Porto, M., Roman, H.E., Vendruscolo, M.: Connectivity of neutral networks, overdispersion, and structural conservation in protein evolution. J. Mol. Evol. 56, 243–254 (2003)
3. Bastolla, U., Porto, M., Roman, H.E., Vendruscolo, M.: Lack of self-averaging in neutral evolution of proteins. Phys. Rev. Lett. 89, 208101/1–208101/4 (2002)
4. Bastolla, U., Porto, M., Roman, H.E., Vendruscolo, M.: Statistical properties of neutral evolution. J. Mol. Evol. 57, S103–S119 (2003)
5. Branden, C., Tooze, J.: Introduction to Protein Structure, 2nd edn. Garland publishing, New York (1999)
6. Johnstone, R.A., Wichern, D.W.: Applied Multivariate Statistical Analysis, 5th edn. Prentice Hall, Englewood Cliffs (2002)

7. Kyte, J., Doolittle, R.F.: A Simple Method for Displaying the Hydropathic character of a Protein. J. Mol. Biol. 157, 105–132 (1982)
8. Lindeberg, T.: Scale-Space Theory in Computer Vision. The Kluwer International Series in Engineering and Computer Science. Kluwer Academic Publishers, Dordrecht (1994)
9. Miyazawa, S., Jernigan, R.L.: Self-consistent estimation of inter-residue protein contact energies based on an equillibrium mixture approximation of residues. Proteins: Structure and Molecular Principles 34, 49–68 (1999)
10. Opper, M., Winther, O.: From Naive Mean Field Theory to the TAP Equations. The MIT Press, Cambridge, Massachusetts London, England (2002)
11. Porto, M., Roman, H.E., Vendruscolo, M., Bastolla, U.: Prediction of site-specific amino acid distributions and limits of divergent evolutionary changes in protein sequences. Mol. Biol. Evol. 22, 630–638 (2005)
12. Fornasari, M.S., Parisi, G., Echave, J.: Site-specific amino acid replacement matrices from structurally constrained protein evolutioin. Mol. Biol. 19, 352–356 (2002)
13. Thorvaldsen, S., Flå, T., Willassen, N.P.: Extracting molecular diversity between populations through sequence alignments. In: Oliveira, J.L., Maojo, V., Martín-Sánchez, F., Pereira, A.S. (eds.) ISBMDA 2005. LNCS (LNBI), vol. 3745, pp. 317–328. Springer, Heidelberg (2005)
14. Wall, M.E., Rechtsteiner, A., Rocha, L.M.: Singular Value Decomposition and Principal Component Analysis. In: Berrar, D.P., Dubitzky, W., Granzow, M. (eds.) A Practical Approach to Microarray Data Analysis, pp. 91–109. Kluwer, Norwell, MA (2003)
15. Tang, C.: Simple Models of the Protein Folding problem. Physica A 31, 288 (2000)
16. Moelbert, S., Emberly, E., Tang, C.: Correlation between sequence hydrophobicity and surface-exposure pattern of database proteins. Protein Science 13, 752–762 (2004)

APMA Database for Affymetrix Target Sequences Mapping, Quality Assessment and Expression Data Mining

Yuriy Orlov[1], Jiangtao Zhou[1], Joanne Chen[2], Atif Shahab[1,2],
and Vladimir Kuznetsov[1,*]

[1] Genome Institute of Singapore, 60 Biopolis Street, Genome,
138672 Singapore
[2] Bioinformatics Institute, 30 Biopolis Street, Matrix,
138671 Singapore
kuznetsov@gis.a-star.edu.sg

Abstract. We have developed an online database APMA (Affymetrix Probe Mapping and Annotation) for interactive presentation, search and visualization of Affymetrix target sequences mapping and annotation <http://apma.bii.a-star.edu.sg>. APMA contains revised genome localization of the Affymetrix U133 GeneChip initial (target) probe sequences. We designed APMA to use it as a filter before data analysis and data mining so that noise expression signals, false correlations and false gene expression patterns can be reduced. Discrepancies found in probeset annotation and target sequence mapping account for up to 30% of probesets, including about 25% of Affymetrix probesets derived from target sequences overlapped interspersed repeats and 1.8% of original target sequences with erroneous orientation of the sequences. 86% of U133 target sequences passed our quality-control filtering.

Keywords: Affymetrix U133, database, target sequences, cross-hybridization, mapping, genome repeats, errors, classification, recognition, data mining.

1 Introduction

The increasing growth of the microarray researches demands high quality standards for microarray expression databases, description and annotation of probes and genes. One of the key problem facing microarray experiments is insufficient reliability of expression measurements due to sub-optimal probe design. The problem could originate from poor gene identification by the probe sequences, whose design may not consider the actual complexity of the human transcriptome. Poor quality control (QC) of microarray probes can also generate many hard statistical problems at data analysis level, starting from selection of differentially expressed genes and ending by identification of co-expressed and co-regulated genes.

One of the widely accepted microarray technologies is provided by Affymetrix Corporation (http://www.affymetrix.com). Our goal was to develop an algorithm and

---

* Corresponding author.

J.C. Rajapakse, B. Schmidt, and G. Volkert (Eds.): PRIB 2007, LNBI 4774, pp. 166–177, 2007.
© Springer-Verlag Berlin Heidelberg 2007

software for quality control and filtering of Affymetrix target sequences. We organized and stored the results of this work in the APMA (Affymetrix Probe Mapping and Annotation) database. Such database suggested is of reasonable practical interest of the users of Affymetrix microarrays.

In situ synthesized oligonucleotide Affymetrix GeneChip uses a set (the so called probeset) of 11-20 oligonucleotide probes, each 25 bases long, to represent a gene or a gene transcript. The perfect match probe comes together with a mismatch probe designed to measure non-specific cross-hybridization. The expression level for a gene is a summary of the signal from the entire probeset. Affymetrix uses ~150-450 nt initial (target) sequences of genes for probes (and whole probeset) location.

The problem of accurate Affymetrix target sequence annotation is related to the complexity of multiple "gene models" including unverified ESTs from public datasets. Reported re-identification of genes may affect 30-50% of probesets [1,2]. Recent papers [3,4,5,6] report re-evaluation of Affymetrix microarray probes using BLAST comparison of probe sequences to the complete human genome. In some cases, multiple probesets can specifically target a single genic sequence coding for protein. In other cases, however, a probeset is capable of hybridizing to more than one transcript (and provide uncertainty in transcript detection) [7].

Selection of original target sequences is one of the key steps of probe design process. There are several basic quality control criteria for verification of the target sequence. The sequences should: (1) detect a unique locus in human genome, (2) match a single transcript without mutations (correct mapping); (3) correspond to the sequence from the transcribed strand of the genome at the locus (correct strand orientation of target sequence); (4) not overlap with any other non-gene sequence that could cross-hybridize or even be independently transcribed (segmental duplications, interspersed repeats); (5) correspond to mature RNA (not intronic sequences that are spliced).

Unfortunately, these basic criteria have not been well controlled. Perhaps, this is the case because transcript databases are incomplete, contain erroneous sequences, and undergo continual growth and change. Figure 1 shows examples of poor designed Affymetrix target sequences.

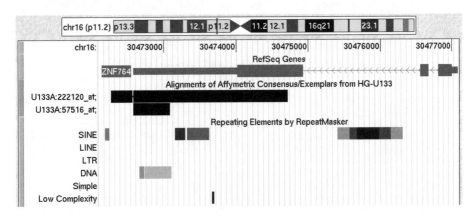

Fig. 1. Examples of problematic Affymetrix target sequences. Target sequence A.222120_at corresponds to ZNF764 gene and contains SINE and DNA repeat elements. Target sequence A.57516_at falls completely (95%) into repeat element (Charlie 8, DNA/MER1 type). The last probeset evidently has great potential for cross-hybridization.

Target sequence A.222120_at maps only one exon of ZNF764 gene (Fig. 1). It contains repeats (SINE and Alu) that could cause cross hybridization of probes. Target sequence A.57516_at overlap DNA/MER1 repeat by 95% and definitely has great potential for cross-hybridization and biased measuring of expression signal for this gene.

In our previous studies [8,9], we have developed software for automatic annotation and quality control of U133A and U133B targets sequences. In this work, we focus on data basing, QC of *target sequences* U133 Plus 2.0 GeneChip and using these tools for characterization of erroneous patterns in target sequences and expression data sets. We use BLAT program for target sequence mapping to check correspondence of probesets to annotated genes. We develop online database for interactive presentation, search, filtering and visualization of Affymetrix U133 Plus 2.0 target sequences and their mapping and annotation. The database collects information on erroneous probesets and provides flexible filters for pre-processing expression data. Finally we use several large cancer cell expression data sets to estimate quality of unreliable target sequences and corresponding probesets.

2 Methods

Affymetrix sequence data for the U133A and U133B GeneChips were downloaded from the NetAffx web site (http://www.affymetrix.com/analysis/index.affx). These sequences, intended to represent genes, are referred to as initial target sequences of the Affymetrix probesets. We used BLAT search at 90% similarity level to match each Affymetrix target sequence to the genome. Then, we annotated overlaps with exonic region(s) of RefSeq, mRNA and spliced EST variants on the NCBI Build 35 and 36.1 (hg17 and hg18) assemblies. Example of target sequences annotation in APMA is in Figure 2.

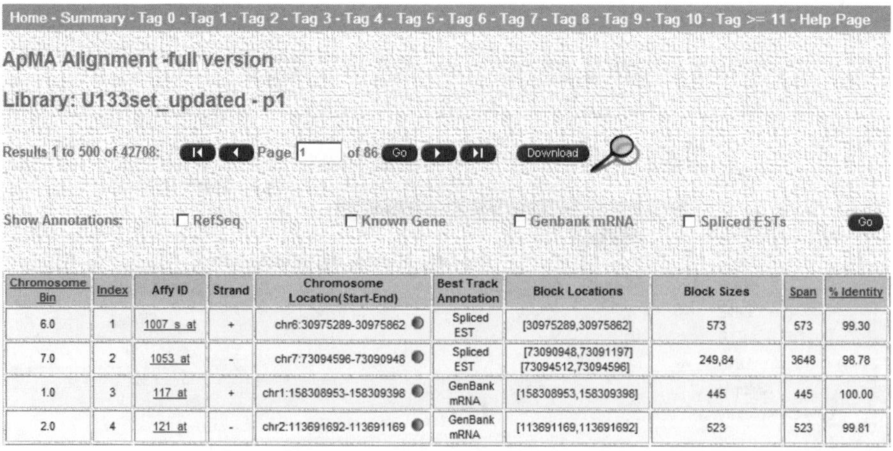

Home - Summary - Tag 0 - Tag 1 - Tag 2 - Tag 3 - Tag 4 - Tag 5 - Tag 6 - Tag 7 - Tag 8 - Tag 9 - Tag 10 - Tag >= 11 - Help Page

ApMA Alignment -full version

Library: U133set_updated - p1

Results 1 to 500 of 42708:   Page 1   of 86   Download

Show Annotations: ☐ RefSeq   ☐ Known Gene   ☐ Genbank mRNA   ☐ Spliced ESTs   Go

Chromosome Bin	Index	Affy ID	Strand	Chromosome Location(Start-End)	Best Track Annotation	Block Locations	Block Sizes	Span	% Identity
6.0	1	1007_s_at	+	chr6:30975289-30975862	Spliced EST	[30975289,30975862]	573	573	99.30
7.0	2	1053_at	-	chr7:73094596-73090948	Spliced EST	[73090948,73091197] [73094512,73094596]	249,84	3648	98.78
1.0	3	117_at	+	chr1:158308953-158309398	GenBank mRNA	[158308953,158309398]	445	445	100.00
2.0	4	121_at	-	chr2:113691692-113691169	GenBank mRNA	[113691169,113691692]	523	523	99.81

Fig. 2. APMA database interface (http://apma.bii.a-star.edu.sg)

We mapped Affymetrix probesets to gene sequence blocks based on the initial target sequences, not based on the individual 25-mers in the probe sets.

We checked for exonic repetitive elements using RepeatMasker. We constructed a table of repeats classified by family and repeat types (DNA, LTR, LINE, SINE, simple and low complexity repeats, etc.) indicating length of the Affymetrix target sequence covered by the each type of repeats.

Large fraction of Affymetrix target sequences maps to a transcript on an opposite strand. Substantial numbers of mRNAs and ESTs in cis-antisense loci represent natural anti-sense transcripts (NAST) derived from the opposite strand of the given (usually protein coding) gene [10]. In order to distinguish the Affymetrix target sequences matching NAST from the Affymetrix target sequences having wrong orientation at non-NAST loci, we developed a pipeline and constructed a local United Sense-Antisense Pairs (USAP) database [11]. The database annotates and classifies SA pairs by three annotation tracks (RefSeq, mRNA and EST sequences) and stores the information about SA genes supported by Affymetrix target sequences.

Expression data. To study functional usefulness of the problematic probes, we analyzed the expression patterns of Affymetrix probesets in 249 primary breast tumors (NCBI Gene Expression Omnibus (GEO) http://www.ncbi.nlm.nih.gov/geo/; data sets GSE4922). The cancer samples were split into groups by histologic grades corresponding to aggressiveness of breast cancer [12,13]. In addition, we used U133A&B expression data from several normal and cancerous human brain tissues (GEO data sets GDS1962), and expression profiles representing lung cancer cell lines (GEO ID: GSE5816). MAS5 normalization was applied [14]. Then we performed global mean normalization to ln(500), which provides better consistency for a large fraction of expressed genes across microarrays.

Software. Our database interface is developed in Perl. For group comparison, Mann–Whitney U-test statistics were used for continuous variables and one-sided Fisher's exact test used for categorical variables (Statistica-6 and StatXact-6 software). We have also used SAM 3.1 (Statistical Analysis of Microarrays) software [15] to estimate the number of differentially expressed genes defined by Affymetrix probesets.

3   APMA Database and Statistical Assessment of Probesets Quality

The results of mapping (chromosome coordinates, orientation, details of overlapping with exons and repeats etc.) were stored in a local database associated with unique Affymetrix probesets ID (http://apma.bii.a-star.edu.sg/). The database has convenient user interface, search engine and visualization tools referring to external (Santa Cruz) and local (Singapore) versions of UCSC Genome Browser. The search engine allows to find annotation of Affymetrix ID by gene name or accession number. The interface is shown in Figure 3.

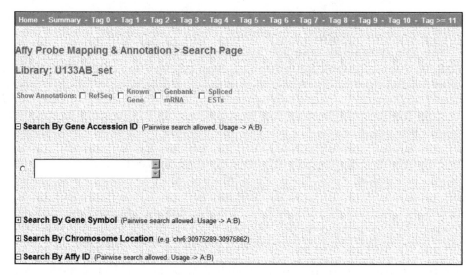

Fig. 3. Interface of the database search menu (http://apma.bii.a-star.edu.sg)

3.1 Statistics of Problematic Groups of Target Sequences

We believe that target sequences of purportedly human microarray probes which, by BLAT, are completely absent in the human genome (sequences to which we hereafter refer to as Tag0 sequences) and target sequences which match multiple loci in the genome (called Tag2, Tag3, etc. based on the number of their BLAT-matched loci) are sources of cross-hybridization effects in gene identification and should be excluded from analysis of microarray experiments. We checked BLAT mappings for all 44,692 sequences on U133A and B microarrays, except service and control probesets [16].

We found: (i) 1212 (2.7%) initial target sequences which do not match any location in the human genome (Tag0 or mismatched sequences, see Tab. 1); (ii) 42708 (95.5%) target sequences with a single reliable mapping (Tag1: reliable target sequences); (iii) 772 target sequences (1.7%) with multiple locations in the human genome (Tag2+). Tag2+ is defined as sum of Tag2, Tag3, Tag4,..., etc. Tag 0 and Tag2+ might cause noise and/or cross-hybridization signals.

Table 1. Statistics of Affymetrix target sequence matches in human genome

#matches	#Affymetrix ID	Percentage	Cumulative %
tag0	1212	2.71	2.71
tag1	42708	95.56	98.27
tag2	450	1.01	99.28
tag3	129	0.29	99.57
tag4	67	0.15	99.72
tag5+	126	0.28	100
Total	**44692**	**100**	**100**

Tag0 sequences are related mostly to mRNA and EST, but not to genomic DNA. These sequences were associated with poorly-designed target sequences, poorly-annotated transcripts, and with nonhuman sequences which were mistakenly labeled as "human" in the GenBank. For instance, some of Tag0 were classified as "xeno-sequence /nonhuman" (mouse, cow, pathogens, rat etc; 224340_at is mouse c-Myc with extra TGA insertion; 217283_at strongly maps mouse short stature homeobox; 217255_at 100% is cow SQSTM1). Other probesets belong to small groups of poorly-defined sequences (for instance, 222196_at falls to random (not assembled) chromosome parts).

Standard assignment of Affymetrix target sequences to genome provided by UCSC Genome Browser using default BLAT parameters either does not define all target sequences or just skips them without any reference. Location of probesets could correspond to the mapping of genes, but the latter maybe not unique. (For example, target sequence for probeset 208303_s_at falls onto different chromosomes in hg18: X, Y following the mapping of CRLF2 (cytokine receptor-like factor 2 isoform 1). The CDS end of the gene is not complete. Another example is 207353_s_at probeset mapped to the unassembled part of chromosome 4 (chr4_random).

We identified multiple genome locations of some extraordinary redundant probes (tag11+). For instance, probeset 81737_at has 22 different locations in human genome; probeset 213089_at has more 11 hits to human genome.

3.2 Repeats in Tag1 Target Sequences

Surprisingly, about 25% of target sequences are covered by mobile elements (repeats) abundant in the human genome such as Alu, LINE and LTR (Tab. 2).

They might serve as a significant source of erroneous detection of expressed genes and cross-hybridization signals.

Table 2. U133 Affymetrix target sequences containing genome repeats

Set of repeats	Repeat class	# in U133A and U133B	# in U133 additional set
Simple repeats	Simple repeat, Low complexity	3233	468
Short transposons (<300 bp)	DNA, SINE/Alu, SINE/MIR	4347	1578
Long transposons (>300 bp)	LINE/CR1, LINE/L1, LTR/ ERV1/ERVK/ERVL/MaLR	5420	1915
Non-transposons and satellites	Other, RNA, rRNA, scRNA, Satellite, snRNA, srpRNA	80	31

3.3 Inversely Oriented Target Sequences

We consider an Affymetrix target sequence as inversely oriented if it matches the opposite strand to any RefSeq, mRNA, or EST-supported gene. If a target sequence matches also any RefSeq or mRNA in the same strand then this sequence may refer to natural antisense transcripts (NAST), but not annotation errors. We developed a pipeline to distinguish annotation errors from sequences matching natural antisense transcripts. We considered a target sequence as misoriented relative to the intended gene (presented by RefSeq or mRNA ID) if:

1) it is aligned perfectly in complete genomic coordinates, block by block (the allowed shift is no more than 8 bp except for the leftmost and rightmost block) to the transcript mapped to the opposite DNA strand;

2) the number of blocks RefSeq/mRNA blocks mapped to the genome was greater than one;

3) there no any RefSeq gene in the same strand;

4) if there are several perfectly matching mRNA transcripts in both strands target sequence matches the majority of GenBank mRNAs in wrong orientation, while there are none or only a single mRNA perfectly matching the Affymetrix target sequence blocks on the same strand.

In total, 810 (1.8%) Affymetrix target sequences were defined as misoriented target sequences. This set was identified by manual curation and automatic comparison of blocks of Affymetrix target sequences with exons of RefSeq or mRNA sequences in opposite strand (Tab. 3). The number of Affymetrix target sequences misoriented relative to intended transcripts is larger than previously reported by Harbig et al. [1].

3.4 Classification of Different Categories of Problematic Affymetrix Target Sequences

Tab. 3 shows the statistics of different categories of poorly-defined Affymetrix target sequences found using hg18 Assembly: Tag0, multiple genome matching Tag2+ (Tag2, Tag3, Tag4 and others) targets sequences, misoriented target sequences and the target sequences covered by genome repeats.

This table shows that only about 86% (38511/44692) U133A&B target sequences could be useful in expression analysis. Our pipeline identified 13260 Affymetrix target sequences matching SA gene pair loci. These target sequences match the natural SA transcripts and should not be excluded from the analysis.

Additionally, we have identified 810 erroneously oriented Affymetrix target sequences, which should be excluded from functional (expression) analysis.

Table 3. Joint classification of problematic Affymetrix GeneChip U133A&B target sequences

Target sequences groups	Non-redundant # of probesets	%
Total # of non-Tag1 sequences, including:	1984	4.43
Tag0	1212	2.71
Tag2+	772	1.72
Total # of misoriented target sequences	810	1.81
Total # of target sequences overlapped with repeats including:	3387	7.57
overlap 80-100% of target sequence length	761	1.7
Total # of useful Tag1 sequences	38511	86.16
TOTAL # of Affymetrix target sequences	**44692**	**100**

3.5   Comparison of Mean Gene Expression Levels Detected by Different Classes of Problematic Target Sequences

We compared average gene expression levels in the groups of problematic target sequences: tag0, multiple loci matching, misoriented relative to given gene, and target sequences covered by repeats by 40-60%, 60-80%, 80-100% of target sequence length (in non-overlapping intervals of percents, i.e. [40;60), [60;80) and [80-100]) (Fig. 4). We used a large set of expression data of genetically and clinically well-separated breast cancer sub-types [12] for analysis of statistical parameters of the probesets. We designated Affymetrix probesets derived from target sequences without any complication or covered by genome repeats by less than 20% to 40% of target sequence length as "Normal". Fig. 4 shows strong negative trend of the mean values of hybridization signal from Normal to misoriented target sequences.

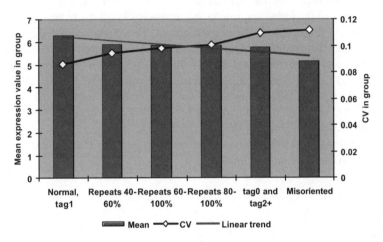

Fig. 4. Population average expression signal of probesets associated with problematic groups of Affymetrix target sequences. Columns present mean values in histologic Grade I of breast cancer samples for the probesets groups, line with diamonds presents corresponding coefficient of variation (CV).

Misoriented and multiple-matching target sequences provide the poorest probesets in comparison with other problematic sequence groups. This trend is exhibited by the lower average expression signal and by the larger coefficient of variation (CV).

We found that simple repeats and low-complexity sequences do not affect the ability of probesets to discriminate tumor-type specific signals [9]. However, as a general trend, target sequences with more genome repeats have progressively worsening proportions among differentially expressed genes in cancer tissue type comparison, especially for longer repeats (LTR and LINE) and for larger sequence span coverage of the target sequences.

Comparison of the numbers and values of correlation coefficients of probesets derived from multiple matching target sequences with random samples from Normal group reveals similarly poor quality of these problematic groups. Our analysis of expression data for cancer samples reveals that larger number of genome loci for the target sequence correlates with 1) higher expression noise (defined by CV-value), 2) lower average signal level, and 3) higher number of spurious positive correlations. This is what we would expect due to nonspecific hybridization signals.

3.6   Comparison of U133A, U133B and Additional to U133 Plus2.0 GeneChips

Figure 5 shows that the averages of signal intensity values for brain cancer cell samples differ for probesets from Normal group and probesets from "problematic" group. Problematic probesets have lower signals and therefore are enriched in the left (noisy-like) part of the empirical signal intensity frequency distribution (Fig. 5). Inversely, the signal value of Normal probesets is much enriched in the right part of the empirical signal intensity frequency distribution.

We have observed that the microarrays U133A and U133B show a markedly different quality of the target sequences and, respectively, of the hybridization signals of the probesets presented on these microarrays. Tab. 5 shows that the fraction of target sequences that passed our QC (quality control, i.e. Tag1, correct orientation on chromosome, and repeat coverage is less than 40% of target sequence length) on microarray U133A is larger in comparison to microarray U133B. In general, microarray U133A is better annotated and as we have showed exhibits higher expression level of genes than microarray U133B. We have observed 89.3% Normal target sequences for microarray U133A and 83% of such sequences for microarray U133B. Additional set (9983 probesets) to microarray U133 Plus 2.0 exhibits 79.6% of such non-problematic target sequences).

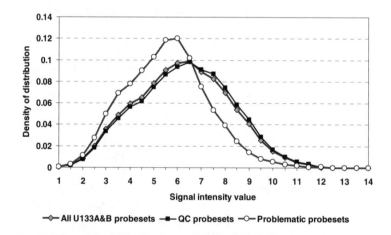

Fig. 5. Comparison of the histograms of signal intensity value for unfiltered (diamond), Normal filtered (square) and problematic (circle) probesets demonstrate systematic shift of the frequency of signal intensity value. Normal probesets filtered out based on quality control (QC) criteria show relatively higher signals. MAS5 normalized and log-transformed signals of brain cancer cell data set were used.

Table 4. Target sequences passed QC for U133A, U133B and additional to U133 Plus 2.0 microarrays

	# target sequences	# target sequences passed QC	target sequences passed QC, %
Common for A and B chip	100	98	98.0
U133A	22115	19753	89.3
U133B	22477	18660	83.0
Additional to U133 Plus 2.0	9983	7942	79.6

There are many examples of significant distinguishing expression levels of probesets, which corresponding to the same gene but are expressed differently on these three U133 sets. The poor-quality target sequences used to design the probesets (designed often using incomplete mRNAs or unreliable EST sequences) should be excluded from the gene expression analysis. Our database allows making this procedure automatically.

4   Discussion and Conclusion

Careful analysis of microarray probe design should be an obligatory component of MicroArray Quality Control (MACQ) project [17] initiated by the FDA USA in order to provide quality control tools to researchers of gene expression profiles and to translate the microarray technology from bench to bedside. In particular, identifying and filtering of unreliable target sequences are important data preprocessing steps before any analysis of microarray expression data. Such search and making decision strategy may provide essential improvement in selection of differentially expressed genes, gene clustering and pattern recognition of genetic and clinical subtypes, and in construction of realistic co-regulatory expression networks.

In this study, we have i) revised genome localization of the Affymetrix U133 GeneChip initial target sequences, ii) evaluated the impact of erroneous and poorly annotated target sequences on the quality of gene expression data and iii) developed an online database for interactive presentation, search and visualization of Affymetrix target sequences mapping and annotation. This DB contains revised genome localization of the Affymetrix U133 GeneChip initial (target) probe sequences.

In many cases, the spurious correlations can lead to serious erroneous interpretation of the microarray results, as was shown in [18]. In performing an analysis of overlaps of Affymetrix target sequences with repeat elements, we have quantitatively demonstrated that the number of positive correlation coefficients between genes in such a type of problematic target sequence group increases as repeat coverage increases. These extra *false* correlations in the groups do not correspond to real gene co-regulation but solely to bad design of target sequences. Similarly, Tag2+ and Tag0 can be also a significant source of spurious correlations of signals from probesets (and representative genes) on microarrays. Nevertheless researchers use such correlations without suitable quality control. Moreover, some gene discovery methodologies, such as hierarchical clustering, principal component analysis and gene-networking use the correlation coefficient of expression signal values between probesets as *basic* information. Other analyses, for example, general linear models, also are ultimately based on correlation-like principles.

Multiple-locus, nonhuman, misoriented, and nonspecific targets sequences are a significant attribute of the U133 GeneChip probesets. The ability of probesets to hybridize to more than one gene product can lead to false positives when analyzing gene expression data. The apparent artifacts in the data exist because the original target sequence annotations do not accurately correspond to the transcripts. Identification and removal of inaccurate target sequences can significantly improve specificity of GeneChip technology.

We summarize that about 14% of U133 A&B Affymetrix probesets have been designed based on erroneous target sequences. This fraction can be further classified as follows:

2.7% of target sequences do not reliably match any location in the human genome;

Another 1.7% of the sequences have multiple locations (up to 10 times and more);

About 7.5% of the remaining Affymetrix target sequences overlap repeat elements abundant in the human genome completely including target sequences located in transposons or over more than 40% of the target sequence length, yielding noisy expression signal;

1.8% of Affymetrix probesets have wrong orientation relative to the transcript they are alleged to detect.

The concrete number of the filtered out problematic Affymetrix probesets could be refined depending on the stringency of criteria. Despite numerous wrongly designed and poorly annotated target sequences, we argue that Affymetrix U133 GeneChip could show reproducible and quantitative hybridization signals. However, about 14% of these signals need filtering based on our expression analysis criteria, genome re-annotation and statistical methods described in this paper. We suggest restricting data analysis and data mining of Affymetrix U133 probesets within the Normal artifact-free Tag1 probes with minimal repeat content.

In conclusion, we suggest that the development of APMA DB and computational QC tools could be used as an integrative filter system to be applied before data analysis and data mining in order to reduce noise expression signals, false correlations and false gene expression patterns.

Acknowledgments. Authors are grateful to Joshy George, Leonard Lipovich, Yong How Choong, Caleb Khor, Leong Cheok and Chuah Yuxin for help in processing of Affymetrix data and discussions as well as to A*STAR for support of the research.

References

1. Harbig, J., Sprinkle, R., Enkemann, S.A.: A sequence-based identification of the genes detected by probesets on the Affymetrix U133 plus 2.0 array. Nucleic Acids Res. 33(3), e31 (2005)
2. Okoniewski, M.J., Miller, C.J.: Hybridization interactions between probesets in short oligo microarrays lead to spurious correlations. BMC Bioinformatics 7, 2761 (2006)
3. Mecham, B.H., Wetmore, D.Z., Szallasi, Z., Sadovsky, Y., Kohane, I., Mariani, T.J.: Increased measurement accuracy for sequence-verified microarray probes. Physiol Genomics 18, 308–315 (2004)

4. Gautier, L., Moller, M., Friis-Hansen, L., Knudsen, S.: Alternative mapping of probes to genes for Affymetrix chips. BMC Bioinformatics 5, 111 (2004)
5. Leong, H.S., Yates, T., Wilson, C., Miller, C.J.: ADAPT: a database of affymetrix probesets and transcripts. Bioinformatics 21, 2552–2553 (2005)
6. Dai, M., Wang, P., Boyd, A.D., Kostov, G., Athey, B., Jones, E.G., Bunney, W.E., Myers, R.M., Speed, T.P., Akil, H., Watson, S.J., Meng, F.: Evolving gene/transcript definitions significantly alter the interpretation of GeneChip data. Nucleic Acids Res. 33, e175 (2005)
7. Stalteri, M.A., Harrison, A.P.: Interpretation of multiple probe sets mapping to the same gene in Affymetrix GeneChips. BMC Bioinformatics 15, 8–13 (2007)
8. Orlov, Y.L., Zhou, J.T., Lipovich, L., Yong, H.C., Li, Y., Shahab, A., Kuznetsov, V.A.: A comprehensive quality assessment of the Affymetrix U133A&B probesets by an integrative genomic and clinical data analysis approach. In: Kolchanov, N.A. (ed.) Proceedings of the Fifth International Conference on Bioinformatics of Genome Regulation and Structure, Novosibirsk, Inst. of Cytology&Genetics, vol. 1, pp. 126–129 (2006)
9. Orlov, Y.L., Zhou, J., Lipovich, L.L., Shahab, A., Kuznetsov, V.A.: Quality assessment of the Affymetrix U133A&B probesets by target sequence mapping and expression data analysis. In: Silico Biol. (in press)
10. Zhang, Y., Liu, X.S., Liu, Q.R., Wei, L.: Genome-wide in silico identification and analysis of cis natural antisense transcripts (cis-NATs) in ten species. Nucleic Acids Res. 34, 3465–3475 (2006)
11. Kuznetsov, V.A., Zhou, J.T., George, J., Orlov, Y.L.: Genome-wide co-expression patterns of human cis-antisense gene pairs. In: Kolchanov, N.A. (ed.) Proceedings of the Fifth International Conference on Bioinformatics of Genome Regulation and Structure, Novosibirsk, Inst. of Cytology&Genetics, vol. 1, pp. 90–93 (2006)
12. Ivshina, A.V., George, J., Senko, O.V., Mow, B., Putti, T.C., Smeds, J., Lindahl, T., Pawitan, Y., Hall, P., Nordgren, H., Wong, J.E., Liu, E.T., Bergh, J., Kuznetsov, V.A., Miller, L.D.: Genetic reclassification of histologic grade delineates new clinical subtypes of breast cancer. Cancer Res. 66, 10292–10301 (2006)
13. Chua, A.L.-S., Ivshina, A.V., Kuznetsov, V.A.: Pareto-Gamma Statistics reveals global rescaling in transcriptomes of low and high aggressive breast cancer phenotypes. In: Rajapakse, J.C., Wong, L., Acharya, R. (eds.) PRIB 2006. LNCS (LNBI), vol. 4146, pp. 49–59. Springer, Heidelberg (2006)
14. MAS 5.0 algorithm. Statistical Algorithms Description Document. Santa Clara, CA: Affymetrix, Inc. (2002), http://www.affymetrix.com/support/technical/whitepapers/sadd-whitepaper.pdf
15. Tusher, V.G., Tibshirani, R., Chu, G.: Significance analysis of microarrays applied to the ionizing radiation response. Proc. Natl. Acad. Sci. USA 98, 5116–5121 (2001)
16. Liu, G., Loraine, A.E., Shigeta, R., Cline, M., Cheng, J., Valmeekam, V., Sun, S., Kulp, D., Siani-Rose, M.A.: NetAffx: Affymetrix probesets and annotations. Nucleic Acids Res. 31, 82–86 (2003)
17. Shi, L., Reid, L.H., Jones, W.D., et al.: The MicroArray Quality Control (MAQC) project shows inter- and intra-platform reproducibility of gene expression measurements. Nat. Biotechnology 24, 1151–1161 (2006)
18. Lu, X., Zhang, X.: The effect of GeneChip gene definitions on the microarray study of cancers. Bioessays 28, 739–746 (2006)

Ensemble of Dissimilarity Based Classifiers for Cancerous Samples Classification

Ángela Blanco, Manuel Martín-Merino[1], and Javier de las Rivas[2]

[1] Universidad Pontificia de Salamanca
C/Compañía 5, 37002, Salamanca, Spain
ablancogo@upsa.es, mmartinmac@upsa.es
[2] Cancer Research Center of Salamanca (CIC)
Salamanca, Spain
jrivas@usal.es

Abstract. DNA Microarray technology allow us to identify cancerous tissues considering the gene expression levels across a collection of related samples.

Several classifiers such as Support Vector Machines (SVM), k Nearest Neighbors (k-NN) or Diagonal Linear Discriminant Analysis (DLDA) have been applied to this problem. However, they are usually based on Euclidean distances that fail to reflect accurately the sample proximities. Several classifiers have been extended to work with non-Euclidean dissimilarities although none outperforms the others because they misclassify a different set of patterns.

In this paper, we combine different kind of dissimilarity based classifiers to reduce the misclassification errors. The diversity among classifiers is induced considering a set of complementary dissimilarities for three different type of models. The experimental results suggest that the algorithm proposed helps to improve classifiers based on a single dissimilarity and a widely used combination strategy such as Bagging.

1   Introduction

DNA Microarray technology allow us to monitor the expression levels of thousands of genes simultaneously across a collection of related samples. This technology has been applied particularly to the prediction of different type of cancer with encouraging results [12].

A large variety of machine learning techniques have been proposed to this aim such as Support Vector Machines (SVM) [10], k Nearest Neighbors [9] or Diagonal Linear Discriminant Analysis (DLDA) [9]. However the algorithms considered in the literature rely frequently on the use of the Euclidean distance that fails often to reflect accurately the proximities among the sample profiles [8,16,19]. The classifiers mentioned above have been extended to work with non-Euclidean dissimilarities [22]. In spite of this, the resulting algorithms misclassify a different set of patterns and fail to reduce significantly the errors. This can be explained because each dissimilarity reflects different features of the data and they induce different type of errors.

J.C. Rajapakse, B. Schmidt, and G. Volkert (Eds.): PRIB 2007, LNBI 4774, pp. 178–188, 2007.

Several authors have pointed out that combining non-optimal classifiers can help to reduce particularly the variance of the predictor [17,24]. In order to achieve this goal, different versions of the classifier are usually built by sampling the patterns or the features [5]. Nevertheless, in our application, this kind of resampling techniques reduce the size of the training set. This may increase the bias of individual classifiers and the error of the combination [24].

In this paper we build the diversity of classifiers considering three different kinds of models such as SVM, k-NN and DLDA. The diversity is increased considering a set of complementary dissimilarities for each model. The classifiers induced will take advantage of the whole sample avoiding the bias introduced by resampling techniques such as Bagging. In order to incorporate non-Euclidean dissimilarities the base classifiers are modified in an appropriate way. Finally, the classifiers are aggregated using a voting strategy [17]. The method proposed has been applied to the prediction of different type of cancer using the gene expression levels with remarkable results.

This paper is organized as follows. Section 2 introduces the dissimilarities considered to build the diversity of classifiers. Section 3 comments how the classifiers can be extended to work from a dissimilarity matrix. In section 4 we present our combination strategy. Section 5 illustrates the performance of the algorithm in the challenging problem of gene expression data analysis. Finally, section 6 gets conclusions and outlines future research trends.

2   Dissimilarities for Gene Expression Data Analysis

An important step in the design of a classifier is the choice of a proper dissimilarity that reflects the proximities among the objects. However, the choice of a good dissimilarity is not an easy task. Each measure reflects different features of the data and the classifiers induced by the dissimilarities misclassify frequently a different set of patterns. Therefore no dissimilarity outperforms the others.

In this section, we comment shortly the main differences among several dissimilarities proposed to evaluate the proximity between cellular samples considering the gene expression levels. For a deeper description and definitions see [8,16,11].

The Euclidean distance evaluates if the gene expression levels differ significantly across different samples. An interesting alternative is the cosine dissimilarity. This measure will become small when the ratio between the gene expression levels is similar for the two samples considered. It differs significantly from the Euclidean distance when the data is not normalized by the L_2 norm.

The correlation measure evaluates if the expression levels of genes change similarly in both samples. Correlation based measures tend to group together samples whose expression levels are linearly related. The correlation differs significantly from the cosine if the means of the sample profiles are not zero. This measure is sensitive to outliers. The Spearman rank dissimilarity is less sensitive to outliers because it computes a correlation between the ranks of the gene expression levels. An alternative measure that helps to overcome the problem of outliers is the Kendall-τ index which is related to the Mutual Information probabilistic measure [11].

Due to the large number of genes, the sample profiles are codified in high dimensional and noisy spaces. In this case, the dissimilarities mentioned above are affected by the 'curse of dimensionality' [1,20]. Hence, most of the dissimilarities become almost constant and the differences among dissimilarities are lost [15]. To avoid this problem, it is recommended to reduce the number of features before computing the dissimilarities.

3   Dissimilarity Based Classifiers

Classical Support Vector Machines (SVM) [25] and Diagonal Linear Discriminant Analysis (DLDA) [9] are not able to work directly from a dissimilarity matrix. In this section, the classical SVM algorithm is extended to work from a dissimilarity matrix by defining a kernel of dissimilarities. Next DLDA is adapted following a different approach by embedding the patterns in a Euclidean space.

The SVM algorithm looks for a linear hyperplane $f(\boldsymbol{x}; \boldsymbol{w}) = \boldsymbol{w}^T\boldsymbol{x}$ that maximizes the margin $\gamma = 2/\|\boldsymbol{w}\|^2$. γ determines the generalization ability of the SVM. The slack variables ξ_i allow to consider classification errors. The figure 1 illustrates the meaning of the SVM parameters.

The hyperplane that minimizes the prediction error is given by the following optimization problem [25]:

$$\text{minimum}_{w,\{\xi_i\}} \quad <\boldsymbol{w}, \boldsymbol{w}> +C\sum_{i=1}^{n}\xi_i^2 \tag{1}$$
$$\text{subject to} \quad y_i(<\boldsymbol{w}, \boldsymbol{x}_i> +b) \geq 1 - \xi_i \quad i = 1,\ldots,n$$
$$\xi_i \geq 0 \quad i = 1,\ldots,n$$

where C is a regularization parameter that achieves a balance between the empirical error and the complexity of the classifier. The optimization problem can

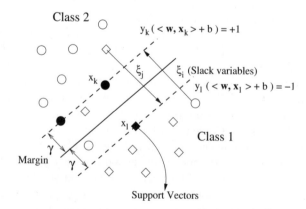

Fig. 1. Scheme of the hyperplane generated by the SVM algorithm for a non-linearly separable problem

be solved efficiently in dual space and the discriminant function can be expressed exclusively in terms of scalar products,

$$f(\boldsymbol{x}) = \sum_{\alpha_i > 0} \alpha_i y_i \langle \boldsymbol{x}, \boldsymbol{x}_i \rangle + w_0 \qquad (2)$$

The SVM algorithm can be easily extended to the non-linear case substituting the scalar products by a Mercer kernel [25].

Non-Euclidean dissimilarities can be incorporated into the SVM algorithm by defining a kernel of dissimilarities [22,23]. Next we detail the idea.

Let d be a dissimilarity [6] and $R = \{p_1, \ldots, p_n\}$ a subset of representatives drawn from the training set. Define the mapping $D(z, R) : \mathcal{F} \to \mathbb{R}^n$ as:

$$D(z, R) = [d(z, p_1), d(z, p_2), \ldots, d(z, p_n)] \qquad (3)$$

This mapping define a dissimilarity space where feature i is given by $d(., p_i)$.

The set of representatives R determine the dimensionality of the feature space. The choice of R is equivalent to select a subset of features in the dissimilarity space. Due to the small number of training samples in our application we have considered the whole sample as a representative set. It has been suggested in the literature that selecting a smaller subset of representatives does not help to improve the resulting classifier [22].

Once the patterns have been represented in the dissimilarity space, a kernel of dissimilarities can be defined as:

$$K_{ij} = \langle D(\boldsymbol{x}_i, R), D(\boldsymbol{x}_j, R) \rangle \qquad (4)$$

where $\langle ., . \rangle$ denotes the scalar product in the feature space. Thus, for the linear SVM the kernel matrix is written as $K = DD^T$. This matrix is positive definite and keeps the nice properties of the optimization problem in the original SVM algorithm.

The DLDA is a variant of the Linear Discriminant Analysis (LDA) that considers diagonal and constant covariance matrices along the classes [9]. However, in order to apply this technique, a vectorial representation of the data should be obtained. To this aim, we follow the approach of [22]. First, the dissimilarities are embedded into an Euclidean space such that the inter-pattern distances reflect approximately the original dissimilarity matrix. Next, the test points are added to this space via a linear algebra operation. Finally the DLDA is applied considering the vectorial representation obtained.

We comment briefly the mathematical details of the embedding operation.

Let $D \in \mathbb{R}^{n \times n}$ be the dissimilarity matrix made up of the object proximities for the training set. A configuration in a low dimensional Euclidean space can be found via a metric multidimensional scaling algorithm (MDS) [6] such that the original dissimilarities are approximately preserved. Let $X = [\boldsymbol{x}_1 \ldots \boldsymbol{x}_n]^T \in \mathbb{R}^{n \times p}$ be the matrix of the object coordinates for the training patterns. Define $B = XX^T$ as the matrix of inner products which is related to the dissimilarity matrix via the following equation:

$$B = -\frac{1}{2} J D^{(2)} J, \qquad (5)$$

where $J = I - \frac{1}{n}\mathbf{1}\mathbf{1}^T \in \mathbb{R}^{n \times n}$ is the centering matrix, I is the identity matrix and $D^{(2)} = (\delta_{ij}^2)$ is the matrix of the square dissimilarities for the training patterns. If B is positive semi-definite, the object coordinates in the low dimensional Euclidean space $\mathbb{R}^k$ can be found through a singular value decomposition [6,13]:

$$X_k = V_k \Lambda_k^{1/2} \,, \tag{6}$$

where $V_k \in \mathbb{R}^{n \times k}$ is an orthogonal matrix with columns the first k eigen-vectors of XX^T and $\Lambda_k = diag(\lambda_1 \ldots \lambda_k) \in \mathbb{R}^{k \times k}$ is a diagonal matrix with λ_i the i-th eigenvalue. Several dissimilarities introduced in section 2 generate inner product matrices B non semi-definite positive. Fortunately, the negative values are small in our application and therefore can be neglected [6] without losing relevant information about the data.

Once the training patterns have been embedded into a low dimensional Euclidean space, the test pattern can be added to this space via a linear projection [22]. Next we comment briefly the derivation.

Let $X_k \in \mathbb{R}^{n \times k}$ be the object configuration for the training patterns in $\mathbb{R}^k$ and $X_n = [\boldsymbol{x}_1 \ldots \boldsymbol{x}_s]^T \in \mathbb{R}^{s \times k}$ the matrix of the object coordinates sought for the test patterns. Let $D_n^{(2)} \in \mathbb{R}^{s \times n}$ be the matrix of the square dissimilarities between the s test patterns and the n training patterns that have been already projected. The matrix $B_n \in \mathbb{R}^{s \times n}$ of inner products among the test and training patterns can be found as:

$$B_n = -\frac{1}{2}(D_n^{(2)} J - U D^{(2)} J) \,, \tag{7}$$

where $J \in \mathbb{R}^{n \times n}$ is the centering matrix and $U = \frac{1}{n}\mathbf{1}^T \mathbf{1} \in \mathbb{R}^{s \times n}$. The derivation of equation (7) is detailed in [22]. Since the matrix of inner products verifies

$$B_n = X_n X_k^T \tag{8}$$

then, X_n can be found as the least mean-square error solution to (8), that is:

$$X_n = B_n X_k (X_k^T X_k)^{-1} \,, \tag{9}$$

Given that $X_k^T X_k = \Lambda_k$ and considering that $X_k = V_k \Lambda_k^{1/2}$ the coordinates for the test points can be obtained as:

$$X_n = B_n V_k \Lambda_k^{-1/2} \,, \tag{10}$$

which can be easily evaluated through simple linear algebraic operations.

4   Combination of Dissimilarity Based Classifiers

In this section we introduce our ensemble of classifiers to reduce the errors and comment briefly the related work.

Our method builds the diversity of classifiers considering three different kind of models such as SVM, k-NN and DLDA. To increase the diversity among

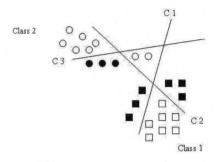

Fig. 2. Aggregation of classifiers using a voting strategy. Bold patterns are misclassified by a single hyperplane but not by the combination.

classifiers we have considered several dissimilarities introduced in section 2. Each dissimilarity reflects different features of the data and the resulting classifiers will produce different errors. Thus, the combination will improve the performance of classifiers based on single dissimilarity [5,18]. Besides, the diversity of classifiers is generated considering the whole training sample. In this way, we avoid to reduce the size of the training set which may induce bias in the individual classifiers. Notice that the combination strategies are not able to reduce the bias of single classifiers [24].

Figure 2 shows in an intuitive way how the combination of classifiers reduces the misclassification errors. For instance bold patterns are assigned to the wrong class by one classifier but using a voting strategy the patterns will be assigned to the right class.

Hence, our combination algorithm proceeds as follows: First, the set of complementary dissimilarities introduced in section 2 are computed. As we mentioned earlier each classifier incorporates the dissimilarities in a different way. For the SVM algorithm, the kernel of dissimilarities is computed and the optimization problem is solved in the usual way. k-NN is able to work directly from a dissimilarity matrix but to avoid the 'curse of dimensionality' and to increase the diversity among dissimilarities it is recommended to reduce previously the number of features. For the DLDA algorithm, the dissimilarities should be embedded in an Euclidean space via a Multidimensional Scaling algorithm. The ensemble of classifiers is aggregated by a standard voting strategy [17]. The diagram 1 shows the steps of the algorithm.

A related technique to combine classifiers is the Bagging [5,3]. This method generates a diversity of classifiers considering several bootstrap samples as training sets. Next, the classifiers are aggregated using a voting strategy. Nevertheless there are three important differences between Bagging and the method proposed in this section.

First, our method generates the diversity of classifiers by considering the whole sample. Bagging trains each classifier using around 63% of the training set. In our application the size of the training set is very small and neglecting part of the patterns may increase the bias of each classifier. It has been suggested in

---

Algorithm 1. Aggregation of classifiers based on multiple models and dissimilarities

1: For each measure compute the dissimilarity matrix
2: Compute the kernel of dissimilarities using equation (4) for the SVM algorithm
3: Embed each dissimilarity into an Euclidean space through equation (6) for DLDA algorithm
4: Train the classifiers for each dissimilarity
5: Combine the different models using a voting strategy
6: Evaluate the ensemble by ten-fold cross-validation. Test points are embedded for DLDA algorithm using equation (10)
7: End

---

the literature that Bagging does not help to reduce the bias [24] and so, the aggregation of classifiers will hardly reduce the misclassification error.

A second advantage of our method is that it is able to work directly with a dissimilarity matrix.

Finally, the combination of several dissimilarities avoids the problem of choosing a particular dissimilarity for the application we are dealing with. This is a difficult and time consuming task.

5   Experimental Results

In this section, the ensemble of classifiers proposed is applied to the identification of cancerous samples using Microarray gene expression data.

Three benchmark gene expression datasets have been considered. The first one consisted of 72 bone marrow samples (47 ALL and 25 AML) obtained from acute leukemia patients at the time of diagnosis [13]. The RNA from marrow mononuclear cells was hybridized to high-density oligonucleotide microarrays produced by Affymetrix and containing 6817 genes. The second dataset consisted of 49 samples from breast tumors [26], 25 classified as positive to estrogen receptors (ER+) and 24 negative to estrogen receptors (ER-). Those positive to estrogen receptors require a different treatment. The RNA of breast cancer cells were hybridized to high-density oligonucleotide microarrays produced by Affymetrix and containing 7129 genes. Finally the third dataset consists of 40 tumor and 22 normal colon samples, analyzed with an Affymetrix oligonucleotide array complementary to more than 6,500 human genes. The number of genes was reduced in the original dataset to 2000 [2].

Due to the large number of genes, samples are codified in a high dimensional and noisy space. Therefore, the dissimilarities are affected by the 'curse of dimensionality' and the correlation among them becomes large [20]. To avoid this problem and to increase the diversity among dissimilarities we have reduced aggressively the number of genes using the standard F-statistic [11]. The number of genes considered for SVM and DLDA are 14% while for k-NN the number of genes kept is 3% because this technique is more sensible to noise.

The dissimilarities have been computed without normalizing the variables because as we have mentioned in section 2 this operation may increase the correlation among them.

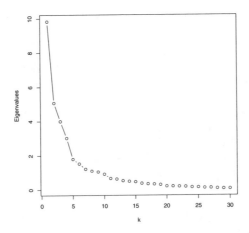

Fig. 3. Curve of eigenvalues for the Multidimensional Scaling algorithm and the χ^2 dissimilarity

The algorithm chosen to train the Support Vector Machines is C-SVM. The C regularization parameter has been set up by ten fold-crossvalidation [21,4]. We have considered linear kernels in all the experiments because the small size of the training set in our application favors the overfitting of the data. Consequently error rates are smaller for linear kernels than for non linear ones.

The number of neighbors for k-NN algorithm is estimated by cross-validation.

Before applying DLDA the dissimilarities should be embedded in an Euclidean space using a Multidimensional Scaling algorithm. An important parameter is the dimensionality of this space generated by the first eigen-vectors of the inner product matrix (5). The number of eigenvectors considered is determined by the curve of eigenvalues.

Figure 3 shows the eigenvalues for the breast cancer data and the χ^2 dissimilarity. The first eleven eigenvalues account for 85% of the variance. Therefore, they preserve the main structure of the data.

The algorithms have been evaluated considering the global errors and the false negative errors. Both have been estimated by ten-fold cross-validation which gives good experimental results for the problem at hand [21].

Table 1 shows the experimental results for the best single classifier for each technique. Table 2 compares the method proposed with Bagging, introduced in section 3. Both, Bagging and the best classifiers based on a single dissimilarity for each model have been taken as a reference.

From the analysis of tables 2 and 2, the following conclusions can be drawn:

- The dissimilarity that minimizes the error depends strongly on the classifier and on the particular dataset considered. No dissimilarity outperforms the others for a wide range of models and datasets. Hence the choice of a proper dissimilarity is not an easy task for human experts.

Table 1. Empirical results for the best single classifier for each technique

Technique	Datasets	Error %	False negative %
SVM (Correlation)	Golub	6.94%	2.77%
SVM (Tau)	Breast	6.12%	2.04%
SVM(Correlation)	Colon	14.5%	6.45%
K-NN (Tau)	Golub	1.38%	1.38%
K-NN(Tau)	Breast	8.16%	2.04%
K-NN (Cosine)	Colon	12.9%	4.83%
DLDA (Tau)	Golub	2.77%	1.38%
DLDA(Spearman)	Breast	8.16%	2.04%
DLDA (Euclidean)	Colon	11.29%	4.83%

Table 2. Empirical results for the combination of classifiers. The Bagging technique has been taken as reference.

Technique	Datasets	Error %	False negative %
	Golub	8.33%	6.94%
Bagging (SVM)	Breast	6.12%	2.04%
	Colon	12.9%	4.83%
	Golub	5.55%	5.55%
Bagging (k-NN)	Breast	14.28%	6.12%
	Colon	14.51%	9.67%
	Golub	6.94%	4.16%
Bagging (DLDA)	Breast	14.28%	2.04%
	Colon	11.29%	3.22%
	Golub	1.38%	1.38%
Combination	Breast	4.08%	2.04%
	Colon	11.2%	3.22%

- The combination strategy proposed outperforms significantly the misclassification errors of the best single classifiers. In particular, the ensemble of classifiers improves significantly the SVM algorithms for the three problems considered. False negative errors are particularly reduced in Golub and Colon datasets. We also report that our method improves the best k-NN classifier for Breast and Colon that are the most complex according to the literature. Finally, DLDA is also improved for Golub and Breast Cancer.
- The ensemble of classifiers proposed improves a widely used combination algorithm such as Bagging. Both kind of errors are particularly reduced for Golub and Breast Cancer. This result supports the idea that our algorithm performs better than the resampling techniques when the sample size is small.

6   Conclusions and Future Research Trends

In this paper, we have proposed an ensemble of classifiers based on a diversity of models and dissimilarities. Our approach aims to reduce the misclassification error of classifiers based solely on a single measure. The algorithm has been applied to the classification of cancerous samples using gene expression data.

The experimental results suggest that the method proposed improves the misclassification error of classifiers based on a single dissimilarity. We also report that our method compares favorably with a widely used combination algorithm such as Bagging.

As future research trends, we will try to extend the method proposed to improve clustering algorithms.

Acknowledgment

This work has been partially supported by the Junta de Castilla y León grant PON05B06.

References

1. Aggarwal, C.C.: Re-designing distance functions and distance-based applications for high dimensional applications. In: Proc. of the ACM International Conference on Management of Data and Symposium on Principles of Database Systems (SIGMOD-PODS), vol. 1, pp. 13–18 (March 2001)
2. Alon, U., Barkai, N., Notterman, D.A., Gish, K., Ybarra, S., Mack, D., Levine, A.J.: Broad patterns of gene expression revealed by clustering analysis of tumor and normal colon tissues probed by oligonucleotide arrays. Proc. Nat'l. Acad. Sci. USA 96, 6745–6750 (1999)
3. Bauer, E., Kohavi, R.: An empirical comparison of voting classification algorithms: Bagging, boosting, and variants. Machine Learning 36, 105–139 (1999)
4. Braga-Neto, U., Dougherty, E.: Is cross-validation valid for small-sample microarray classification? Bioinformatics 20(3), 374–380 (2004)
5. Breiman, L.: Bagging predictors. Machine Learning 24, 123–140 (1996)
6. Cox, T., Cox, M.: Multidimensional Scaling, 2nd edn. Chapman & Hall/CRC Press, New York (2001)
7. Cristianini, N., Shawe-Taylor, J.: An Introduction to Support Vector Machines and Other Kernel-Based Learning Methods. Cambridge University Press, Cambridge (2000)
8. Drăghici, S.: Data Analysis Tools for DNA Microarrays. Chapman & Hall/CRC Press, New York (2003)
9. Dudoit, S., Fridlyand, J., Speed, T.: Comparison of discrimination methods for the classification of tumors using gene expression data. Journal of the American Statistical Association 97, 77–87 (2002)
10. Furey, T., Cristianini, N., Duffy, N., Bednarski, D., Schummer, M., Haussler, D.: Support vector machine classification and validation of cancer tissue samples using microarray expression data. Bioinformatics 16(10), 906–914 (2000)
11. Gentleman, R., Carey, V., Huber, W., Irizarry, R., Dudoit, S.: Bioinformatics and Computational Biology Solutions Using R and Bioconductor. Springer, Heidelberg (2006)
12. Golub, T., Slonim, D., Tamayo, P., Huard, C., Gaasenbeek, M., Mesirov, J., Coller, H., Loh, M., Downing, J., Caligiuri, M., Bloomfield, C., Lander, E.: Molecular classification of cancer: Class discovery and class prediction by gene expression monitoring. Science 286(15), 531–537 (1999)

13. Golub, G.H., Loan, C.F.V.: Matrix Computations, 3rd edn. Johns Hopkins university press, Baltimore, Maryland, USA (1996)
14. Guyon, I., Weston, J., Barnhill, S., Vapnik, V.: Gene selection for cancer classification using support vector machines. Machine Learning 46, 389–422 (2002)
15. Hinneburg, C.C.A.A., Keim, D.A.: What is the nearest neighbor in high dimensional spaces? In: Proc. of the International Conference on Database Theory (ICDT), pp. 506–515. Morgan Kaufmann, Cairo, Egypt (2000)
16. Jiang, D., Tang, C., Zhang, A.: Cluster analysis for gene expression data: A survey. IEEE Transactions on Knowledge and Data Engineering 16(11) (November 2004)
17. Kittler, J., Hatef, M., Duin, R., Matas, J.: On combining classifiers. IEEE Transactions on Neural Networks 20(3), 228–239 (1998)
18. Kuncheva, L.I.: Combining Pattern Classifiers. John Wiley, New Jersey (2004)
19. Martín-Merino, M., Muñoz, A.: Self organizing map and sammon mapping for asymmetric proximities. Neurocomputing 63, 171–192 (2005)
20. Martín-Merino, M., Noz, A.M.: A new sammon algorithm for sparse data visualization. In: International Conference on Pattern Recognition (ICPR), pp. 477–481. IEEE Press, Cambridge (UK) (2004)
21. Molinaro, A., Simon, R., Pfeiffer, R.: Prediction error estimation: a comparison of resampling methods. Bioinformatics 21(15), 3301–3307 (2005)
22. Pekalska, E., Paclick, P., Duin, R.: A generalized kernel approach to dissimilarity-based classification. Journal of Machine Learning Research 2, 175–211 (2001)
23. Schölkopf, B., Smola, A.: Learning with Kernels. MIT Press, Cambridge, USA (2002)
24. Valentini, G., Dietterich, T.: Bias-variance analysis of support vector machines for the development of svm-based ensemble methods. Journal of Machine Learning Research 5, 725–775 (2004)
25. Vapnik, V.: Statistical Learning Theory. John Wiley & Sons, New York (1998)
26. West, M., Blanchette, C., Dressman, H., Huang, E., Ishida, S., Spang, R., Zuzan, H., Olson, J., Marks, J., Nevins, J.: Predicting the clinical status of human breast cancer by using gene expression profiles. PNAS 98(20) (September 2001)

Gene Expression Analysis of Leukemia Samples Using Visual Interpretation of Small Ensembles: A Case Study

Gregor Stiglic[1], Nawaz Khan[2], Mateja Verlic[1], and Peter Kokol[1]

[1] University of Maribor, FERI, Smetanova 17, 2000 Maribor, Slovenia
[2] School of Computing Science, Middlesex University, The Burrough, Hendon,
London NW4 4BT, UK
{gregor.stiglic,kokol}@uni-mb.si,
N.X.Khan@mdx.ac.uk

Abstract. Many advanced machine learning and statistical methods have recently been employed in classification of gene expression measurements. Although many of these methods can achieve high accuracy, they generally lack comprehensibility of the classification process. In this paper a new method for interpretation of small ensembles of classifiers is used on gene expression data from real-world dataset. It was shown that interactive interpretation systems that were developed for classical machine learning problems also give a great range of possibilities for the scientists in the bioinformatics field. Therefore we chose a gene expression dataset discriminating three types of Leukemia as a testbed for the proposed Visual Interpretation of Small Ensembles (VISE) tool. Our results show that using the accuracy of ensembles and adding comprehensibility gains not only accurate but also results that can possibly represent new knowledge on specific gene functions.

Keywords: gene expression analysis, machine learning, decision trees.

1 Introduction

Gene expression analysis is a novel technique that in contrast to measurement of a single gene transcription enables measurement of all genes in an organism at once. Finding combinations of genes whose expression levels distinguish different groups of diseases is a complex task that is usually solved by different machine learning or statistical algorithms. While most of the algorithms gain very accurate results in classification of gene expression samples, there is still very limited number of algorithms that can offer a good interpretation of the results that were gained using advanced machine learning techniques.

Methods like bagging, boosting and random forests, which combine decisions of multiple hypotheses, also called ensemble methods, are some of the strongest existing machine learning methods. Ensemble methods are learning algorithms that build a set of classifiers which are used to classify new instances by combining their predictions. It was shown that ensembles clearly outperform the single classifiers in terms of classification accuracy [1-5].

One of the main drawbacks of ensemble classifiers is weak comprehensibility of the produced classification models. Many times it is possible to convert all single

J.C. Rajapakse, B. Schmidt, and G. Volkert (Eds.): PRIB 2007, LNBI 4774, pp. 189–197, 2007.

models from an ensemble to a set of rules, but such rule sets quickly become too complex to be comprehensible. Main scheme for such methods is rule extraction, that is, symbolic rules are extracted from the 'black-box' model. Most usual method is simple rule extraction from all components of a classification model that is followed by aggregation of the extracted rules. One of first such systems was presented by Setiono in [5], where the neural network is pruned and the outputs of hidden units are discretized. The rule extraction algorithm is executed iteratively for each sub-network constructed from hidden units with many outputs. Sometimes this process can be even simpler – e.g. when working with decision tree (DT), rules can be extracted directly from the branches of a tree.

Another option when improving the comprehensibility of classification process is introduction of classification visualization. One of the first papers where visualization of high-dimensional classifiers is presented was written by Melnik [6], where visual interpretation of neural networks is described. An extensive work in visualization of multiple and single DTs that also includes their interpretation was done by Urbanek in [7]. He presents a tool for interactive visual interpretation of DT forests. Another paper by Frank and Witten [8] presents a technique that uses a two-dimensional visualization based on class probability estimates. All above mentioned papers suggest that visual interpretation of classification models is worth further research to help both experts and non-experts understand the most accurate classification techniques.

Above mentioned examples demonstrate use of visual interpretation in classical machine learning problems, while it should also be mentioned that there were some experiments that combine visualization and microarray classification process. A study that uses Support Vector Machines and tries to interpret the results using visualization was presented by Caragea et al. [9]. A similar study in terms of visualization of microarray data to interpret results of classification was conducted by Lee et al. [10]. Their tool called GeneGobi is mostly based on statistical instead of machine learning methods. Another tool was developed by Curk et al. [11] where visualization is used for setting the experiments and interpretation of results, which represents a major simplification of experimental process in microarray analysis.

The following sections of this paper present a case study where a novel Visual Interpretation of Small Ensembles (VISE) method [12] is used on a microarray dataset discriminating three types of Leukemia that was initially presented by Armstrong et al. [13]. In contrast to experiments described in [12] another version of VISE tool was used where DTs are generated based on bagging instead of boosting DTs. Section 2 contains a presentation of virtual interpretation of small ensembles. It is followed by a section describing the experimental settings and results. Section 4 presents a validation study by providing an interpretation of the results in the context of rule sets and then by comparing the proposed adaptations with the combined and simple DTs for leukemia grouping. In the last section, the main contribution of this paper is summarized and several issues for future works are indicated.

2 Interpretation Tool

Usually as the number of classifiers in ensemble increases it means an increase of complexity and decrease of comprehensibility, assuming that single models combined in an ensemble are comprehensible models (e.g. DTs or a set of rules). This paper demonstrates a novel tool for visual interactive interpretation of ensembles consisting of three DTs. It is based on idea that a small ensemble can increase the accuracy and still keep the complexity of the ensemble as low as possible. To ensure the diversity of induced DTs is high enough we use a simple variant of bagging [14] technique for building DTs. Training set is split into three equal parts, where the first DT is generated from the first two thirds, the second from the last two thirds and the last tree from first and last third of the examples in training set. Default pruning settings are used to achieve lower complexity levels of generated DTs. All DTs used are standard C4.5 trees as implemented in Weka environment [15]. The same environment was used as a core for the developed small ensembles interpretation tool.

Main screen of the VISE (Visual Interpretation of Small Ensembles) tool is presented in Fig. 1. Primary DT window can be seen on the left hand side of the screen, while on the opposite side the other two DTs are displayed in smaller windows. Each of the trees on the right side can be magnified and transferred to the main window by switching the main and one of the two side windows containing simplified visualization of the tree. Bottom of the screen contains a set of rules that are extracted from the above trees in an interactive way. Interaction is an integral part of the tool; therefore user is allowed to select branches of trees that he is interested in, either by decision at the terminal node of the branch or by features (i.e. nodes) that are included in the branch. The first interactive step is selection of a significant branch (according to expert's opinion) in a tree, which is followed by automatic extraction of the rule from this branch and all the rules that could possibly contribute to the decision from the remaining two trees.

First step is followed by automatic extraction of rules that can be done in two ways:

1. Using the training set examples, a single or a group of branches is selected (and rules are extracted from them) which contain the examples that were used when the selected branch was built.
2. In case there are too few examples in the selected branch, we artificially create the examples whose attribute values correspond to the selected branch and label them using a robust and accurate ensemble (in our case we use random forests ensemble consisting of 100 DTs)

This way user is able to observe which rules (i.e. DT branches) could possibly vote against decision of the main DT. Using this knowledge we are able to understand how and why an ensemble would vote differently in case of using a single DT for specific samples that fit in the selected branch of the tree.

For each small ensemble we can also get the quick accuracy estimation using 10-fold cross-validation.

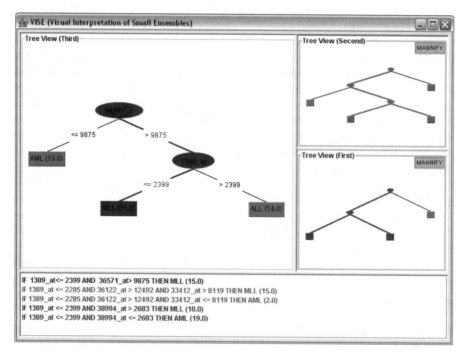

Fig. 1. Main screen of VISE tool

The informative value of resulting rules is marked by their color that represents their origin and by their decision class. The following section demonstrates usage of the tool on a gene expression dataset discriminating three types of Leukemia.

3 Experimental Settings and Results

This section highlights the details of our study and key findings that were obtained by applying the VISE tool to Leukemia microarray dataset. In the original research by Armstrong et al. [13] clustering algorithms revealed that lymphoblastic leukemias with MLL translocations can clearly be separated from conventional acute lymphoblastic and acute myelogenous leukemias. The same dataset consisting of 72 tissue samples, each of them containing 12582 gene expression measurements was used in our experiment. In the original study a dataset was split in a training set containing 57 samples and testing set with another 15 samples. In our study all 72 samples (24 ALL, 20 MLL, 28 AML) were used in a single dataset, while 10-fold cross validation was used for accuracy estimations. Basic DT that was used to extract rules from a small ensemble of three DTs is presented in Figure 2 where number in parentheses indicates that all examples from training set were correctly classified.

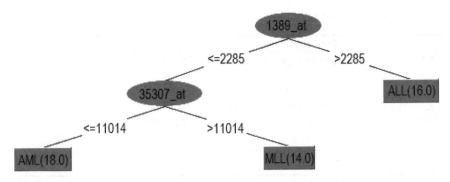

Fig. 2. Primary decision tree induced by VISE

Rules that were directly extracted from small ensemble are presented in Table 1. All rules extracted from primary DT are displayed in bold and are followed by rules that are fired in other two DTs using the corresponding samples from the selected primary tree branch. When evaluating the accuracy of decision trees that were built using Leukemia dataset [13] it was indicated that three decision trees together reached an average 10-fold cross validation accuracy rate of 90.5% compared to 84.2% that was achieved by single decision trees.

For easier understanding and rule interpretation gene id to gene description mappings are presented in Table 2. When interpreting the results from VISE tool it should be noticed that among the rules fired in secondary DTs it is possible to find rules that are voting against the rule extracted from primary DT. Those rules could also be called opposing rules and should be taken into consideration when interpreting results. In our case there are two genes that are included in such rules – i.e. genes with identification numbers 41503_at and 38046_at.

Table 1. Rules fired for each branch in the primary DT

AML Branch
IF 35307_at NOT EXPRESSED AND 1389_at NOT EXPRESSED THEN AML
IF 1389_at NOT EXPRESSED AND 38046_at NOT EXPRESSED THEN AML
IF 41503_at EXPRESSED AND 1389_at NOT EXPRESSED THEN MLL
IF 41503_at NOT EXPRESSED THEN AML
MLL Branch
IF 35307_at EXPRESSED AND 1389_at NOT EXPRESSED THEN MLL
IF 1389_at NOT EXPRESSED AND 38046_at EXPRESSED THEN MLL
IF 1389_at NOT EXPRESSED AND 38046_at NOT EXPRESSED THEN AML
IF 41503_at EXPRESSED AND 1389_at NOT EXPRESSED THEN MLL
ALL Branch
IF 1389_at EXPRESSED THEN ALL
IF 1389_at EXPRESSED THEN ALL
IF 41503_at NOT EXPRESSED THEN AML
IF 41503_at EXPRESSED AND 1389_at EXPRESSED THEN ALL

Table 2. Gene descriptions for easier interpretation of results in Table 1

Gene ID	Description
35307_at	Homo sapiens mRNA for GDP dissociation inhibitor beta
1389_at	Human common acute lymphoblastic leukemia antigen (CALLA) mRNA, complete cds
38046_at	Homo sapiens mRNA for Prer protein
41503_at	Homo sapiens mRNA for KIAA0854 protein, complete cds

4 Interpretation of Results

This section provides an expert evaluation of results and shows the differences between traditional gene expression analysis techniques and VISE tool in terms of results interpretation. Evaluation is based on rules that were extracted from DTs and are presented in Table 1.

GDP dissociation inhibitor (GDI) is a protein that controls the GDP-GTP exchange reactions. GTP-binding proteins involve in trafficking of molecules between cellular organelles. GDIs slow the rate of dissociation of GDP and release GDP from membrane-bound Rabs [16]. The GDI beta gene is vulnerable to inversion/deletion mutation and may cause leukemia. The association of GDI and its expression involving cellular transport have been reported by many researchers, for example [17] and [18]. It is evident from many researches that GDI expression is responsible for chronic myelogenous leukemia.

Common acute lymphocytic leukemia antigen (metallo endopeptidase; neutral endopeptidase) is an important cell surface marker in the diagnosis of human acute lymphocytic leukemia (ALL) [19]. It is present on leukemic cells of pre-B phenotype, which represent 85% of cases of ALL. Yagi et al. [20] and Fasching et al. [21] have suggested that the specific antigen receptor may be present at birth in some patients with ALL, suggesting a prenatal origin for the leukemic clone. They also have showed that some patients with ALL characterized by specific translocations have been demonstrated to have cells showing the translocation at the time of birth. This is because Lymphoblasts antigen receptors are unique to a particular patient. Sheikh et al. [22] has reported of peripheral blood lymphocytosis caused by CD23, CD25 in addition to CD5 and CD10. The expression of antigens for ALL have been reported my many researchers. For example, Ogawa et al. [23], Cutrona et al. [24] and Shipp [25] have reported the close correlation between expression of CD10/neutral endopeptidase and tumor development.

Red protein (RER protein; IK factor; cytokine IK) involves in the negative regulatory pathway of constitutive MHC Class II antigens expression. It expressed at similar levels in fetal and adult tissues in developmental stage. A lower expression of mRna for the protein may lead to fetal brain placenta COT 25-normalized squamous cell carcinoma, B cell metastatic chondrosarcoma and colon tumor.

Transcription factor ZHX2 involves in transcription factor activities and regulates the transcription [26, 27]. The irregular expression of mRNA may lead to lymphoma, B-cell lymphatic leukemia and lung and spleen lymphoma.

The rules above, although, show the direct association of GDI and lymphoblastic leukemia antigen to the ALL and AML, some of the features of leukemia exhibit a mixed type of leukemia, for example, MLL. The morphological features and immunophenotypic profile of the leukemia is not readily classifiable and may be influenced by some other expressions, for example, expression of Prer proteins and Transcription factors. The importance of these genes that influence the classification of leukemia cannot be ignored.

5 Conclusions and Future Work

From the previous section it is evident that results obtained from VISE tool can reveal potential new knowledge and make interpretation of results a simple task for bioinformatics experts. It was shown that in most cases it is enough to select a few crucial genes that are sufficient for improvement of classification accuracy. But a step further enables extraction of additional rules and significant genes that can be decisive for comprehensibility of classification results.

Another important aspect of VISE tool is the interactiveness of the classification process. It enables interaction with the expert in a way where it can be specified which rules (i.e. DT branches) are important for him and does not rely only on automatic feature selection like most of other methods.

As usual in the gene expression research we should emphasize that all the results are obtained from datasets containing a low number of samples. The increase of datasets that will provide us with more samples in the future also brings some new challenges. We can expect more complex classifiers which will also be more accurate. Therefore one of the main aims for the future is reduction of produced classifiers when working with many of them at once as it is the case in ensembles of classifiers.

References

1. Bauer, E., Kohavi, R.: An empirical comparison of voting classification algorithms: Bagging, boosting and variants. Machine Learning 36(1/2), 525–536 (1999)
2. Dietterich, T.G.: An experimental comparison of three methods for constructing ensembles of decision tress: Bagging, boosting and randomization. Machine Learning 40(2), 139–158 (2000)
3. Freund, Y., Schapire, R.E.: Experiments with a new boosting algorithm. In: Proceedings of the 13th International Conference on Machine Learning, pp. 148–156. Morgan Kauffman, San Francisco (1996)
4. Kuncheva, L., Whitaker, C.: Measures of Diversity in Classifier Ensembles and Their Relationship with the Ensemble Accuracy. Machine Learning 51, 181–207 (2003)
5. Hall, L.O., Bowyer, K.W., Banfield, R.E., Bhadoria, D., Kegelmeyer, W.P., Eschrich, S.: Comparing Pure Parallel Ensemble Creation Techniques Against Bagging. In: The Third IEEE International Conference on Data Mining, Melbourne, Florida, pp. 533–536 (November 2003)

6. Melnik, O., Pollack, J.B.: Theory and scope of exact representation extraction from feed-forward networks. Cognitive Systems Research 3(2) (2002)
7. Urbanek, S.: Exploring Statistical Forests. In: Proc. of the 2002 Joint Statistical Meeting, Mira DP (2002)
8. Frank, E., Hall, M.: Visualizing Class Probability Estimators. In: Proceedings of the European Conference on Principles and Practice of Knowledge Discovery in Databases, Cavtat, Croatia (2003)
9. Caragea, D., Cook, D., Honavar, V.: Visual Methods for Examining Support Vector Machine Results, ISU Technical Report (December 2005)
10. Lee, E.K., Cook, D., Wurtele, E., Kim, D., Kim, J., An, H.: GENEGOBI: Visual Data Analysis Aid Tools for Microarray Data. In: Computational Statistics 2004 Symposium (COMPSTAT 04) (2004)
11. Curk, T., Demsar, J., Xu, Q., Leban, G., Petrovic, U., Bratko, I., Shaulsky, G., Zupan, B.: Microarray data mining with visual programming. Bioinformatics 21(3), 396–398 (2005)
12. Stiglic, G., Mertik, M., Podgorelec, V., Kokol, P.: Using Visual Interpretation of Small Ensembles in Microarray Analysis. In: Proceedings of Computer Based Medical Systems, Salt Lake City, UT, USA (2006)
13. Armstrong, S.A., Staunton, J.E., Silverman, L.B., Pieters, R., den Boer, M.L., Minden, M.D., Sallan, S.E., Lander, E.S., Golub, T.R., Korsmeyer, S.J.: MLL translocations specify a distinct gene expression profile that distinguishes a unique leukaemia. Nat. Genet. 30(1), 41–47 (2002)
14. Breiman, L.: Bagging predictors. Machine Learning 24(2), 123–140 (1996)
15. Witten, I.H., Frank, E.: Data Mining: Practical machine learning tools with Java implementations. Morgan Kaufmann, San Francisco (2005)
16. Bachner, D., Sedlacek, Z., Korn, B., Hameister, H., Poustka, A.: Expression patterns of two human genes coding for different rab GDP-dissociation inhibitors (GDIs), extremely conserved proteins involved in cellular transport. Hum. Mol. Genet. 4(4), 701–708 (1995)
17. Cutrona, G., Tasso, P., et al.: CD10 is a marker for cycling cells with propensity to apoptosis in childhood ALL. Br. J. Cancer 86(11), 1776–1785 (2002)
18. Fasching, K., Panzer, S., Haas, O.A., et al.: Presence of clone-specific antigen receptor gene rearrangements at birth indicates an in utero origin of diverse types of early childhood acute lymphoblastic leukemia. Blood 95(8), 2722–2724 (2000)
19. Kawata, H., Yamada, K., Shou, Z., Mizutani, T., Yazawa, T., Yoshino, M., Sekiguchi, T., Kajitani, T., Miyamoto, K.: Zinc-fingers and homeoboxes (ZHX) 2, a novel member of the ZHX family, functions as a transcriptional repressor. Biochem. J. 373(Pt 3), 747–757 (2003)
20. Ogawa, H., Iwaya, K., Izumi, M., Kuroda, M., Serizawa, H., Koyanagi, Y., Mukai, K.: Expression of CD10 by stromal cells during colorectal tumor development. Hum. Pathol. 33(8), 806–811 (2002)
21. Sheikh, S.S., Kallakury, B.V., Al-Kuraya, K.A., Meck, J., Hartmann, D.P., Bagg, A.: CD5-negative, CD10-negative small B-cell leukemia: variant of chronic lymphocytic leukemia or a distinct entity? Am. J. Hematol. 71(4), 306–310 (2002)
22. Shipp, M.A., Tarr, G.E., Chen, C.Y., Switzer, S.N., Hersh, L.B., Stein, H., Sunday, M.E., Reinherz, E.L.: CD10/neutral endopeptidase 24.11 hydrolyzes bombesin-like peptides and regulates the growth of small cell carcinomas of the lung. Proc. Natl. Acad. Sci. USA 88(23), 10662–10666 (1991)
23. Shisheva, A., Sudhof, T.C., Czech, M.P.: Cloning, characterization, and expression of a novel GDP dissociation inhibitor isoform from skeletal muscle. Mol. Cell Biol. 14(5), 3459–3468 (1994)

24. Strausberg, R.L., Feingold, E.A., et al.: Generation and initial analysis of more than 15,000 full-length human and mouse cDNA sequences. Proc. Natl. Acad. Sci. USA 99(26), 16899–16903 (2002)
25. Toyoda, M., Nakamura, M., Makino, T., Kagoura, M., Morohashi, M.: Sebaceous glands in acne patients express high levels of neutral endopeptidase. Exp. Dermatol. 11(3), 241–247 (2002)
26. Weitzdoerfer, R., Stolzlechner, D., Dierssen, M., Ferreres, J., Fountoulakis, M., Lubec, G.: Reduction of nucleoside diphosphate kinase B, Rab GDP-dissociation inhibitor beta and histidine triad nucleotide-binding protein in fetal Down syndrome brain. J. Neural Transm. Suppl. 61, 347–359 (2001)
27. Yagi, T., Hibi, S., Tabata, Y., et al.: Detection of clonotypic IGH and TCR rearrangements in the neonatal blood spots of infants and children with B-cell precursor acute lymphoblastic leukemia. Blood 96(1), 264–268 (2000)

Ant-MST: An Ant-Based Minimum Spanning Tree for Gene Expression Data Clustering

Deyu Zhou, Yulan He, Chee Keong Kwoh, and Hao Wang

School of Computer Engineering, Nanyang Technological University
Nanyang Avenue, Singapore 639798
{zhou0063,asylhe,asckkwoh,wang0046}@ntu.edu.sg

Abstract. We have proposed an ant-based clustering algorithm for document clustering based on the travelling salesperson scenario. In this paper, we presented an approach called Ant-MST for gene expression data clustering based on both ant-based clustering and minimum spanning trees (MST). The ant-based clustering algorithm is firstly used to construct a fully connected network of nodes. Each node represents one gene, and every edge is associated with a certain level of pheromone intensity describing the co-expression level between two genes. Then MST is used to break the linkages in order to generate clusters. Comparing to other MST-based clustering approaches, our proposed method uses pheromone intensity to measure the similarity between two genes instead of using Euclidean distance or correlation distance. Pheromone intensities associated with every edge in a fully-connected network records the collective memory of the ants. Self-organizing behavior could be easily discovered through pheromone intensities. Experimental results on three gene expression datasets show that our approach in general outperforms the classical clustering methods such as K-means and agglomerate hierarchical clustering.

Keywords: gene expression data, clustering, ant-based clustering, minimum spanning tree.

1 Introduction

Microarrays enable biologists to study genome-wide patterns of gene expressions in any given cell type, at any given time, and under any given set of condition. Using these arrays can generate large amounts of data, potentially capable of providing fundamental insights into biological processes ranging from gene function to cancer, ageing and pharmacology [1]. Even partial understanding of the available information can provide helpful clues. For example, co-expressions of novel genes may provide leads to the function of many genes for which information is not available currently.

Clustering is a fundamental technique in exploratory data analysis and pattern discovery, aiming at extracting underlying cluster structures. Cluster analysis is concerned with multivariate techniques that can be used to create groups

J.C. Rajapakse, B. Schmidt, and G. Volkert (Eds.): PRIB 2007, LNBI 4774, pp. 198–205, 2007.
© Springer-Verlag Berlin Heidelberg 2007

amongst the observations, where there is no *a priori* information regarding the underlying group structure. Clustering of the genes on the basis of the tissues can be used to search for groups of gene that might be regulated together. Dozens of clustering algorithm exist in the literature and a number of *ad hoc* clustering procedures have been applied to microarray data. Available methods can be categorized broadly as being hierarchical such as agglomerative hierarchical clustering (AHC) [2, 3] or non-hierarchical such as k-means clustering [4] and clustering through Self-Organizing Maps [5]. A major limitation of hierarchical methods is their inability to determine the number of the clusters. The limitation of k-means methods is their high computational complexity.

The concepts and properties of graph theory make it very convenient to describe clustering problems by means of graphs [6]. Nodes of a weighted graph correspond to data points in the pattern space and edges reflect the proximities between each pair of data points. Approaches based on minimum spanning trees have been proposed for clustering gene expression data [7]. Minimum spanning tree (MST), a concept from the graph theory, is used for representing multi-dimensional gene expression data. Based on the representation, gene expression data clustering problem is converted to a tree partitioning problem. Advantages of using this method have been described and demonstrated as follows [7]: 1) the simple structure of a tree facilitates efficient implementations of rigorous clustering algorithm; 2) clustering based on MST does not depend on detailed geometric shape of a cluster; 3) inter-data relationship is greatly simplified in MST representation and no essential information for clustering is lost.

We have proposed an ant-based clustering algorithm for document clustering based on the traveling salesperson (TSP) scenario [8]. It not only has the traits of self-organization and robustness, but also can generate optimal number of clusters without incorporating any other algorithms such as K-means or AHC. In [8], to break the linkages of the fully connected network in order to generate clusters, average pheromone strategy is used. The average pheromone of all the edges is computed at first and then edges with pheromone intensity less than the average pheromone will be removed form the network. Nodes will then be separated by their connecting edges to from clusters. In this paper, we investigate using the method based on minimum spanning trees (MST) to break the linkages in order to generate clusters. The reasons behind are: 1) the method based on MST has been proven efficient in the domain of gene expression clustering, 2) and it has strong mathematical foundation.

Our proposed approach called Ant-MST consists of two steps. First, a fully connected network of nodes is generated using the ant-based clustering method. Then the linkages is broken based on MST in order to generate clusters. It uses pheromone intensity to measure the similarity between two genes instead of using Euclidean distance or correlation distance. Pheromone intensities associated with every edge in a fully-connected network records the collective memory of the ants. Self-organizing behavior could be easily discovered through pheromone intensities.

The rest of the paper is organized as follows. Section 2 presents the Ant-MST approach. Experimental results on three gene expression datasets are discussed in section 3. Finally, section 4 concludes the paper and outlines the future work.

2   Ant-MST: An Ant-Based Minimum Spanning Tree

2.1   Ant-Based Clustering

The Ant Colony Optimization (ACO) algorithm belongs to the natural class of problem solving techniques which is initially inspired by the efficiency of real ants as they find their fastest path back to their nest when sourcing for food. An ant is able to find this path back due to the presence of pheromone deposited along the trail by either itself or other ants. An open loop feedback exists in this process as the chances of an ant taking a path increases with the amount of pheromone built up by other ants.

Early approaches in applying ACO to clustering are to first partition the search area into grids. A population of ant-like agents then move around this 2D grid and carry or drop objects based on certain probabilities so as to categorize the objects. However, this may result in too many clusters as there might be objects left alone in the 2D grid and objects still carried by the ants when the algorithm stops. Therefore, Some other algorithms such as k-means are normally combined with ACO to minimize categorization errors. More recently, variants of ant-based clustering have been proposed, such as using inhomogeneous population of ants which allow to skip several grid cells in one step, representing ants as data objects and allowing them to enter either the active state or the sleeping state on a 2D grid. Existing approaches are all based on the same scenario that ants move around in a 2D grid and carry or drop objects to perform categorization.

We have proposed an ant-based clustering algorithm for document clustering based on the travelling salesperson (TSP) scenario [8]. The advantages of our ant-based clustering approach are: 1) It does no rely on a 2D grid structure. 2) It can generate optimal number of clusters without incorporating any other algorithms such as k-means or AHC. 3) When compared with both the classical document clustering algorithms such as K-means and AHC and the Artificial Immune Network (aiNet) based method, it shows improved performance when tested on the subsets of 20 Newsgroup data[1]. Here, we investigate the ant-based clustering algorithm for gene expression data analysis.

2.2   Minimum Spanning Trees

The concept of minimum spanning trees (MSTs) is from graph theory. For a connected and undirected graph G, a spanning tree of the graph G, T is a subgraph which is a tree and connects all the vertices together. A single graph can have many different spanning trees. If we assign a weight to each edge,

---

[1] http://people.csail.mit.edu/jrennie/20Newsgroups/

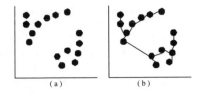

Fig. 1. 2D representation of a set of gene expression data (a) and its corresponding MST (b)

Table 1. Three objective functions and their corresponding clustering algorithms

Method	Objective Function	Procedure
Removing longest edges (MST-R)	Partition an MST into K subtrees so that the total edge-distance of all the K subtrees is minimized	Find the $K-1$ longest MST-edges, cut them and get a K-clustering achieving the global optimality of the objective function.
Iterative clustering (MST-I)	Partition an MST T into K subtrees $\{T_i\}_{i=1}^{K}$ to optimize: $\sum_{i=1}^{K} \sum_{d \in T_i} Dist(d, \text{center}(T_i))$ where d is the data point in the T_i and center(T_i) is dependent on the distance measure.	Start with an arbitrary K-partitioning of the tree and iteratively do the following until converging. For each pair of adjacent clusters, go through all tree edges within the merged cluster to find an edge which globally optimizes the 2-partitioning of the merged cluster and then cut the edge.
Global optimization (MST-G)	Partition the tree T into K subtrees and select K representatives $d_1, \ldots, d_K \in D$ to optimize $\sum_{i=1}^{K} \sum_{d \in T_i} Dist(d, d_i)$	Use dynamic programming to find the K representatives

and use this to assign a weight to a spanning tree by computing the sum of the weights of the edges in that spanning tree, a minimum spanning tree or minimum weight spanning tree is then a spanning tree with weight less than or equal to the weight of every other spanning tree.

We can use an MST to represent a set of gene expression data and their significant inter-data relationship. An example of a set of expression data and its corresponding MST is given in 1. In this example, the weight between two node is calculated using Euclidean distance. There are also other ways to measure the distance between two gene expression profiles such as correlational distance and mahalanobis distance. An MST of a weighted graph can be found by a greedy method, such as the classical Kruskal's algorithm [9].

After finding an MST T for a weighted graph, we can partition T into K subtrees, for some specified integer $K > 0$. These K subtrees correspond to K clusters. Since different clustering problems need different objective functions to achieve best performance, three objective functions and their corresponding procedures [7] are presented in Table 1.

2.3   Gene Expression Clustering Based on Ant-MST

We propose Ant-MST, an ant-based minimum spanning tree, for gene expression clustering. Given N genes $g_i, i = 1, \ldots, N$ and their expression profile $E_i = \langle a_{i1}, a_{i2}, \ldots, a_{im} \rangle, i = 1, \ldots, N$, we want to cluster these genes into several categories based on similarities between their expression profiles. Figure 2 describes our algorithm in details.

---

1. **Initialization.**
   N genes corresponds to N points in the graph. N genes are connected by $\frac{1}{2}N \times (N-1)$ edges.
   For every edge (i, j), set an initial value $\tau_{ij}(t)$ for pheromone intensity.
   Place m ants randomly on the N points.
2. **Construct a fully connected network of nodes G**
   The fully connected network of nodes is built using the ant-based clustering algorithm. Details can be found in [8]. Each edge is associated with a pheromone intensity τ.
3. **Build an MST T for the connected graph G**
   Initially, set T contain an edge with the smallest
   pheromone intensity in the G, remove the edge from G.
   Do the following iteratively Until all vertices are
   connected by the selected edges:
      add the edge with the smallest pheromone intensity
      in the G
      make sure that no cycle is formed.
   EndLoop
4. **Partition T into K subtrees**
   There are three methods to perform the partition which have been presented in Table 1.

Fig. 2. Gene expression clustering algorithm based on Ant-MST

3   Experimental Results

3.1   Setup

After the investigation of the suitability of various datasets in Stanford Genomic Resource Database[2], three datasets were chosen to evaluate the performance of our algorithms.

The dataset I is a subset of gene expression data in the yeast Saccharomyces cerevisiae (SGD)[3], which is commonly known as baker's or budding yeast. A set of 68 genes with each gene having 79 data points is chosen.

The dataset II is a temporal gene expression dataset in response of human fibroblasts to serum[4]. It consists of 517 genes and each gene has 18 data points.

---

[2] http://genome-www.stanford.edu/
[3] http://www.yeastgenome.org/
[4] http://genome-www.stanford.edu/serum/

In this dataset, genes are listed according to their cluster order along with their Gene bank Accession number and Clone IDs. Gene names with the SID prefix are not sequence verified. The expression changes are given as the ratio of the expression level at the given time-point to the expression level in serum-starved fibroblasts.

The dataset III is the rat central nervous system development dataset[5]. It is obtained by researchers using the method of reverse transcription-coupled PCR to study the expression levels during rat central nervous system development.

3.2   Results

Rand index [10] is used to evaluate the performance of the clustering algorithm. It is a metric to measure the similarity between two clusters which contain exactly the same data objects. In our experiments, rand index is used to measure the number of pair-wise agreements of resultant clusters from our algorithms and the "expert" classes, normalized by the total number of pair-wise combinations.

The expression of Rand Index is as following:

$$R(M, N) = \frac{a + d}{a + b + c + d} \tag{1}$$

Where M is the number of "expert" classes, N is the number of clusters to be evaluated, a is "true positive pairs", it is the number of pairs with same class label of "expert class" that are assigned into the same cluster, d is "true negative pairs", it is the number of pairs with different class label that are assigned into different cluster, b is "false negative pairs", it is the number of pairs of the same "expert" class label that are assigned to different clusters, c is "false positive pairs", it is number of pairs of different "expert" class label that are assigned to the same clusters. Rand index lies between 0 and 1; a high value indicates high the degree of agreements of resultant clusters and "expert" classes. Table 2 shows the detailed "expert" information of these datasets.

Table 2. Statistics on experimental data

Dataset	Gene Cluster					
	A	B	C	D	E	F
I	28	17	15	8	-	-
II	305	43	7	162	-	-
III	27	20	21	17	21	6

Table 3 lists the experimental results based on the different objective functions, MST-R, MST-I and MST-G as shown in Table 1, and on different datasets. Results using the classical clustering algorithms such as Agglomerative Hierarchical Clustering (AHC) and K-means are also presented.

---

[5] http://www.arclab.org/node_pages/265.html

Table 3. Comparison of experimental results on different algorithms

	Rand Index		
Methods	Dataset I	Dataset II	Dataset III
MST-R	0.910	0.541	0.293
MST-I	**0.936**	0.682	0.582
MST-G	0.923	**0.811**	0.568
AHC	0.803	0.628	0.575
K-means	0.701	0.565	**0.676**

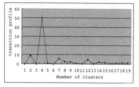 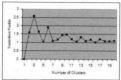 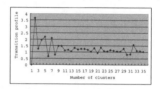

Fig. 3. Transition profile diagram of dataset I   **Fig. 4.** Transition profile diagram of dataset II   **Fig. 5.** Transition profile diagram of dataset III

It can be observed from Table 3 that the performance of clustering algorithm based on MST is better than that of AHC and K-means on dataset I and II. The rand index value achieved is 93.6% by MST-I on dataset I and 81.1% by MST-G on dataset II. However, the rand index values obtained using the MST-based methods on dataset III are lower than that of K-means with MST-I slightly outperforming AHC. The probably reason of better performance of K-means on dataset III is that the exact cluster number 6 was preset by the user while in practice it is hard to predict the correct cluster number.

The MST-based methods are able to calculate the optimal number of clusters automatically based on the transition profile values. Figure 3, 4, 5 are the transition profile diagrams for dataset I, II and III respectively. In the transition profile diagram, the x-axis represents the number of cluster, while the y-axis represents transition profile values. The highest transition profile value indicates the optimal number of clusters. It can be observed from Figure 3 that the optimal number of clusters in dataset I is 4, which is same as the actual number of clusters as can be found in Table 2. While for dataset II, the optimal number of clusters is 3 as shown in Figure 4. This is slightly different from the actual cluster number 4. Figure 5 reveals that the optimal number of clusters in dataset III is 3 which is different from the actual cluster number 6. This also explains the worse performance of MST-based methods in dataset III.

4   Conclusions and Future Work

In this paper, we have presented a clustering algorithm Ant-MST for gene expression data clustering. It consists of two stages. First construct a fully connected network of nodes using the ant-based clustering algorithm and then build an MST from the fully connected graph and partition it into K clusters. Experimental results on three different datasets have been presented to illustrate its feasibility and efficiency. In future work we will continue on the enhancement of the gene expression data clustering component and conduct a large scale of experiments to evaluate the system performance.

References

1. Baldi, P., Brunak, S.: Bioninformatics: The machine learning approach (2001)
2. Eisen, M.B., Spellman, P.T., Brown, P.O., Botstein, D.: Cluster analysis and display of genome-wide expression patterns. Proceedings of the National Academy of Sciences of the United States of America 95(14), 14863–14868 (1998)
3. Wen, X., Fuhrman, S., Michaels, G.S., Carr, D.B.: Large-scale temporal gene expression mapping of central nervous system development. Proceedings of the National Academy of Sciences of the United States of America 95(1), 334–339 (1998)
4. Herwig, R., Poustka, A.J., Mller, C., Bull, C.: Large-scale clustering of cdna-fingerprinting data. Genome Research 9(11), 1093–1105 (1999)
5. Tamayo, P., Slonim, D., Mesirov, J., Zhu, Q., Kitareewan, S.: Interpreting patterns of gene expression with self-organizing maps: Methods and application to hematopoietic differentiation. Proceedings of the National Academy of Sciences of the United States of America 96(6), 2907–2912 (1999)
6. Xu, R., Wunsch II, D.: Survey of clustering algorithms. IEEE Transactions on Neural Networks 16(3), 645–678 (2005)
7. Xu, Y., Olman, V., Xu, D.: Clustering gene expression data using a graph-theoretic approach: an application of minimum spanning trees. Bioinformatics 18(4), 536–545 (2002)
8. He, Y., Hui, S.C., Sim, Y.: A Novel Ant-Based Clustering Approach for Document Clustering. In: Asia Information Retrieval symposium, pp. 537–544. Springer, Heidelberg (2006)
9. Aho, A.V., Hopcroft, J.E., Ullman, J.D.: The design and analysis of computer algorithms (1974)
10. Rand, W.M.: Objective criteria for the evaluation of clustering methods. Journal of the American Statistical Association 66, 622–626 (1971)

Integrating Gene Expression Data from Microarrays Using the Self-Organising Map and the Gene Ontology

Ken McGarry*, Mohammad Sarfraz, and John MacIntyre

School of Computing and Technology, University of Sunderland,
St Peters Campus, St Peters Way, SR6 ODD, UK
ken.mcgarry@sunderland.ac.uk

Abstract. The self-organizing map (SOM) is useful within bioinformatics research because of its clustering and visualization capabilities. The SOM is a vector quantization method that reduces the dimensionality of original measurement and visualizes individual tumor sample in a SOM component plane. The data is taken from cDNA microarray experiments on Diffuse Large B-Cell Lymphoma (DLBCL) data set of Alizadeh. The objective is to get the SOM to discover biologically meaningful clusters of genes that are active in this particular form of cancer. Despite their powers of visualization, SOMs cannot provide a full explanation of their structure and composition without further detailed analysis. The only method to have gone someway towards filling this gap is the unified distance matrix or U-matrix technique. This method will be used to provide a better understanding of the nature of discovered gene clusters. We enhance the work of previous researchers by integrating the clustering results with the Gene Ontology for deeper analysis of biological meaning, identification of diversity in gene expression of the DLBCL tumors and reflecting the variations in tumor growth rate.

1 Introduction

Microarrays are an exciting and recent technological breakthrough that has enabled the detailed analysis of cellular activity and condition [1]. Recent work has highlighted how components of metabolic pathways can be identified and how the protein targets of drug treatment can be determined using expression profiles [2]. Microarray technology can deliver an extremely detailed analysis of cellular activity and condition [3]. Recent work has highlighted how components of metabolic pathways can be identified and how the protein targets of drug treatment can be determined using expression profiles for example Alizadeh et al [4] discovered a new sub-class of cancer with implications for clinical treatment. Microarray experiments are producing unprecedented quantities of genome data, the management and analysis of this data is starting to receive greater attention [5]. However, there is no one technique that appears to be superior, either for data management or data analysis.

---

* Corresponding author.

J.C. Rajapakse, B. Schmidt, and G. Volkert (Eds.): PRIB 2007, LNBI 4774, pp. 206–217, 2007.
© Springer-Verlag Berlin Heidelberg 2007

Microarrays have been used extensively for gene expression analysis and geno-typing [6]. Expression analysis seeks to uncover the activity level of certain genes and groups of genes. This is of vital importance in drug discovery where not only are the anticipated effects on the target genes must be confirmed but also for any side-effects on non-target genes must to be monitored. Genotyping seeks to discover and identify many of the mutations within a single gene and can be used for the screening of individuals for particular diseases [7]. Obtaining such information at an early stage will lead to to improved clinical treatment [8].

Microarrays are small glass slides or chips that contain many thousands of genes (strands of DNA) formed as spots which are laid out in a regular grid-like structure. The genes are selected by scientists from gene libraries, and because of their microscopic size they must be located on the glass substrate by auto-mated robotic equipment. The selected genes are usually chosen because they are deemed important for the particular biological process to be investigated. The microarrays are then introduced to the biological samples (DNA that have been labeled by fluorescent materials), which then bind to the original DNA placed on the glass substrate. The microarray image is then scanned and digitised by a laser system. Image processing software is used to reveal the intensity of the fluorescent labels and depending on the type of microarray, their colour. The intensity of the spot is proportional to the level and activity at which the genes are being expressed. Colour, where applicable, is used to identify sample and control populations.

The starting point for any microarray experiment is to define the biological question to answer [9]. For example, a scientist may wish to pursue the hypothesis that a certain number of specific genes are active (up-regulated) in a particular type of cancer and if treated with a particular drug should be inactive (down-regulated). The choice of microarray must also be made, often Affymetrix Gene chips are used in parallel with cDNA microarrays [10].

This paper is concerned with analyzing gene expression data generated from microarrays. We use the self-organizing map (SOM) because of its clustering and visualization capabilities. SOM is a vector quantization method that reduces that simplifies and reduces the dimensionality of original measurement and visualizes individual tumor sample in a SOM component plane. The data is taken from cDNA microarray experiments on Diffuse Large B-Cell Lymphoma (DLBCL) data set of Alizadeh [4]. Diffuse Large B-Cell Lymphoma is the most prevalent lymphoid cancer in adults and accounts for 30-40% of cancers, unfortunately, 50% patients cannot be cured.

The remainder of this is paper is structured as follows; section two discusses the details of the new microarray technology and the problems inherent in the data they generate for machine learning researchers; subsections deal with the characteristics of the SOM that make it suitable for bioinformatic work and the gene ontology system which enables the representation and processing of information about gene products and functions. Section three describes the data, experimental setup and preprocessing issues specific to the microarray data and the experimental results, finally section four presents the conclusions.

2   The Biological Basis of Microarray Technology

Figure 1 shows the internal structure of a typical microarray, the substrates can be glass slides, plastic slides or membranes where the cDNA can be deposited. They have a regular matrix structure, each spot corresponds to a gene sequence. The same gene sequences are usually repeated elsewhere on the chip for reasons of precision and accuracy. Several thousand genes may be placed on an individual chip, the cost of running microarray experiments is directly related to the number of genes per chip.

Although the process of creating microarrays and the analysis of the resultant data is fraught with difficulties their essential operation is relatively straightforward to understand. A set of DNA sequences stored in libraries that correspond to specific genes selected by scientists for their experiment are transferred or *spotted* onto a glass slide by robots. Cell cultures are taken from the patients (a sample and a control) and each is labelled by a fluorescent dye, usually red for the sample population and green for the control population. These cultures are then introduced to the microarray and allowed to bind or *hybridise* with their complementary target cDNA sequences on the chip. The more active a gene is, the more mRNA it should produce and so the intensity and colour of the spot corresponding to that gene ought to appear greater than non-active genes. If the control population is in greater quantity then it will appear green, if the sample population is in greater quantity then it will appear red, if the spot is yellow

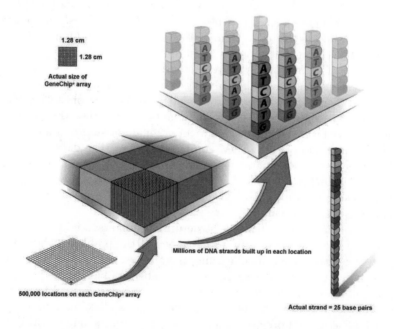

Fig. 1. Each spot is composed of millions of cDNA strands, diagram courtesy of Affymetrix Corporation

then both populations are expressed in equal quantities, if the spot is black then no hybridisation has occurred.

The basic idea behind microarray analysis is to examine the intensities of the spots which is an indirect indication of the level of expression of the genes. The expression levels are often compared against biologically related samples to see which genes are differentially expressed. This can be displayed as a ratio between the sample and control genes, there are disadvantages to using only expression ratios for data analysis. The ratios can help determine important relationships between genes but they also remove information relating to the absolute gene-expression levels. The information pertaining as to wether a gene is up- or -down regulated appears differently when using ratios; i.e. a up- factor of 2 have a value of 2 while those genes that are down-regulated by 2 have a value of -0.5 [11]. Transforming the data using a Log2 base produces a more intuitive range of values, see figure 2. This is a simple way to compare the two channels. Points that are above the diagonal in this plot correspond to genes that have higher expression levels in the sample than in the sample.

Typically, the first and most commonly used technique is to normalise the data, this manipulates the hybridisation intensities to balance them in order to make meaningful comparisons [12]. Normalisation usually needs to be applied because of various problems with experimental bias such as background intensities of the microarrays are not uniform, also differences can occur between pen-tips/print-tips, or blocks. These must be compensated for by normalizsation, hopefully the information will be available to normalise each block separately. Normalisation of data means that weaker signals are amplified, this could

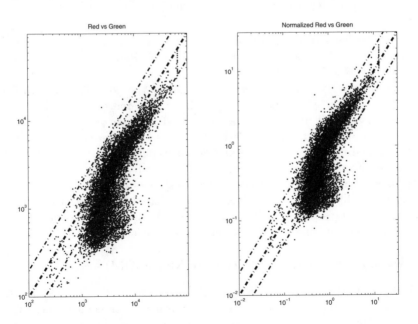

Fig. 2. Comparison of Normalisation of intensity data

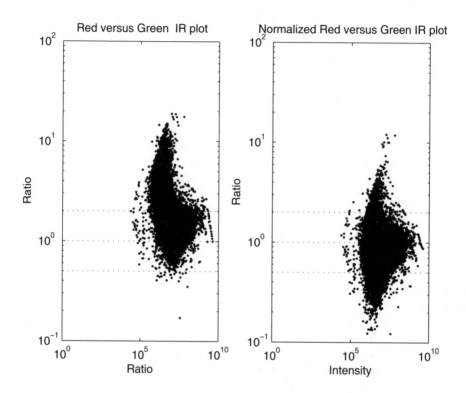

Fig. 3. Comparison of Normalisation of intensity/ratio data

mean they are related to important cellular activity that is expressed in small quantities of cDNA or perhaps could just be noise. Replicates, are one way of determining such effects.

It is also useful to plot the log_2 ratios against the intensity for each spot. Figure 3 shows how such a plot can highlight the difference.

Typically, the first and most commonly used technique is to normalise the data, this manipulates the hybridisation intensities to balance them in order to make meaningful comparisons. Normalisation usually needs to be applied because of various problems with experimental bias such as background intensities of the microarrays are not uniform. Normalisation of data means that weaker signals are amplified, this could mean they are related to important cellular activity that is expressed in small quantities of cDNA or perhaps could just be noise. Replicates, are one way of determining such effects.

2.1   Kohonen Self-Organising Feature Map (SOM)

The Kohonen SOM consists of a simple architecture. Since its initial introduction by Kohonen several improvements and variations have been made to the training algorithm. The SOFM consists of two layers of neurons, the input and output layers. The input layer presents the input data patterns to the output layer and

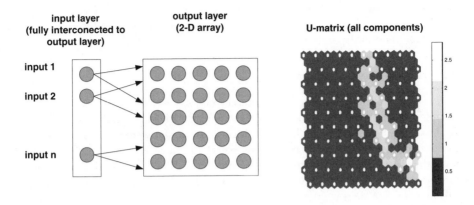

Fig. 4. Architecture of SOM, showing a regular grid of neurons. The U-matrix technique calculates the weighted sum of all Euclidean distances between the weight vectors for all output neurons. The resulting values can be used to interpret the clusters learned by the SOM. Each white dot represents a neuron and the colours represents different values of the weights, a distinct boundary is formed forming two large clusters.

is fully interconnected. The output layer is usually organised as a 2-dimensional array of units which have lateral connections to several neighbouring neurons. The architecture is shown in Figure 4.

Each output neuron by means of these lateral connections is effected by the activity of its neighbours. The activation of the output units according to Kohonens original work is by equation 1. The modification of the weights is given by equation 2 :

$$O_j = F_{min}(d_j) = F_{min}(\sum_i (X_i - W_{ji})^2) \tag{1}$$

$$\Delta W_{ij} = O_j \eta (X_i - W_{ji}) \tag{2}$$

where:
$O_j =$ activation of output unit, $X_i =$ activation value from input unit, $W_{ji} =$ lateral weights connecting to output unit, $d_j =$ neurons in neighbourhood, $F_{min} =$ unity function returning 1 or 0, $\eta =$ gain term decreasing over time.

The lateral connections enable the SOM to learn "competitively", this means that the output neurons compete for the classification of the input patterns. During training the input patterns are presented to the SOM and the output unit with the nearest weight vector will be classed as the winner.

The Kohonen self-organising feature map (SOM) is a neural network which is unsupervised technique that represents multi-dimensional patterns into 2-dimensional form for visualisation [13]. It also has the important feature of topological preservation i.e. clusters that are close to each other represent patterns that are very similar. The SOM is often used to group microarray gene expression data into related clusters, for example Kaski selected a subset of 1551 yeast genes of known functional classes [14,15]. Since neural networks are not amenable to internal scrutiny (they are known as black boxes), Kaski was interested in determining the internal representation by using U-matrix analysis to show *how* the SOM partitioned the boundaries between the clusters.

2.2   The Gene Ontology

The use of ontologies is increasingly perceived as a way forward to overcome the complexity of biological information, for comprehensive introductions see [16]. A substantial amount of biological information is hierarchial in nature and the inter-relationships between the various pieces of knowledge can be meaningfully formalized, structured and represented by an ontology. One should not confuse Gene Ontology with a database of gene sequence or with a catalogue of gene product, rather than it gives us an idea of how gene product behaves at cellular level. It is not a way to bring together all the available biological datasets. The authors of GO have tried to provide a practically useful framework for keeping track of biological annotations which are applied to gene products.

GO is divided in to three disjoint term hierarchies, which are cellular component, biological process and molecular function. A cellular component is just a component of a cell with a condition that it is a part of large object, which might be a gene product or anatomical structure. A biological process is defined in GO as:"A phenomenon marked by changes that lead to a particular result, mediated by one or more gene product" [17]. Biological process terms can be quite specific

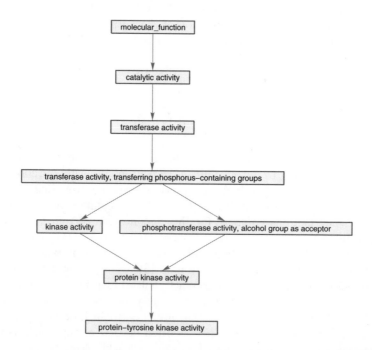

Fig. 5. Gene Ontology identifies gene JNK3 as active in protein-tyrosine kinase activity

(glycolysis) or very general (apotosis). Molecular function and biological process terms are clearly closely interrelated. Molecular Function describes activities at molecular level, like that of binding activities or catalytic activities, In GO it represent activities rather than molecules or complexes that perform the action, and do not specify the context in which action take place.

3 Experimental Results

The work of Alizadeh is often cited as a clustering success, whereby the authors were able to identify a new sub-class of cancer [4]. The novel variety was revealed through hierarchial clustering of tumors DLBCL (diffuse large B-cell lymphoma) data. The authors identified two distinct groups that were highly correlated with patient survival rates (40% of patients respond well to conventional treatment), these patients showed *germinal centre B-like DLBCL* stages of expression. This implied a major breakthrough for the treatment for this variety of cancer as the 60% of patients who succumbed to the disease showed *activated B-like cells* stages of expression. Sources of experimental data: All the data used in this study including survival data of lymphoma patients was obtained from the web supplement of the publication of Alizadeh available at http:/llmpp.nih.gov/lymphoma/data.shtml.

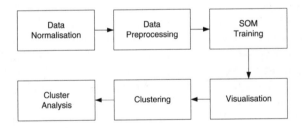

Fig. 6. Experimental setup and process

3.1 Data, Experimental Setup and Preprocessing

The various stages involved are highlighted in figure 6. The fluorescent intensity of each gene was tested and if greater than 1.4 times the local background were considered well measured. The ratio values were log-transformed (base 2) and stored in a table (rows, individual cDNA clones; columns, single mRNA samples). The Alizadeh data was preprocessed by the Lowess function with zero-norm with linear models and kernel methods. Each feature was given mean zero value and standard deviation was reduced to one. After cleaning the data that is removing all those which were under expressed and any bad measurement in the data, the original data set of 4026 genes was reduced to 3535 genes from 96 samples.

Figure 7 shows the U-Matrix of DLBCL entire data set, the individual clusters are quite well differentiated. The name of the genes superimposed over the

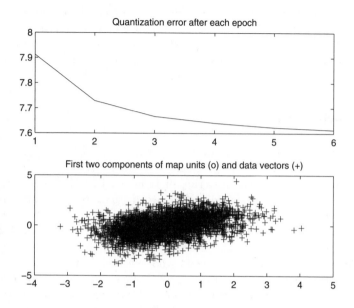

Fig. 7. Training run on DLBCL data

map unit so it is very easy to observe and analysis which genes are part of a particular cluster. The expression data can be judged by the colours, predominately reddish colour implies that a particular gene is highly expressed. The bluer the colour implies that a particular gene is less expressed. Despite their powers of visualization, SOMs cannot provide a full explanation of their structure and composition without further detailed analysis. The only method to have gone someway towards filling this gap is the *unified distance matrix* or U-matrix technique of Ultsch [18]. Further U-matrix research involving the analysis of individual component features was undertaken by Kaski [19]. Recent work by Malone makes explicit the contribution of each variable in the cluster to be assessed for characterising the cluster and can be expressed in rule format [20].

A deeper analysis of the SOM component plane (figure 8) reveals 42 DLBCL samples and three DLBCL lines (OCILy3, OCILy10 and OCILy1), the topology of the SOM is 20x15 and the colour scale of component plane represent the mean ratio in each map node.

Through the proposed approach applied above one can directly observe gene expression patterns of different lymphomas sub types i.e. DLBCL, CLL and FL, as it can be seen by the figures above that there are four prominent clusters identified in DLBCL 4, 2, 9 and the large group of clusters of 1 and 12 a short summary of the genes included in these cluster are listed table 1. After selecting the genes in the second subset file, the annotations have to be extracted from the ontology website. The particular information of interest for humans is gene-association-goa-human. It contains up to date annotations of Homo Sapiens, the more interesting genes were tested to get their ancestor list and also their root graph.

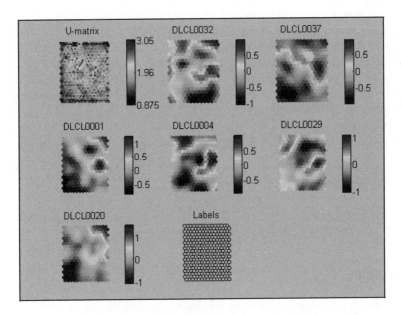

Fig. 8. Umatrix and SOM component planes

Table 1. Important DLBCL Genes clustered by the SOM

GeneName	GO ID	Description
TP73L	GO:0045892	Tumor protein p73-like
JNK3	GO:0004713	Catalysis of ATP and a protein tyrosine.
LYSp100	GO:0006952	Defense/immunity protein activity
RAD50	GO:0030674	physically linking the bound proteins or complexes to each other
CD44	GO:0016337	The attachment of one cell to another cell via adhesion molecules
GADD34	GO:0030968	Results in changes in the regulation of transcription and translation
CD5	GO:0025383	Involvement in DLBCL tumor progression

The DLBCL data was applied to the Gene Ontology to look at the significance of interesting genes and Gene Ontology terms that are used in the micro array. For the ontology study the data used all 3535 genes, first we applied K-means clustering was done to select only interesting genes during this all under expressed genes were removed, the total number of gene were reduced to 1157, than clustering was done into 4 sets. The difficulty of course is accurately identifying "interesting" genes.

4   Conclusions

We have demonstrated the use of Self Organising Map as a tool for analysis of gene expression data. The approach taken in our paper for the analysis of gene expression data were consistent with results originally published. However, the aim of this study was to demonstrate the visualization capabilities of SOM with the original data. We also integrated the Gene Ontology with the discovered clusters of genes, which provides additional domain knowledge regarding gene function and common biological pathways. Finally in this study, the theoretical and practical approach of analysis of gene expression data of human Diffuse Large B cell Lymphoma have been discussed using SOM. We conclude that the SOM provides an excellent perfect platform for visualization and analysis of microarray data, and it will be very useful in extracting biologically meaningful information, when combined with domain knowledge such as the Gene Ontology.

Acknowledgements

This work was part supported by a Research Development Fellowship funded by HEFCE and the Biosystems Informatics Institute (Bii).

References

1. Berkum, N., Holstege, F.: Dna microarrays raising the profile. Current Opinions in Biotechnology 12(1), 48–52 (2001)
2. Soinov, L., Krestyaninova, M., Brazma, A.: Towards reconstruction of gene networks from expression data by supervised learning. Genome Biology 4(1), 1–10 (2003)
3. Sherlock, G.: Analysis of large-scale gene expression data. Current Opinion in Immunology 12, 201–205 (2000)
4. Alizadeh, A., Eisen, M., Davis, R., Ma, C.: Distinct types of diffuse large B-cell lymphoma identified by gene expression profiling. Nature 403, 503–511 (2000)
5. Kuo, P., Kim, E., Trimarchi, J., Jenssen, T., Vinterbo, S., Ohno-Machado, L.: A primer on gene expression and microarrays for machine learning researchers. Journal of Biomedical Bioinformatics 37, 293–303 (2004)
6. Huges, T., et al.: Functional discovery via a compendium of expression profiles. Cell 102, 109–126 (2000)
7. Lu, Y., Han, J.: Cancer classification using gene expression data. Information Systems 28, 242–268 (2003)
8. Peterson, C., Ringer, M.: Analyzing tumor gene expression profile. Artificial Intelligence in Medicine 28(1), 59–74 (2003)
9. Moreau, Y., Aerts, S., Moor, B.D., DeStrooper, B., Dabrowski, M.: Comparison and meta-analysis of microarray data: from the bench to the computer desk. Trends in Genetics 19(10), 570–577 (2004)
10. Kuo, P., Jenssen, T., Butte, A., Ohno-Machado, L., Kohane, I.: Analysis of matched mRNA measurements from two different microarray technologies. Bioinformatics 18(3), 405–412 (2003)
11. Quackenbush, J.: Computational analysis of microarray data. Nature Reviews Genetics 2, 418–427 (2001)
12. Quackenbush, J.: Microarray data normalisation and transformation. Nature Genetics Supplement 32, 496–501 (2002)
13. Kohonen, T., Oja, E., Simula, O., Visa, A., Kangas, J.: Engineering applications of the self-organizing map. Proceedings of the IEEE 84(10), 1358–1383 (1996)

14. Kaski, S., Nikkilä, J., Törönen, P., Castrén, E., Wong, G.: Analysis and visualization of gene expression data using self-organizing maps. In: Proceedings of NSIP-01, IEEE-EURASIP Workshop on Nonlinear Signal and Image Processing 2001, Baltimore, USA (2001)
15. Nikkila, J., Kaski, S., Toronen, P., Castren, E., Wong, G.: Analysis and visualization of gene expression data using self-organizing maps. Neural Networks 8(9), 953–966 (2002)
16. Bard, J., Rhee, S.: Ontologies in biology: design applications and future challenges. Nature Reviews Genetics 5, 213–222 (2004)
17. Ashburner, M.: Gene ontology: tool for the unification of biology. Nature Genetics 25, 25–29 (2000)
18. Ultsch, A., Siemon, H.P.: Kohonens self organizing feature maps for exploratory data analysis. In: Proceedings of the International Neural Network Conference, pp. 305–308 (1990)
19. Kaski, S.: Dimensionality reduction by random mapping: Fast similarity computation for clustering. In: Proceedings of IJCNN'98, International Joint Conference on Neural Networks, Piscataway, NJ, vol. 1, pp. 413–418 (1998)
20. Malone, J., McGarry, K., Bowerman, C., Wermter, S.: Rule extraction from kohonen neural networks. Neural Computing Applications Journal 15(1), 9–17 (2006)

Order Preserving Clustering by Finding Frequent Orders in Gene Expression Data

Li Teng and Laiwan Chan

Department of Computer Science and Engineering,
The Chinese University of Hong Kong, Hong Kong

Abstract. This paper concerns the discovery of Order Preserving Clusters (OP-Clusters) in gene expression data, in each of which a subset of genes induce a similar linear ordering along a subset of conditions. After converting each gene vector into an ordered label sequence. The problem is transferred into finding frequent orders appearing in the sequence set. We propose an algorithm of finding the frequent orders by iteratively Combining the most Frequent Prefixes and Suffixes (CFPS) in a statistical way. We also define the significance of an OP-Cluster. Our method has good scale-up property with dimension of the dataset and size of the cluster. Experimental study on both synthetic datasets and real gene expression dataset shows our approach is very effective and efficient.

1 Introduction

In gene expression dataset, each row stands for one gene and each column stands for one condition. Traditional methods for pattern discovery in gene expression matrices are based on clustering genes (conditions) by comparing their expression levels in all conditions (genes). However, general understanding of cellular processes leads us to expect subsets of genes to be coregulated and coexpressed only under certain experimental conditions. Recent research works [1-8],focus on discovering such local patterns embedded in high dimensional gene expression data.

Order preserving clustering (OP-Clustering)[9] is one discipline of looking for submatrices in which value the rows induce the same linear ordering in the columns. In former study of gene expression profiles, researchers regard there are certain stages for genes. They use on or off to stand for the state of gene. There could be more than two stages. The idea of stages encourages us to measure the similarities by comparing the condition orders between two genes. Therefore we expect the data to contain a set of genes and a set of conditions such that the genes are identically ordered on the conditions. By finding the hidden order and genes that support it, we can potentially find the different stages shared by the genes. Figure 1 shows an example of the order preserving subsequences in two data sequences. In Figure 1(a) there is no obvious trend between the two sequences. However, if we only consider columns [c d e j] as shown in Figure 1(b), the two subsequences show strong coherence on these four columns. The

J.C. Rajapakse, B. Schmidt, and G. Volkert (Eds.): PRIB 2007, LNBI 4774, pp. 218–229, 2007.

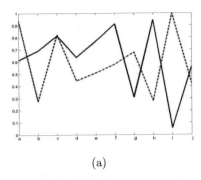
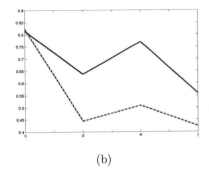

(a)   (b)

Fig. 1. An example of the Order Preserving Subsequences in two sequences.(a) Original data, (b) the OP subsequences.

two subsequences show the same ascending pattern if we rearrange the columns as [j d e c].

Problem Statement. Given an n×m gene expression dataset A, where $G = \{g_1, g_2, \ldots, g_n\}$ is the set of genes (also the rows), $C = \{c_1, c_2, \ldots, c_m\}$ is the set of conditions (columns). Order Preserving Cluster (OP-Cluster) $OPC = (P, Q)$ is a submatrix of A, where $P \subseteq G$, $Q \subseteq C$, that all the genes in P share the same linear order on the conditions in Q.

That means there is a permutation of the conditions in Q, after which all the genes in P show the ascending patterns. An embedded OP-Cluster can be identified by the order of conditions which involve in. Here we call the permutation of conditions the order of the OP-Cluster and the subset of genes the supports for that order. From this point of view the work of finding OP-Clusters is related to the so called sequential pattern mining in some other literature [10][11][12].

In this paper we propose a model of OP-Clustering on numerical dataset. After converting the numerical dataset into a sorted label sequence set. We present a heuristic method of finding the frequent orders by combining the most frequent prefix and suffix iteratively in a statistical way. The structure of the paper is as follow. In section 2 some related work is discussed. The OP-Clustering algorithm (CFPS) is presented in section 3. The experimental results on both synthetic dataset and real gene expression dataset are presented in section 4 and we draw the conclusion in section 5.

2   Related Work

The concept of Order-Preserving SubMatrix (OPSM) was first introduced by Ben-Dor et al.[9] and they also proved the OPSM problem was NP-hard in the worst case. By discovering a subset of genes identically ordered among a subset of conditions they focused on the coherence of the relative order of the conditions rather than the coherence of actual expression levels. A stochastic model was developed by them to discover the best row supported submatrix given a fixed size of conditions. For OPSM the quality of their resulted cluster is very sensitive to some given parameters and the initial selection of partial models.

Liu et al.[13] proposed an exhaustive bicluster enumeration algorithm to discover all significant OP-Clusters. Their pattern discovery algorithm is heavily inspired in sequential pattern min-ing algorithms [14]. Mining sequential pattern was first introduced in the work of Agrawal et al.[10], and most methods in this area are based on breadth-first or depth-first search. Liu et al. used a compact tree structure to store the identities of the rows and the sequences sharing the same prefixes. A depth-first algorithm was devised to discover the OP-Clusters with a user-specified threshold. The drawback is that the construction of the OPC-Tree is very time consuming and it needs excessive memory resources. For large dataset pruning techniques have to be taken before it can be effective.

Bleuler S. and Zitzler Z.[15] used a scoring function which combines the mean squared residue score with the OPSM concept. It allows the arbitrarily scale and degree of orderedness required for a cluster. They proposed an evolutionary algorithm framework that allows simultaneous clustering over multiple time course experiments. A local search procedure based on a greedy heuristic was applied to each individual before evaluation. Their method is still time consuming and they can only find non-overlapping clusters in one run.

Most existing methods based on exhaustive searching and need excessive computation and memory cost. In our work, a heuristic method based on statistical information is proposed to find the OP-Clusters by finding the frequent orders in the sequence set which comes from the original numerical dataset.

3   Algorithm

In this section we present the algorithm. Firstly the main procedures of the algorithm is introduced in section 3.1. Then some details of the algorithm are given in section 3.2-3.4. Examples are used to to illustrate the algorithm.

3.1   A Top Down Algorithm to Find Frequent Orders

We divide the model into the following three phases. In Step 1 numerical dataset is converted into sequence set. This processing takes only once in the first run. In Step 2 and 3 the frequent orders is constructed iteratively. These two steps would be repeated when necessary.

Step 1. Sorting Phase. Each condition is identified with a label. Then each gene expression vector is sorted in non-decreasing order. According to the sorting result each gene expression vector is converted into an ordered label sequence. The order of each label in a sequence stands for the ranking of the corresponding condition in the particular gene expression vector. Now the whole dataset is converted to a sequences set.

The conditions with very close values are grouped to form an order equivalent group in which we make no no difference on their relative order. That means

ID	a	b	c	d	e	f
1	4	5	5	3	8	1
2	3	4	5	2	7	8
3	5	3	7	4	9	1
4	8	1	5	4	3	7

ID	Sequence					
1	f	d	a	(c b)	e
2	d	a	b	c	e	f
3	f	b	d	a	c	e
4	b	e	d	c	f	a

(a)   (b)

Fig. 2. (a) Original numerical data matrix, (b) sequences of column labels of the data matrix

conditions in order equivalent group could have exchangeable orders. The strict order requirement is too sharp in OP-Clustering scheme and orders between close values would be meaningless. Since order between some conditions might be disturbed by sampling processing especially when noises exist. We define a threshold θ. When difference between two conditions is smaller than θ, we group them together to form an order equivalent group. θ is relative to the magnitude of dataset. Figure 2 shows an example the sorting phase. "()" means order equivalent group. Before going on we give some notations.

Definition 1. For any two labels, no matter they are adjacent or not in a sequence with no repeating labels, the one that comes before is called the prefix and the other one is called the suffix. E.g. in sequence "fdecab", "e" is prefix of "c", "a" and "b". Also it is suffix of "f" and "d".

Definition 2. For two label sequences x and y, if all labels of x are contained in y and any two labels of x have the same order as they have in y then we say x is a subsequence of y or x appears in y. E.g. "dca" is a subsequence of "fdecab".

Definition 3. Sequence x is different from sequence y when any of the following two cases occurs, 1. There are labels in x which do not appear in y, 2. The common labels of the two sequences have different orders. E.g. sequence "adcdf" and "acbed" are different from sequence "abcde", while "abe" is not.

An embedded OP-Cluster can be identified by first finding the order of its columns, which is the frequent order appearing in the sequence set. The idea of frequent order is similar to the frequent pattern in the sequential pattern mining problem [16][11][12]. However, in our work it is more complicated than conventional sequential pattern mining problem. The original sequences we are handling have the same length. Different labels appear once and only once in each sequence. The identities of the genes associated with each frequent order have to be recorded in order to determine the genes involved in an OP-Cluster. While conventional sequential mining algorithms only keep the number of appearance of frequent subsequences but not their identities.

Step 2. Counting Phase. We scan the label sequences and construct an order matrix O which counts the occurrence of any label being prefix or suffix of any other labels. In the example there are 6 labels. Then order matrix O is a 6×6 matrix as Figure 3 shows. Each row/column stands for one label. $O(a, b)$ stands for the occurrence frequency of "a" being prefix of "b" (or "b" being suffix of "a") in all the rows. Suppose the original numerical data matrix has n rows an m columns then O is a $m \times m$ matrix and $O(a, b) + O(b, a) = n + z$ (z is the number of order equivalent groups which contain "a" and "b").

O	a	b	c	d	e	f	
a	0	2	3	0	3	1	9
b	2	0	4	2	4	2	14
c	1	1	0	0	3	2	7
d	4	2	4	0	3	2	15
e	1	0	1	1	0	2	5
f	3	2	2	2	2	0	11
	11	7	14	5	15	9	

Fig. 3. Finding the frequent prefix and suffix combination of [d e] from order occurrence matrix by picking out the row (column) with the maximum accumulated occurrence frequencies

Order matrix shows occurrence frequencies of any length-2 subsequences. Suppose original dataset is randomly constructed, orders between any two labels would be evenly distributed in the rows. That means the possibility of occurrence of "a" before "b" would be the same as that of "b" before "a". However if an OP-Cluster exists in data matrix some labels would have much higher frequencies of being prefix/suffix then the other labels in the global way. This statistical information is useful to find the frequent orders shared by a significant number of gene vectors.

Step 3. Finding the most frequent prefix and suffix. We accumulate the total number of occurrences of each label being prefix (suffix) of another labels in all rows. It is the summation of the corresponding row (column) of order matrix O. The most frequent prefix (suffix) is the label with the maximum number of accumulated occurrences of being prefix (suffix). In the example of Figure 3 "d" and "e" is the most frequent prefix and suffix respectively. We combine the most frequent prefix and suffix to form an initial frequent order candidate which has higher occurrence frequency in the sequences set. So [d e] is the seed for growing the frequent order in this example.

Only the supporting rows for this order is kept. Other rows which do not support the order is removed. And we extract subsequence between the current prefix and suffix from the remaining rows. Figure 4(a) shows the 3 subsequences

ID	Subsequences				
1		d	a	(c b)	e
2	d	a	b	c	e
3			d	a	c	e
4						

O'	a	b	c	d	e	f	
a	0	2	3	0	0	0	5
b	0	0	2	0	0	0	2
c	0	1	0	0	0	0	1
d	0	0	0	0	0	0	0
e	0	0	0	0	0	0	0
f	0	0	0	0	0	0	0
	0	3	5	0	0	0	

(a)   (b)

Fig. 4. (a) Set of the extracted subsequences, (b) updated order matrix

"acb", "abc" and "ac". Then we go back to step 2 and update order matrix O based on the remain subsequences as shown in Figure 4(b). In the second run [a c] is found as the new prefix and suffix combination. Since order [d e] is supported by all remaining rows and "d" is the first prefix ("e" is the last suffix) of all labels in the subsequences, [a c] is put between the former prefix and suffix to form an enlarged frequent order of [d a c e]. Step 2 and 3 could be repeated when the minimum support (will be discussed later) is satisfied. By iteratively finding the most frequent prefix (suffix) and combining them we enlarge frequent order candidate step by step. Due to the noise in data we could miss some prefix or suffix of target frequent order during iterations. Enhancement can be made by repeating the whole procedures on the remaining rows and all columns.

3.2   Minimum Support for the Frequent Order

Some researchers choose an exact value as the minimum support While we use $p(t)$ as the minimum support criterion,

$$p(t) = \frac{1 - p_{ini}}{ln(t)\_max} ln(t) + p_{ini} \tag{1}$$

where t stands for the current iteration. In the first iteration $p(1) = p_{ini}$ and $p(t) \leq 1$. $p(t)$ increases with the iteration. t_max stands for the maximum value of t. If original dataset has m columns, the maximum value of the iteration would be $m/2$.

We compute the portion of number of supports for enlarged frequent order to number of current remaining rows. Iteration goes on only when the value is bigger than $p(t)$. Since we delete the non-supporting rows after each iteration, number of remaining rows keeps decreasing. If non-frequent sub-prefix or sub-suffix is added to the frequent order candidate number of remaining supports would decrease by a large number. In this way we reject non-frequent orders. To loosen the minimum support requirement one can use $p(t)^2$ instead.

ID	Sequences					
1	f	d	a	c	b	e
2	d	a	b	e	c	f
3	c	b	d	a	f	e
4	b	e	d	c	f	a

O'	a	b	c	d	e	f	
a	0	2	1	0	0	1	4
b	0	0	0	0	0	0	0
c	0	1	0	0	0	0	1
d	0	0	0	0	0	0	0
e	0	0	0	0	0	0	0
f	0	0	0	0	0	0	0
	0	3	1	0	0	1	

(a)   (b)

Fig. 5. (a) Updated sequence set (the gray elements construct an OP-Cluster), (b) updated occurrence matrix

3.3 Find Proper Single Label When No Frequent Combination of Prefix and Suffix Exists

When target frequent order has an odd number of labels, it can not be discovered by including both prefixes and suffixes to the frequent order iteratively. The hidden OP-Cluster in Figure 6(a)has frequent order of [d a e]. [d e] would be found in the first interation and [a b] in the second iteration. While for order [d a b e] there are only two supports. That could lead to a non-frequent order. So when no significant prefix and suffix combination could be found, we count the occurrence frequencies of all labels in remaining subsequences. Column with the maximum number of appearance would be checked to see whether it could be added into the candidate order. So "a" is added to form the frequent order [d a e] and result in a 3×3 OP-Cluster when satisfying the minimum support.

3.4 Find Multiple OP-Clusters

We repeat the whole algorithm and whenever we find a new frequent order candidate we compare it to all existing frequent orders. If the candidate is not different from the existing frequent orders (see Definition 3), we ignore it and update the occurrence matrix O then repeat until new candidate which is different from the existing ones is found. Theoretically as iteration goes on, all label combinations could be tested. But frequent orders would be processed first and non-frequent sequences would be rejected at very early stage without much processing. This scheme reduces a large proportion of computation cost. There is rarely multiply operation in our algorithm. The major cost is on computing matrix O by counting the accumulated occurrence. Cost for the whole algorithm is hard to estimate since number of iteration varies a lot for different datasets. In step 1 the sorting requires time in $O(nm)$ and only take place for once. Calculation of matrix O in the first run is an $O(m^2n)$ effort. This cost drops dramatically in following iterations since number of supporting rows and length of subsequences decrease a lot. Space complexity of the whole algorithm is $O(nm)$ which is very small comparing to most of the existing methods.

4   Experiments

The algorithm is tested with both synthetically generated datasets and real gene expression datasets for effectiveness and efficiency. The experiments are implemented with MATLAB and executed on a computer with a 3.2 GHz and 0.99 GB main memory.

Size of OP-Cluster, which means number of columns and rows involved, is the measurement for its significance. Also length of potential frequent order and number of its supports are two critical factors that affect the performance of our algorithms. However, the best OP-Cluster is hard to define since shorter orders are supported by much more genes. For n rows with length m, the probability of finding at least k supports for any order with length m is,

$$P(n, m, k) = \sum_{i=k}^{n} (\frac{1}{m!})(\frac{m! - 1}{m!})^{n-i} C_n^i \qquad (2)$$

P decreases with the increasing of m and k. P could be used to measure the significance of an OP-Cluster. The smaller P is, the more significant the OP-Cluster is.

4.1   Synthetic Data

We generate the synthetic data in this way: Firstly, random datasets were generated. Then OP-Clusters were inserted manually into the datasets. In the following experiment each case has been implemented for 10 times. We get the average value of all runs.

Sometimes instead of finding the exact manually inserted OP-Cluster we find OP-Clusters which overlap with the inserted one. These OP-Clusters could be formed by chance when other conditions were included in the frequent order. Suppose a $k \times l$ OP-Cluster was inserted and an OP-Cluster, which has p rows and q columns in common with the inserted one, is found. The found OP-Cluster could have more conditions than the inserted one. In that case it may have fewer supports than the inserted one. We define the accuracy to be,

$$accuracy = (\frac{p}{k} + \frac{q}{l})/2 \qquad (3)$$

When the exact OP-Cluster is found, the accuracy is 1. In other cases the accuracy is proportional to the volume of the overlapping part between the embedded OP-Cluster and the found one.

Test For Effectiveness. Number of supporting rows and length of frequent order relative to the size of dataset are the two important factors which effect the

Table 1. Result on varying the size of the dataset and the inserted OP-Cluster

	100×30	200×30	500×30
20×15	100%	100%	96.5%
40×10	100%	100%	100%
200×15	–	100%	100%

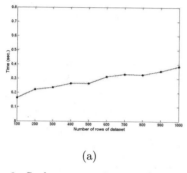

(a)

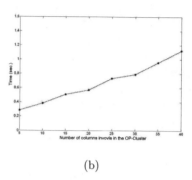

(b)

Fig. 6. Scale-up experiments. Response time V.S. (a) number of rows of dataset, (b) number of columns involved in OP-Cluster.

performance. To test the effectiveness we run the algorithm with combinations of different sizes of datasets and OP-Clusters. For the OP-Clusters we also change the ratio of rows to columns.

Table 1 shows the result. Rows of the table show the size of inserted cluster and columns show the size of the original dataset. Average accuracy of each case of 10 runs was shown. Our method works very well in nearly all cases, especially when size of inserted OP-Cluster is significant to size of original dataset. 20×15 OP-Cluster was exactly found for 9 out of the 10 runs when it is inserted into a 500×30 data matrix. And in another case our algorithm finds a 6×16 OP-Cluster, which covers all the columns involved in the inserted OP-Cluster.

Test For Scalability. The scale-up properties of the algorithm were analyzed by varying size of dataset and embedded cluster. We report the time cost for the first run. Figure 6(a) shows the result when varying the number of rows of dataset. A 50×20 OP-Cluster was inserted into datasets which have 50 columns but increasing number of rows from 100 to 1000. Figure 6(b) shows the result when increasing the number of columns involve in the OP-Cluster. An OP-Cluster with 200 supports was inserted into 1000×50 datasets. The number of conditions involve was increased from 5 to 40. Our algorithms got high precision almost in all cases and scale linearly with size of dataset and size of OP-Cluster.

4.2 Microarray Data

In addition to simulated datasets, we run our algorithm on the breast tumor dataset from Chen et al. [17]. This dataset has 3,226 genes and 22 tissues. Among

Table 2. Comparison of result significance with OPSM algorithm and OPC-Tree algorithm. (N/A: not available)

Number of tissues	Number of Max supporting rows		
	CFPS	OPSM	OPC-Tree
4	771	347	690
5	142	N/A	N/A
6	124	42	126
7	32	8	47

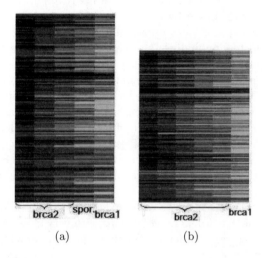

(a)   (b)

Fig. 7. (a) The largest 5-tissue OP-Cluster, (b) the largest 6-tissue OP-Cluster

the 22 tissues, there are 7 brca1 mutations, 8 brca2 mutations and 7 sporadic breast tumors. Our experiments demonstrate the power of finding biological OP-Clusters in the gene expression data. We compared our result with that from the OPSM algorithms of Ben-Dor et al. [9] and OPC-Tree of Liu et al. [13].

Firstly, we report the significance of our clusters in table 2 with comparisons with other two algorithms. Our clustering algorithm generates much more significant clusters than OPSM. It costs only 7.17 seconds for the finding the first 20 OP-Clusters. These OP-Clusters cover all the 22 tissues and 74.3% of the genes. Ben-Dor et al. reported only three clusters with size 4×347, 6×42 and 8×7 respectively. However, our OP-Clustering algorithm was able to find 4 tissues clusters supported by a maximum of 771 genes, which doubles the number of genes of the result of OPSM and also outperformed OPC-Tree. Our algorithm also found 5-tissue clusters with a maximum support of 142 genes. $P(5, 142, 3226) \approx 1.1659e - 55$ that means the cluster has very high significance. 5-tissue clusters were not reported by OPSM and OPC-Tree. The order imposed on these five tissues are interesting, 3 brca2 mutations show lower expression, 1 brca2 mutation in the middle and 1 sporadic breast tumor shows the highest expression. This result suggest these 142 genes has different expression levels in these tissues. Figure 7 shows the largest 5-tissue and 6-tissue OP-Clusters and type of the tissues involved. No OP-Clusters with more than 7 tissues were reported. Since our algorithm is good at finding the statistical majority. Larger clusters would be found first. And we only process the first 20 clusters, small cluster would lose the chance of being found.

5   Discussion

Order preserving clustering has been used in many applications to capture the consistent tendency exhibited by a subset of objects in a subset of dimensions in high dimensional space. We proposed a heuristic approach CFPS which discovers the OP-Clusters by finding the frequent orders using a top-down scheme in a statistical way. The method is easy to use with a low computation and space cost. Few parameter has to be initialized for the algorithm. We define the significance of an OP-Cluster and by it we can discriminate the meaningless OP-Clusters constructed by chance. The algorithm works very well in the experiment. It scale linearly to the size of the dataset and the size of the clusters. For the real gene expression profiles our method outperform OPSM and OPC-Tree in finding significant clusters. But the nature of NP-hardness of this problem implies that there may be sizeable OP-Clusters evading the search by any efficient algorithm.

There are several extensions we can make based on our algorithm. Although we permit exchangeable orders for conditions with very close values, the requirement on exactly the same order is still sharp in some applications, especially when noise or outlier exists. One extension of the current model is to explore similar but not exact the same order among conditions. There are many overlapping OP-Clusters in our result. Further steps could be taken to merge some of the overlapping clusters to form new clusters with more columns.

References

1. Cheng, Y., Churhc, G.: Biclustering of expression data. In: ISMB'00, pp. 93–103. ACM Press, New York (2000)
2. Tanay, A., Sharan, R., Shamir, R.: Discovering statistically significant biclusters in gene expression data. IEEE Transactions on Knowledge and Data Engineering 18, 136–144 (2002)
3. Wang, H., Wang, W., Yang, J., Yu, P.: Clustering by pattern similarity in large data sets. In: ACM SIGMOD Conference on Management of Data'02, pp. 394–405 (2002)
4. Yang, J., Wang, W., Wang, H., Yu, P.: δ-clustering: Capturing subspace correlation in a large data set. In: 18th IEEE Int'l. Conf. Data Eng., pp. 517–528 (2002)
5. Bleuler, S., Prelic, A., Zitzler, E.: An ea framework for biclustering of gene expression data. In: Congress on Evolutionary Computation'04, pp. 166–173 (2004)
6. Cho, H., Dhillon, I.S., Guan, Y., Sra, S.: Minimum sum-squared residue cococlustering of gene expression data. In: Fourth SIAM Int'l. Conf. Data Mining (2004)

7. Teng, L., Chan, L.: Biclustering gene expression profiles by alternately sorting with weighted correlation coefficient. In: IEEE International Workshop on Machine Learning for Signal Processing'06 (2006)
8. Madeira, S., Oliveira, A.: Biclustering algorithms for biological data analysis: A survey. IEEE/ACM Transactions on Computational Biology and Bioinformatics 1, 24–45 (2004)
9. Ben-Dor, A., Chor, B., Karp, R., Yakhini, Z.: Discovering local structure in gene expression data: The order-preserving submatrix problem. In: RECOMB'02. ACM Press, New York (2002)
10. Agrawal, R., Srikant, R.: Mining sequential patterns. In: 11th International Conference on Data Engineering, pp. 3–14 (1995)
11. Han, J., Pei, J., Yin, J.: Mining frequent frequent patterns without candidate generation. In: ISMB'00 ACM SIGMOD Conference on Management of Data'02, pp. 1–12 (2000)
12. Pei, J., Han, J., Mortazavi-Asl, B., Wang, J., Pinto, H., Chen, Q., Dayal, U., Hsu, M.: Mining sequential patterns by pattern-growth: The prefixspan approach. IEEE Transactions on Knowledge and Data Engineering 16, 1424–1440 (2004)
13. Liu, J., Yang, J., Wang, W.: Biclustering in gene expression data by tendency. In: IEEE Computational Systems Bioinformatics Conference, pp. 182–193. IEEE Computer Society Press, Los Alamitos (2004)
14. Hipp, J., Guntzer, U., Nakhaeizadeh, G.: Algorithms for association rule mining-a general survey and comparison. SIGKDD Explorations 2, 58–64 (2000)
15. Bleuler, S., Zitzler, E.: Order preserving clustering over multiple time course experiments. In: EvoBIO'05, pp. 33–43 (2005)
16. Agrawal, R., Srikant, R.: Fast algorithms for mining association rules. In: 20th Int'l. Conf. Very Large Data Bases, pp. 487–499 (1994)
17. Chen Y., Radmacher, M., Bittner, M., Simon, R., Meltzer, P.: Gene expression profiles in hereditary breast cancer. NEJM 344, 539–548 (2001).

Correlation-Based Relevancy and Redundancy Measures for Efficient Gene Selection

Kezhi Z. Mao and Wenyin Tang

School of Electrical & Electronic Engineering
Nanyang Technological University
Singapore 639798

Abstract. The gene-label correlation provides an effective measure of the relevancy of a gene. However, this measure evaluates genes on an individual basis, and the gene sets thus obtained may exhibit severe redundancy. In this study, we propose a new correlation heuristic for set-based gene selection, with the goal of alleviating the redundancy problem. The new correlation heuristic consists of two components that account for gene relevancy and redundancy respectively. The relevancy of a gene is evaluated in terms of its correlation with class label on an individual basis, while the redundancy of a gene with respect to a given gene subset is measured by its correlation with a new dimension built upon the gene subset. The new correlation heuristic retains the simplicity of individual gene evaluation and the redundancy handling capacity of set-based gene evaluation. Two different ways of using the relevancy and redundancy measures are presented in this study. One way is the maximization of the ratio of relevancy measure to redundancy measure, and another way is the maximization of the relevancy measure subtracting redundancy measure. Experimental studies on six gene expression problems show that both criteria produce excellent results.

1 Introduction

Gene selection has been an active research area since the birth of the gene microarray technology, and a variety of gene selection algorithms have been proposed. The various gene selection algorithms can be classified into two categories, namely individual gene selection (see for example [8,4,7,15]) and gene subset selection (see for example [14,11,6,10,12,21,20,1]). The two types of gene selection algorithms often serve different purposes. If gene selection is for efficient pattern classification or class prediction, subset-based gene selection should be employed. This is because a gene subset consisting of top individually ranked genes may far from optimal due to the severe redundancy existed. Whatever category a gene selection algorithm belongs to, it involves an evaluation criterion to measure the goodness of an individual gene or a subset of genes. A variety of evaluation criteria have been used in the gene selection algorithms mentioned above, motivated by different considerations. These include t-test, F-test, Fisher ratio, entropy, cross validation error, Bayesian error estimation, loss functions of regression, and support vector machine (SVM) criteria etc.

J.C. Rajapakse, B. Schmidt, and G. Volkert (Eds.): PRIB 2007, LNBI 4774, pp. 230–241, 2007.
© Springer-Verlag Berlin Heidelberg 2007

Correlation measures have also been used for gene evaluation and selection. To minimize gene redundancy, one available correlation measure is the set-based correlation heuristic proposed by [13], where the merit of a feature subset is evaluated using the ratio of the average feature-label correlation to the average feature-feature correlation. Similar measures were also proposed in [3]. Another correltaion-based algorithm is the two-phase relevancy-redundancy analysis proposed by [19], where relevant genes are first selected through individual relevancy analysis, and redundant genes are then removed through Markov blanket-based redundancy analysis. But our experiment studies show that this algorithm could over-prune and the number of genes finally obtained might be insufficient.

In this study, we propose a new correlation heuristic for forward, *i.e.* bottom-up, gene selection. The new correlation consists of two components accounting for relevancy and redundancy respectively. The relevancy of a gene is evaluated individually in terms of its correlation with class label, while the redundancy of a gene with respect to a given gene subset is measured by its correlation with the output of the classifier built upon the gene subset. This way of evaluating redundancy is an outstanding character of the new correlation heuristic. The rationale lies in the fact that the major discriminative information underlying the gene subset is captured by the classifier, and thus the correlation between the candidate gene and the output of the classifier reflects the redundancy of the candidate gene with respect to the gene subset. Two ways of using relevancy and redundancy measures are presented. One is the ratio of relevancy measure to redundancy measure, and another is the relevancy subtracting redundancy. Through maximizing the two criteria, genes with high relevancy and minor redundancy could be selected.

The new correlation heuristic inherits the simplicity of individual gene evaluation and the redundancy handling capacity of set-based evaluation. Experimental studies show that both criteria produce excellent results.

2 Correlation-Based Relevancy and Redundancy Measures for Gene Selection

2.1 Relevancy and Redundancy Measures

Assume there are N training data pairs:

$$\{\mathbf{x}(1), y(1)\}, \{\mathbf{x}(2), y(2)\}, \ldots, \{\mathbf{x}(N), y(N)\}$$

where $y(k)$ denotes the class label of sample k, with value of either $+1$ or -1. $\mathbf{x}(k)$ is the feature vector of sample k consisting of n genes:

$$\mathbf{x}(k) = [x_1(k), x_2(k), \ldots, x_n(k)]$$

The gene-label correlation is defined as the correlation between a gene and the class label:

$$r_{yx_i} = \frac{1}{N-1} \frac{\sum_{k=1}^{N} x_i(k)y(k)}{\sigma_{x_i}\sigma_y} \tag{1}$$

where σ_{x_i} and σ_y denote the standard deviation of gene x_i and class label y respectively. The gene-label correlation reflects the predictive power, or relevancy, of a gene and could be used to identify biologically related genes of certain biological phenomenon of interest. However, the correlation criterion Eqn (1) evaluates genes on an individual basis, without considering correlations between genes. Severe redundancy might exist if it is used to select gene subsets. To achieve good pattern classification results, an ideal gene subset should possess the following properties:

(i) having maximum relevancy;
(ii) having minimum redundancy.

To yield gene subsets with maximum relevancy and minimum redundancy, we can select gene subsets that maximizes the ratio of relevancy measure to redundancy measure or the difference between the two measures [13,3].

In a forward gene selection algorithm, the gene subset is built up step by step, by adding one gene at one step. Assume m genes have already been selected: $s_m = \{x_1, x_2, \ldots, x_m\}$, the objective is to select the next best gene. To select the gene with maximum relevancy and minimum redundancy, we can evaluate and select genes using the following criteria

$$J_1 = \frac{R_{yx_i}}{R_{s_m x_i}} \qquad (2)$$

or

$$J_2 = R_{yx_i} - R_{s_m x_i} \qquad (3)$$

where R_{yx_i} denotes the relevancy measure of gene x_i, and $R_{s_m x_i}$ denotes redundancy measure of gene x_i with respect to gene subset s_m. The gene with the maximum J_1 or J_2 should be selected.

The relevancy of a gene can be easily measured in terms of its correlation with class label as in Eqn (1) or other measures such as Fisher ratio. The major issue here is how to evaluate the redundancy of x_i with respect to the given subset s_m. In [13] and [3], the redundancy is measured in terms of the average correlation between candidate x_i and those in the gene subset selected s_m. Next, we propose a new approach to redundancy evaluation.

2.2 A New Approach to Redundancy Evaluation

The basic idea of the new way of evaluating redundancy of a candidate gene with respect to gene subset s_m is to project data from the m-dimensional space to a new one-dimensional space using a linear transform, and then measure the redundancy of a candidate gene based on its correlation with the new dimension. Assume the linear transform is given by:

$$z_m(k) = \sum_{j=1}^{m} w_j x_j(k) \qquad (4)$$

where w_j, $j = 1, 2 \ldots, m$ are the coefficients of the linear transform. Eqn(4) is such a transform that the major discriminative information underlying the m dimensions, *i.e.* m genes in s_m, is compressed onto z_m. The linear transform that projects data from m-dimensional space to one-dimensional space can be obtained by the support vector machine (SVM) method because the SVM classifier captures the major discriminative power underlying s_m.

The redundancy of x_i with respect to gene subset s_m is measured using the correlation between x_i and z_m. The rationale of the new way of evaluating redundancy can be explained from the point of view of variable selection in multiple regression. Assume the regression of class label on the m features in s_m is as Eqn (4), then the resultant regression error is given by:

$$e(k) = y(k) - z_m(k) \tag{5}$$

The variable to be selected next should have maximum correlation with the regression error. Assume

$$\mathbf{y} = [y(1), y(2), \ldots, y(N)]^T$$

$$\mathbf{e} = [e(1), e(2), \ldots, e(N)]^T$$

$$\mathbf{z_m} = [z_m(1), z_m(2), \ldots, z_m(N)]^T$$

$$\mathbf{x}_i = [x_i(1), x_i(2), \ldots, x_i(N)]^T$$

The correlation between x_i and e, denoted by r_{ex_i} is given by:

$$
\begin{aligned}
r_{ex_i} &= \frac{1}{N-1} \frac{\mathbf{x}_i^T \mathbf{e}}{\sigma_e \sigma_{x_i}} \\
&= \frac{1}{N-1} \frac{\mathbf{x}_i^T \mathbf{y} - \mathbf{x}_i^T \mathbf{z}_m}{\sigma_e \sigma_{x_i}}
\end{aligned}
\tag{6}
$$

where σ_e denote the standard deviation of error signal e. If genes, class label and sample projections on the new dimension are normalised to zero mean and unit standard deviation, Eqn (6) can be written as

$$r_{ex_i} = \frac{1}{\sigma_e}[r_{yx_i} - r_{z_m x_i}] \tag{7}$$

where r_{yx_i} and $r_{z_m x_i}$ denotes the correlations between x_i and class label and the output of the classifier respectively. To ensure the minimum regression error after adding the new feature, selection of the new feature should be based on maximization of r_{ex_i}. A comparison of Eqn (7) with Eqn (3) shows that if the correlation between x_i and class label is used to evaluate the relevancy of x_i, then the redundancy $R_{s_m x_i}$ can be measured using the correlation between gene x_i and the output of the classifier built upon s_m.

The heuristic J_1 and J_2 can be rewritten as:

$$J_1 = \frac{|\mathbf{y}^T \mathbf{x}_i|}{|\mathbf{z}^T \mathbf{x}_i|} \tag{8}$$

$$J_2 = |\mathbf{y}^T \mathbf{x}_i| - |\mathbf{z}^T \mathbf{x}_i| \tag{9}$$

where $|.|$ denotes the absolute value. This is because the correlations can take both positive or negative values.

J_2 actually can be modified by putting a weighting element on the redundancy measure:

$$J_2^* = |\mathbf{y}^T \mathbf{x}_i| - \lambda |\mathbf{z}^T \mathbf{x}_i| \tag{10}$$

where λ denotes the weighting element.

The main characteristic of the present study is that the redundancy of a gene with respect to a gene subset selected is measured using the correlation between the gene and a new dimension built upon the gene subset. An important issue here is how to create the new dimension. As analysed above, the correlation measure Eqn (3) is equivalent to regression error based feature evaluation when the role of the previously selected features is controlled. This suggest that we may control the effect of the previously selected gene subsets when a new dimension is created after a new gene is added. This is briefly described below. A new dimension, named z_2, is first created using x_1 and x_2. Selection of the third gene is based on the correlation criteria where the redundancy of a candidate gene is measured using the correlation between the candidate gene and z_2. After the 3rd gene, say x_3 is selected, a new dimension z_3 is created using x_3 and z_2. In this process, the creation of a new dimension is always done in a 2-dimensional space. And the creation can be based on different approaches such as support vector machine (SVM).

Due to small sample size and very high dimensionality in gene expression data, the training data could be mapped to the class label. Thus, the redundancy measure would approaches the relevancy measure and a zero value of the criterion would be obtained. To overcome this problem, the new dimension created at each step is rotated by an angle. Taking z_{m-1}, x_i and z_m as an example, where z_m is created by z_{m-1} and x_i.

$$z_m(k) = w_{m1} z_{m-1}(k) + w_{m2} x_i(k) \tag{11}$$

Taking the z_{m-1} as an reference, the angle of the new dimension is given by:

$$\alpha = \arctan\left(\frac{w_{m2}}{w_{m1}}\right) \tag{12}$$

After a few genes are selected, the sample projections on z_{m-1} are very close to class labels, and play dominant role in creating z_m. Thus, the value of w_{m1} has a much greater amplitude than w_{m2}, and the angle becomes very small. Hence we have:

$$\alpha \approx \frac{w_{m2}}{w_{m1}} \tag{13}$$

To rotate the new axis, we can reduce the value of w_{m1} to w_{m1}/γ, where $\gamma > 1$. Thus, the new angle is given by:

$$\beta \approx \gamma \frac{w_{m2}}{w_{m1}} = \gamma \alpha \tag{14}$$

The new dimension is usually obtained by optimizing certain criterion. The transform obtained is therefore optimal in the sense of maximum separating margin in support vector machine, maximum class separability in Fisher's linear discriminant analysis, and minimum regression error in least mean square estimation etc. The rotation introduce with deteriorate the optimality, and is therefore can be regarded as a regularization.

Criterion J_1 and J_2^* consist of two components. One component accounts for the relevance of the gene, and another component accounts for the redundancy of the gene with respect to gene subset s_m. The relevance is measured on an individual basis, while the redundancy is measured on a set basis. The merit of this way of evaluating a candidate gene is that it retains the simplicity of individual gene evaluation and the capacity of redundancy handling of set-based gene evaluation.

2.3   The Correlation Criteria-Based Gene Selection Algorithm

The procedure of forward gene selection based on the correlation J_1 and J_2^* is summarized below:

(i) Normalise data including class label to zero mean and unit standard deviation.

(ii) Evaluate the correlation between class label and each of the n genes in the candidate gene pool: $x_1, x_2, \ldots, x_n$. Identify the gene that has the maximum correlation measure, say x_j, add it to the gene subset and remove it from the candidate gene pool. Let $z = x_j$.

(iii) Evaluate the correlation between z and each of the $n - 1$ genes in the candidate gene pool, and calculate J_1 or J_2^* using Eqn (8) or (10). Identify the gene having the maximum measure, say x_i, add it to the gene subset and remove it from the candidate gene pool.

(iv) Train the linear SVM classifier using the genes in the gene subset selected and denote the decision value of classifier for the training samples as z. Normalise z to zero mean and unit standard deviation. Evaluate the correlation between z and each of the $n - 2$ genes in the candidate gene pool, and calculate J_1 or J_2^* using Eqn (8) or (10). Identify the gene having the maximum measure, say x_k, add it to the gene subset and remove it from the candidate gene pool.

(v) Step (iv) is repeated until a stopping criterion, say the number of genes selected, is satisfied.

To identify the $m + 1^{th}$ gene from a candidate gene pool of $n - m$ genes at step $m + 1$, the computations involved include training a linear classifier such as a linear support vector machine (SVM) once and performing $n - m$ vector product in N-dimensional space, where N is the training sample size. Apparently, the computational complexity of the proposed method is very limited.

3   Experimental Studies

In the experiment, the performance of the proposed correlation heuristic was studied. For comparison purpose, the two-phase relevancy-redundancy analysis proposed in [19] and set-based correlation heuristic proposed in [13] were also studied. In addition, the recursive feature elimination (RFE) algorithm [12], which is often considered as a benchmark algorithm, was also studied.

The performance of these gene selection algorithms was evaluated in terms of classification error rate. The study in [2] revealed that error estimation based on cross validation including leave-one-out and repeated k-fold cross validation may exhibit excessive variability. In this study, .632+ bootstrapping [5] was used. In the bootstrap testing, 200 replica were generated to estimate the error rate, and the splits of training and test data in the 200 replica were kept identical during the testing of the gene selection algorithms.

Six gene expression datasets were used to test the performance of the proposed algorithm. The eight datasets are summarized in Table 1:

Table 1. Datasets description

Datasets	Original sources	Genes
Leukaemia	[8]	7129
Breast cancer (ER)	[18]	7129
Breast cancer (LN)	[18]	7129
Lung cancer	[9]	12533
CNS tumour	[16]	7129
Breast cancer	[17]	24481

Each of these datasets was standardized to zero mean and unit standard deviation across genes. Since the dimensionality (*i.e.* the number of genes) of gene expression data is very high, and most of these genes are irrelevant to the discriminant task, a pre-selection procedure was employed to reduce the number of candidate genes to 1000 based on Fisher's ratio, which is an individual gene ranking criterion. All the experiments and comparisons in this work were conducted on the pre-selected data.

The experimental results on the 6 datasets are shown in Figures 1-6 respectively. On each dataset, 4 algorithms were tested, including the recursive feature elimination (RFE), correlation-based feature selection (CFS), and the two new correlation heuristics Eqn (8) and Eqn (9), named as CH1 and CH2 respectively. In the experimental study, the weight λ on the redundancy measure in criterion J_2^* was set to 2, and the weight on slack variable in RFE was set to a wide range of values, as small as 0.001 and as great as 100, but the results were almost identical.

Across the 6 problems, the two-phase relevancy-redundancy analysis produced gene subsets consisting of just a few genes since the Markov blanket principle removed most of the candidate genes while the other 3 algorithms used the number of genes selected as the stopping criterion. As shown in Figures 1-6, the

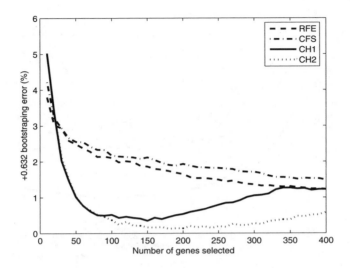

Fig. 1. Comparison of RM and RRM with RFE in Leukaemia problem

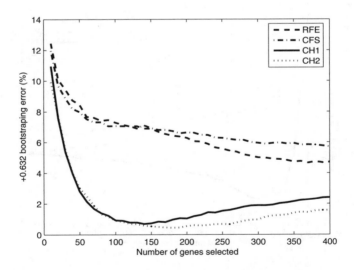

Fig. 2. Comparison of RM and RRM with RFE in Breast Cancer (ER) problem

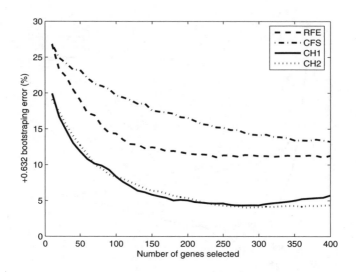

Fig. 3. Comparison of RM and RRM with RFE in Breast Cancer (LN) problem

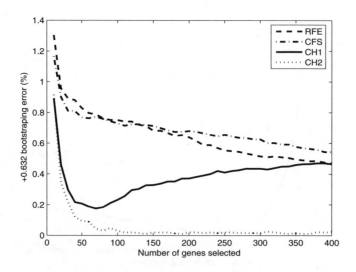

Fig. 4. Comparison of RM and RRM with RFE in Lung Cancer problem

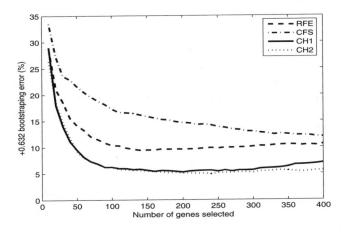

Fig. 5. Comparison of RM and RRM with RFE in CNS Tumor problem

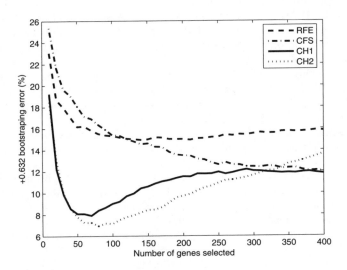

Fig. 6. Comparison of RM and RRM with RFE in Breast Cancer problem

RFE algorithm outperform the CFS algorithm in all the 6 problems. However, CH1 and CH2 outperform both CFS and RFE substantially. The results of CH2 are a bit inferior to those of CH1, this is probably because the introduction of the weight element λ improves the adaptability and flexibility of the correlation heuristic.

4   Conclusions

In this study, we have proposed a new correlation heuristic for efficient gene selection, where relevancy and redundancy components of a gene are considered explicitly in merit evaluation. Two formulae have been presented by different way of combining the two components. The proposed correlation heuristic retains the simplicity of individual gene evaluation and the capacity of redundancy handling of set-based gene evaluation. Experimental studies have shown that the correlation heuristic produces gene subsets leading to excellent classification accuracy.

References

1. Braga-Neto, U., Dougherty, E.R.: Bolstered error estimation. Pattern Recognition 37(6), 1267–1281 (2004a)
2. Braga-Neto, U.M., Dougherty, E.R.: Is cross-validation valid for small-sample microarray classification? Bioinformatics 20(3), 374–380 (2004b)
3. Ding, C., Peng, H.: Minimum redundancy feature selection from microarray gene expression data. In: Proceedings of 2nd IEEE Computer Society Bioinformatics Conference. IEEE Computer Society Press, Los Alamitos (2003a)
4. Dudoit, S., Fridyand, J., Speed, T.P.: Comparison of discrimination methods for the classification of tumors using gene expression data. Journal of the American Statistical Association 97, 77–87 (2002)
5. Efron, B., Tibshirani, R.: Improvements on cross-validation: the.632+ bootstrap method. Journal of the American Statistical Association 92(438), 548–560 (1997)
6. Fan, L., Yang, Y.: Analysis of recursive gene selection approaches from microarray data. Bioinformatics 21(19), 3741–3747 (2005)
7. Furlanello, C., Serafini, M., Merler, S., Jurman, G.: Entropy-based gene ranking without selection bias for the predictive classification of microarray data. BMC Bioinformatics 4(54) (2003)
8. Golub, T., Slonim, D., Tamayo, P., Huard, C., Gaasenbeek, M., Mesirov, J., Coller, H., Loh, M., Downing, J., Caligiuri, M., Bloomfield, C., Lander, E.: Molecular classification of cancer: class discovery and class prediction by gene expression monitoring. Science 286, 531–537 (1999)
9. Gordon, G.J., Jensen, R.V., Hsiao, L.-L., Gullans, S.R., Blumenstock, J.E., Ramaswamy, S., Richards, W.G., Sugarbaker, D.J., Bueno, R.: Translation of microarray data into clinically relevant cancer diagnostic tests using gene expression ratios in lung cancer and mesothelioma. Cancer Research 62 (2002)
10. Guan, Z., Zhao, H.: A semiparametric approach for marker gene selection based on gene expression data. Bioinformatics 21(4), 529–536 (2005)
11. Gui, J., Li, H.: Penalized cox regression analysis in the high-dimensional and low-sample size settings, with applications to microarray gene expression data. Bioinformatics 21(13), 3001–3008 (2005)
12. Guyon, I., Weston, J., Barnhill, S., Vapnik, V.: Gene selection for cancer classification using support vector machines. Machine Learning 46(1-3), 389–422 (2002)
13. Hall, M.: Correlation-based feature selection for discrete and numeric class machine learning. In: Proceedings of Seventeenth International Conference on Machine Learning, San Francisco, CA, USA (2000)

14. Li, Y., Campbell, C., Tipping, M.: Bayesian automatic relevance determination algorithms for classifying gene expression data. Bioinformatics 18(10), 1332–1339 (2002)
15. Liu, X., Krishnan, A., Mondry, A.: Entropy-based gene selection for cancer classification using microarray data. BMC Bioinformatics 6(76) (2005)
16. Pomeroy, S.L.: Prediction of central nervous system embryonal tumour outcome based on gene expression. Nature 415 (2002)
17. van't Veer, Dai, H., van de Vijver, He, Y.D., Hart, A.A., Mao, M., Peterse, H.L., van der Kooy, Marton, M.J., Witteveen, A.T., Schreiber, G.J., Kerkhoven, R.M., Roberts, C., Linsley, P.S., Bernards, R., Friend, S.H.: Gene expression profiling predicts clinical outcome of breast cancer. Nature 415 (2002)
18. West, M., Blanchette, C., Dressman, H., Huang, E., Ishida, S., Spang, R., Zuzan, H., Olson, J.A., Marks, J.R., Nevins, J.R.: Predicting the clinical status of human breast cancer by using gene expression profiles. Proc. Natl. Acad. Sci. USA 98(20), 11462–11467 (2001)
19. Yu, L., Liu, H.: Efficient feature selection via analysis of relevance and redundancy. Journal of Machine Learning Research 5 (2004)
20. Zhang, H.H., Ahn, J., Lin, X., Park, C.: Gene selection using support vector machines with non-convex penalty. Bioinformatics 22(1), 88–95 (2006)
21. Zhou, X., Mao, K.Z.: Ls bound based gene selection for dna microarray data. Bioinformatics 21(8), 1559–1564 (2005)

SVM-RFE with Relevancy and Redundancy Criteria for Gene Selection

Piyushkumar A. Mundra[1] and Jagath C. Rajapakse[1,2]

[1] Bioinformatics Research Center, School of Computer Engineering,
Nanyang Technological University, 50 Nanyang Avenue,
Singapore 639798
[2] Singapore-MIT Alliance, N2-B2C-15, 50 Nanyang Avenue, Singapore
asjagath@ntu.edu.sg

Abstract. This paper introduces a novel gene selection method incorporating mutual information in the support vector machine recursive feature elimination (SVM-RFE). We incorporate an additional term of mutual information based minimum redundancy maximum relevancy criteria along with feature weight calculated by SVM algorithm. We tested proposed method on colon cancer and leukemia cancer gene expression dataset. The results show that the proposed method performs better than the original SVM-RFE method. The selected gene subset has better classification accuracy and better generalization capability.

Keywords: Gene selection, mutual information, minimum redundancy, maximum relevancy, SVM-RFE, cancer classification.

1 Introduction

DNA-microarray has emerged as a very powerful method to analyze gene expression of cells. This high throughput technology enables simultaneous monitoring of expression level of thousand of genes and hence results in a vast pool of data. Detecting differences among the gene expressions can be very useful in disease diagnosis and distinction of specific tumor type. Most of gene expression datasets contain small number of samples and very high number of genes. For accurate classification, it is extremely imperative to select relevant genes. Because it is possible that totally irrelevant genes are selected, the classifier still produces very high classification accuracy.

Broadly, two approaches of gene selections appear in machine learning and bioinformatics literature: the filter and wrapper methods [1-2]. Filter methods are purely based on the statistical correlations and independent of the classifier used. They evaluate the goodness of the feature subset only by intrinsic characteristic of the data. Based on the relation of each single gene with class labels by the calculation of simple statistical measures computed from the empirical distribution, feature ranking is performed. Some of the statistical measures are Shannon-entropy, Euclidean distance, Kolmogorov-dependence, t-score, P-metric, mutual information etc [3].

On the other hand, wrapper methods rank features based on their effect on the classification accuracy. In this method, the feature selected will be highly dependent

J.C. Rajapakse, B. Schmidt, and G. Volkert (Eds.): PRIB 2007, LNBI 4774, pp. 242–252, 2007.

on the classification algorithm used. It is claimed by many authors that wrapper approach obtains better subset of predictive genes than filter approach [3]. Both filter and wrapper methods have their own advantages and disadvantages. Filters are usually simple and have less computational cost but fail to provide a small subset of genes. Wrappers are more computationally complex and gene subset selected may not generalize with other classification algorithm as it is highly dependent on the classification algorithm used in feature ranking. Different wrapper approaches are proposed by various authors [4-10].

Support Vector Machine - Recursive Feature Elimination (SVM-RFE) is one of the most successful wrapper method based algorithm in the feature (gene) ranking and hence reduction in the dimensionality of the dataset [10]. Multiple SVM-RFE (MSVM-RFE) has shown improvement on the classification accuracy over SVM-RFE [7]. Similarly like SVM-RFE, [6] presented Recursive Cluster Elimination (RCE) algorithm. Though SVM-RFE is very powerful method, it does not ensure to select the genes which are maximally relevant to the class and at the same time possesses minimum redundancy among them as feature selected are highly dependent on the weights derived from SVM algorithm.

Maximum gene relevancy and minimum gene redundancy is very important for gene selection as it can result in more balanced coverage of the feature space, capturing broad characteristics of the dataset and improvement in the classification accuracy. Minimum Redundancy Maximum Relevancy (MRMR) algorithm was proposed for maximizing gene relevancy and minimizing the gene redundancy [11-13]. They had ranked all the genes and according to information theoretic criteria and selected the top ranked genes for the classification. Another approach in degree of differential prioritization (DDP) criteria was proposed to strike the balance between relevancy and redundancy [14].

In this paper, we propose a novel hybrid approach to incorporate MRMR criteria in SVM-RFE algorithm itself. Mutual information based additional term will be added along with the SVM-ranking criteria. This additional term will be useful in achieving maximum relevant and minimum redundant gene subset without sacrificing on classification accuracy. The resulting gene subset may represent whole gene expression dataset broadly and may have better generalization capabilities.

The rest of the paper is organized as follows: in Section II, we will review Minimum Redundancy and Maximum Relevancy (MRMR) criteria and SVM-RFE. In Section III, we will propose hybrid algorithm of MRMR with SVM-RFE. Section IV will discuss the experimental procedure to test the algorithm on various gene expression datasets and results. Finally, in Section V we analyzed the results and conclude the paper.

2 Method

2.1 Minimum Redundancy Maximum Relevancy (MRMR) Criteria

Here, this criteria attempts to find the subset of genes having maximal relevancy to target class and least redundant among themselves. If a gene is expressed randomly or uniformly in different class, its relevancy to the respective class will be zero, i.e. its mutual information with the class will be zero. Strongly expressed gene in one class will have larger mutual information with the respective class.

Let $S = \{x_i: i = 1,2,\ldots,n\}$ be subset of dataset where i is the gene and x_i is expression of gene i. Each x_i can be represented as $(x_{i1}, x_{i2},\ldots, x_{iJ})$, where x_{ij} is the expression of ith gene in j^{th} sample. If target classes are $c = \{c_1,c_2,\ldots,c_K\}$, where K is number of class, then $I(c;x_i)$ will quantify the relevance of gene i to the classification task. Thus the by maximizing the total relevance of all genes in subset S should be the maximum relevance criteria:

$$\max \ V_i, \quad V_i = \frac{1}{|S|}\sum_{x_i \in S} I(c; x_i) \tag{1}$$

Here, $|S|$ is number of genes in subset S.

It is possible that the features selected from the maximum relevancy criteria are highly redundant. Removal of features from this redundant set will not affect the discrimination power. The 'minimum redundancy' means select the features in such a way that they are mutually maximally dissimilar in the subset. Mutual information can also be used as a measure to find similarity between two features. Hence, following redundancy removal criteria can be used to achieve mutually exclusive features:

$$\min \ W_i, \quad W_i = \frac{1}{|S|^2} \sum_{x_i,x_j \in S} I(x_i, x_j) \tag{2}$$

It is necessary to optimize both maximum relevancy and minimum redundancy criteria to get the best feature subset. To achieve so, we will need a single objective function which can describe both the criteria. Such simplest objective criteria can be written as,

$$\max\left(\frac{V_i}{W_i}\right) \quad or \quad \max(V_i - W_i) \tag{3}$$

In present work, we have used the quotient objective criteria with SVM-RFE.

2.2 SVM-RFE

To select the genes for accurate cancer classification, SVM-RFE algorithm was proposed by [10]. The algorithm produces nested subset of the genes by backward elimination, starting with all the features and removing one feature in every iteration. Here, the feature removal is based on the SVM ranking criteria, the ith feature with the smallest ranking score $c_i = (w_i)^2$ is eliminated, where w_i is the corresponding weight of ith feature calculated from SVM.

The reason for choosing $c_i = (w_i)^2$ as ranking criteria is the feature removed by this criteria will have least change in the objective function. The objective function in the SVM-RFE is $J = \frac{1}{2}\|w\|^2$. Optimal Brain Damage (OBD) algorithm [15] has explained this effect. It approximates the change in objective function caused by removing the feature by second order Taylor series expansion of the objective function,

$$\Delta J(i) = \frac{\partial J}{\partial w_i} \Delta w_i + \frac{\partial^2 J}{\partial w_i^2} (\Delta w_i)^2 \tag{4}$$

At the optimum, first derivative can be neglected and using $J = \frac{1}{2}\|w\|^2$, equation (4) becomes

$$\Delta J(i) = (\Delta w_i)^2 \tag{5}$$

The SVM-Recursive feature elimination procedure can be described as follows:

Start: Ranked feature set $R = [\ ]$ and selected feature subset $S = [1, 2,..., n]$
    Repeat until all features are ranked
        a) Train linear SVM with feature set S in input variable

        b) Compute the weight vector

$$w = \sum_i \alpha_i y_i x_i$$

        c) Compute the ranking score of features

$$c_i = (w_i)^2$$

        d) Select the feature with smallest ranking score

$$e = \arg\min(c)$$

        e) Update $R = [e,R]; S = S - [e]$

Output: Ranked feature set R.

3 SVM-RFE with MRMR Criteria

The final subset obtained from the SVM-RFE algorithm may contain many redundant genes. Many biologically important genes may have lost because of less weight compare to these redundant features. We propose to integrate MRMR criteria with the weight criteria in SVM-RFE. MRMR criteria will make the subset less redundant and SVM-RFE weight will make sure the selected genes are useful in classification. The final selected gene subset will represent best mutually exclusive genes. The final dataset obtained will represent the whole dataset better than obtained by SVM-RFE alone.

The detailed SVM-RFE algorithm with MRMR criteria is discussed below:

Start: Ranked feature set $R = [\ ]$ and selected feature subset $S = [1, 2,..., n]$

    Repeat until all features are ranked
        a) Train linear SVM with feature set S in input variable and calculate the weight of
           each vector $w_{i,svm}$

$$w_{i,svm} = \sum_i \alpha_i y_i x_i$$

b) Calculate the class relevancy of each feature and mutual information among features using equation (2)

$$w_{i,MI} = \frac{I(c; x_i)}{\frac{1}{|S|} \sum_{x_i, x_j} I(x_i; x_j)}$$

c) Compute the weight vector

$$w_i = \frac{|w_{i,svm}|}{\max(w_{svm})} + \frac{w_{i,MI}}{\max(w_{MI})}$$

d) Compute the ranking score of features

$$c_i = (w_i)^2$$

e) Select the feature with smallest ranking score

$$e = \arg\min(c)$$

f)   Update $R = [e,R]$; $S = S - [e]$

Output: Ranked feature set R.

4   Experiments

4.1   Data

To evaluate the performance of MRMR based SVM-RFE, experiments were carried out on two most popular gene expression dataset, leukemia cancer dataset [16] and colon cancer dataset [17]. For the present study, we had only taken available training data of the leukemia dataset. These datasets were obtained from http://ligarto.org/rdiaz/Papers/rfVS/randomForestVarSel.html [9]. Both the dataset were further divided in two separate training and testing dataset. The details of the dataset are given in Table 1.

Table 1. Sizes of training and test sets, number of gene in two gene expression dataset

Dataset	Training Samples	Testing Samples	Total Number of Genes
Colon	42	20	2000
Leukemia	24	14	3051

4.2   Preprocessing

The dataset was randomly divided into training and testing set with maintaining the class ratio in both the sets. Training dataset was normalized to zero mean and unit variance. These continuous datasets were directly used in SVM-RFE after normalization.

It is difficult to find the mutual information of two continuous features. Hence for the simplicity of calculating mutual information, training dataset was discretized. Discretization will also help in the noise reduction. Mean (μ) and standard deviation (σ) of each individual gene expression variable was used to discretize the observation. Following criteria is then used to categorize the data: Data larger than $\mu + \sigma/2$ will be changed to state 2 ; Data in between $\mu + \sigma/2$ and $\mu - \sigma/2$ will be transformed to state 0 ; Data smaller than $\mu - \sigma/2$ will be transformed to state -2.

4.3 Parameter Estimation

SVMs performances depend upon its two critical hyperparameters, the kernel function and the regularization parameter C. It is imperative to select these parameters carefully. In present study, linear SVMs were used, which require only C parameter to tune. C values were chosen from finite set $\{2^{-20},....,2^{0},...,2^{15}\}$. This set was used for both recursive feature elimination (both from SVM-RFE and hybrid of mutual information and SVM-RFE) and performance evaluation.

To estimate the prediction generalization error, CV can be used. The resulting estimate of generalization error is often used as model selection criteria. Model that has the smallest generalization error are chosen. In k-fold CV, the data instances are divided into k – mutual folds with equal size. Model is trained with k-1 folds and tested on omitted fold. This average testing error, calculated by testing on each fold, represents the generalization error estimate. Another important variant of k-fold CV is 'Leave-one-out' method. In this method, k equals to the number of data instances. Classifier is built with all samples except one and tested on the omitted sample.

As sample size is small and class imbalance prevalent in most of the dataset, we used Matthew's Correlation Coefficient (MCC) with 10 fold cross validation. After each 10 fold CV, we summed the true positive (TP), true negative (TN), false positive (FP) and false negative (FN). These values were used to calculate MCC[1] parameter. MCC will vary between -1 to 1. Higher the MCC value means classifier has high sensitivity and specificity.

To increase the speed of the numerical simulations with both SVM-RFE and proposed hybrid method, we eliminate m features each time when number of features n is large in recursive feature subset S. If $n > 10000$, we choose $m = 100$, if $1000<n<10000$, m will be 10, and if $n < 1000$, $m = 1$.

4.4 Testing

It is necessary to check the validation accuracy of the classifiers as many times classifier fits training data extremely good but their prediction accuracy on unseen data may be very poor. However, the training and testing set of gene expression data are small and test error may not represent the true validation accuracy due to "unfortunate" partition of training and testing sets. To avoid such situation, we merge the training and testing datasets and then partition the total samples again in training

---

[1] $MCC = \dfrac{(TP)(TN) - (FP)(FN)}{\sqrt{(TP + FP)(TP + FN)(TN + FP)(TN + FN)}}$

and testing sets by random sampling. This process is performed 100 times and for each time, classifier is trained on the training set and tested on the corresponding testing set. The test error, sensitivity and specificity were computed for these 100 trials.

The feature ranking is carried out using only the training data. The goodness of feature subset is evaluated using linear SVM classifier trained with ranked genes as input variables. No data discretization was done for testing. For each gene expression dataset and method, we test feature subsets with number of genes ranging from 1 to 100. We take the gene subset with the least average test error as the best feature subset. This gene set is used to calculate the performance of the each method in terms of sensitivity and specificity on each gene expression dataset.

To compare the results of the proposed algorithm with MRMR filtering, we ranked the features using program available at http://research.janelia.org/peng/proj/mRMR/index.htm [13]. We obtained top 100 features from the training dataset for both the gene expression data. The classification accuracy of gene subset is evaluated with SVM algorithm.

In all feature selection methods and testing the classifier, we had used LIBSVM – 2.83 software [18].

4.5 Results

We applied proposed hybrid of mutual information and SVM-RFE on Colon cancer and Leukemia cancer gene expression dataset. To compare our method, we also tested with SVM-RFE and MRMR method. The results are shown in the Tables 2 and 3. The results are shown in terms of number of genes, overall accuracy of the classifier and class-wise accuracy (sensitivity and specificity). In Figs 1-2, average test error of linear SVM classifier on selected gene subsets with SVM-RFE, MRMR + SVM and hybrid method is plotted.

From Table 2-3, it is clear that proposed hybrid of mutual information and SVM-RFE performs better than the SVM-RFE in both Colon cancer and Leukemia gene expression dataset. Apart from accuracy, number of genes in the best subset of both dataset is also small compare to SVM-RFE. Comparing with MRMR method, hybrid method performs better in colon cancer dataset and results are comparable in the Leukemia dataset.

Table 2. Performance of SVM classifier with feature selection by SVMRFE, hybrid of SVM and MI based RFE and MRMR method on Colon Cancer gene expression dataset

	Number of genes	Accuracy (%)	Sensitivity (%)	Specificity (%)
MRMR - SVM	13	87.7 ± 7.12	86.2 ± 8.32	89.13 ± 8.32
SVM-RFE	74	88.5 ± 5.97	85.6 ± 12.19	90.19 ± 7.48
SVM-RFE with MRMR	51	89.3 ± 6.71	85.98 ± 12.49	91.44 ± 7.91

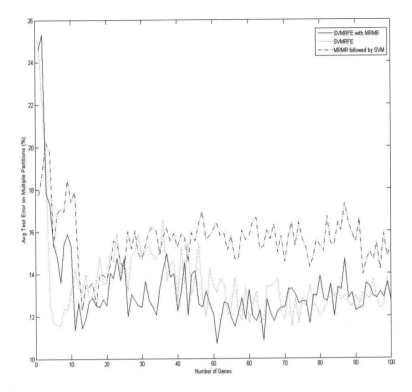

Fig. 1. Average misclassification error of gene subset selected by SVMRFE, hybrid of SVM and MI based RFE and MRMR method on 100 random testing with Colon Cancer gene expression dataset

Table 3. Performance of SVM classifier with feature selection by SVMRFE, hybrid of SVM and MI based RFE and MRMR method on Leukemia Cancer gene expression dataset

	Number of genes	Accuracy (%)	Sensitivity (%)	Specificity (%)
MRMR - SVM	74	97.6 ± 4.19	99.36 ± 3.78	97.04 ± 5.5
SVM-RFE	44	97.13 ± 4.67	100 ± 0	96.13 ± 6.23
SVM-RFE with MRMR	21	97.27 ± 4.75	99.25 ± 2.5	96.63 ± 5.92

5 Discussion and Conclusion

As seen the Table 2-3, the results are better than SVM-RFE both in terms of small number of genes and better classification accuracy. Results are comparable with MRMR based feature ranking. MRMR is basically a type of filter method. Mostly,

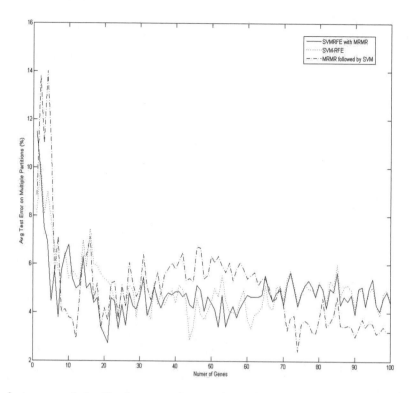

Fig. 2. Average misclassification error of gene subset selected by SVMRFE, hybrid of SVM and MI based RFE and MRMR method on 100 random testing with Leukemia Cancer gene expression dataset

filter method gives large gene subset for comparable classification accuracy with wrapper method. Small number of gene in subset will produce inferior classification accuracy in filter approach. The results with colon cancer and leukemia gene expression dataset show exactly the same nature.

From the result table, we observe that standard deviations of all performance measures (accuracy, sensitivity and specificity) over 100 times training and testing are large. In other words, the variability of single test is large and such test results are not fair performance reference due to possible 'unfortunate' partitioning. When dataset is small, the risk of 'unfortunate' partitioning increases.

In our proposed hybrid method, advantage of both continuous data (in SVM feature weighting) and discrete data (for MRMR) is encoded. Hence, we believe that this ranking method is noise tolerant without affecting the continuous nature of the gene expression data.

As seen from the Figure 1 and 2, hybrid method gave small classification error than SVMRFE and MRMR filtering in most part of gene subset tested. It means this method has better generalization capability. The proposed hybrid method selects the genes based on their effect on classification accuracy and make sure that they are least redundant among themselves. As the gene subset selected is best representing the whole dataset and least redundant, better generalization was expected and hence seen in the results. We also believe that the gene subset selected by this method should give similar classification accuracy with other classifiers.

Finally in conclusion, the gene subset selected by this method represents the broader class characteristics than SVMRFE. It will insure better screening of the dataset and better representation of whole dataset in small number of most relevant and least redundant gene subset.

References

1. Blum, A., Langley, A.: Selection of relevant features and examples in machine learning. Artif. Intell. 97, 245–271 (1997)
2. Kohavi, R., John, G.: Wrappers for feature subset selection. Artif. Intell. 97, 273–324 (1997)
3. Inza, I., Larranaga, P., Blanco, R., Cerrolaza, A.: Filter versus wrapper gene selection approaches in DNA microarray domains. Arti. Intelli. Medicine 31, 91–103 (2004)
4. Rakotomamonjy, A.: Variable selection using SVM criteria. J. Mach. Learn. Res (Special Issue on Variable Selection) 3, 1357–1370 (2003)
5. Ruiz, R., Riquelme, J., Aguilar-Ruiz, J.: Incremental wrapper-based gene selection from microarraydata for cancer classification. Patter. Recog. 39, 2383–2392 (2006)
6. Yousef, M., Jung, S., Showe, L., Showe, M.: Recursive Cluster Elimination (RCE) for Classification and Feature Selection from Gene Expression Data. BMC Bioinfo. 8, 144 (2007)
7. Kai-Bo, D., Rajapakse, J.C., Wang, H., Azuaje, F.: Multiple SVM-RFE for Gene Selection in Cancer Classification With Expression Data. IEEE Trans. Nanobio. 4, 228–234 (2005)
8. Rajapakse, J.C., Kai-Bo, D., Yeo, W.K.: Proteomic Cancer Classification with Mass Spectrometry Data. Am. J. Pharmacogenomics 5, 281–292 (2005)
9. Diaz-Uriarte, R., Andres, S.: Gene Selection and classification of microarray data using random forest. BMC Bioinfo. 7, 3 (2006)
10. Guyon, I., Weston, J., Barhill, S., Vapnik, V.: Gene Selection for Cancer Classification using Support Vector Machines. Machine Learning 46, 389–422 (2002)
11. Ding, C., Peng, H.: Minimum Redundancy Feature Selection from Microarray Gene Expression Data. J. Bioinfo. Compu. Bio. 3, 185–205 (2005)
12. Ding, C., Peng, H.: Minimum Redundancy Feature Selection from Microarray Gene Expression Data. In: Proceed. Second IEEE Comp. System. Bioinfo. Conferen., pp. 523–529. IEEE Computer Society Press, Los Alamitos (2003)
13. Peng, H., Long, F., Ding, C.: Feature Selection Based on Mutual Information: Criteria of Max-Dependency, Max-Relevance, and Min-Redundancy. IEEE Trans. Patt. Anal. Machi. Intell. 27, 1226–1237 (2005)
14. Ooi, C., Chetty, M., Teng, S.: Differential prioritization between relevance and redundancy in correlation-based feature selection techniques for multiclass gene expression data. BMC Bioinfo. 7, 320–339 (2006)
15. LeCun, Y., Denker, J., Solla, S., Howard, R., Jackel, L.: Optimal Brain Damage. In: Touretzky, D. (ed.) Advances in Neural Information Processing Systems II, pp. 598–605. Morgan Kaufmann, San Mateo, CA (1990)

16. Golub, T., Slonim, D., Tamayo, P., Huard, C., Gaasenbeek, M., Mesirov, J., Coller, H., Loh, M., Downing, J., Caligiuri, M., Bloomfield, C., Lander, E.: Molecular Classification of Cancer: Class Discovery and Class Prediction by Gene Expression. Science 286, 531–537 (1999)
17. Alon, U., Barkai, N., Notterman, D., Gish, K., Ybarra, S., Mack, D., Levine, A.: Broad patterns of gene expression revealed by clustering analysis of tumor and normal colon tissues probed by oligonucleotide arrays. PNAS 96, 6745–6750 (1999)
18. Chang, C., Lin, C.: LIBSVM: A Library for Support Vector Machines (2001), http://www.csie.ntu.edu.tw/ cjlin/libsvm

In Silico Expression Profiles of Human Endogenous Retroviruses

Merja Oja

Helsinki Institute for Information Technology,
Helsinki University of Technology, P.O. Box 5400, 02015 TKK, Finland,
and Department of Computer Science, University of Helsinki
merja.oja@tkk.fi

Abstract. Human endogenous retroviruses (HERVs) are remnants of ancient retrovirus infections and now reside within the human DNA. Recently HERV expression has been detected in both normal and diseased tissues. However, the patterns of expression of individual HERV sequences are mostly unknown. In this work we use a generative mixture model, based on hidden Markov models, for estimating the activities of individual HERV sequences from databases of expressed sequences. We determine the relative activities of sixty HERVs from the HML2 group in five human tissues, i.e. we estimate the *expression profile* of each HERV. This allows us to gain insight into HERV function.

1 Introduction

Human endogenous retroviruses (HERVs) are remains of retrovirus infections that occured millions of years ago. They are viruslike DNA sequences that reside within the human genome. HERV sequences form 8% of the human genomic DNA [3,4].

Retroviruses can move and copy their DNA to other locations in the genome. These copying events will eventually yield several mutated versions of the original virus. A group of such sequences is called a HERV group and it may contain hundreds of very similar sequences. Most of the HERV sequences are heavily mutated and/or broken due to genomic rearrangements and have partially lost the typical retroviral structure consisting of 4 genes (gag, pro, pol and env) and two long terminal repeat sequences (LTRs), one at each end of the retrovirus sequence.

In this paper we study the HML2 group because it is the youngest and as such has the largest proportion of full length HERVs and the smallest number of mutations. Thus, it has the most potential for containing active HERVs.

HERVs are interesting for two reasons: they can express viral genes in human tissues and their presence in the genome may affect the function of nearby human genes. Retroviral activity might cause disease; retroviral mRNAs have been detected in schizophrenia, autoimmune diseases and cancer [2,10] although a causal role of HERVs in these conditions is highly uncertain. In addition, a few retroviral genes have adopted functions beneficial to the human host [8].

J.C. Rajapakse, B. Schmidt, and G. Volkert (Eds.): PRIB 2007, LNBI 4774, pp. 253–263, 2007.
© Springer-Verlag Berlin Heidelberg 2007

In this work we study activities of individual HERV sequences in various tissues, i.e. will estimate the *expression profile* of each HERV. The profile contains measurements from several tissues and thus enables us to study the differential expression patterns of individual HERVs. This leads to better understanding of the function of individual HERVs. For example, HERVs that are more/only active in the brain tissue may have functions related to neurodegenerative diseases or to normal brain functions. This profiling approach is widely used in the study of human gene function, see for example [17]. In contrast, the only work that we know of where individual HERVs have been studied in several tissues is [16], where a small set of full-length HERV-K elements (HML2 is a subgroup of HERV-K) were studied using a heuristic method.

We have earlier studied the overall expression of individual HERVs (one expression value for each HERV without distinguishing between different tissues and conditions). In this work we extend the approach to estimation of expression *profiles* over various tissues. Furthermore, we analyze the expression profiles of *individual* HERV sequences. In contrast, most previous studies of HERV expression report activities only for HERV groups (e.g. [13]); the only exceptions we know of are [6] where HERVs are searched from gene mRNAs but activities are not compared across HERVs and [16] mentioned above.

To find evidence of HERV expression, we use a large public database of expressed sequence tags (ESTs). ESTs are short and noisy samples from mRNA sequences. The amount of ESTs originating from a particular HERV is evidence of its activity. However, it is nearly impossible to match an EST sequence to only one HERV sequence: Each EST will match several HERVs very well due to the similarity of the HERV sequences within a HERV group and the noise (sequencing errors) in the ESTs. We have introduced earlier a probabilistic model [11] to handle the uncertainty in EST to HERV matching. In the methods section we describe how this model can be used for estimating HERV expression profiles. The expression profiles for the HERV sequences of the HML2 group are presented in the results section.

2 Methods

In [11] a generative mixture model, based on hidden Markov models, for estimating the activities of individual HERV sequences from ESTs was introduced. Below we briefly describe this model and then move on to describe how it can be used when the aim is to estimate expression profiles instead of overall expression values.

The hidden Markov mixture model is a generative model for the set of EST sequences. It is designed to mimic the actual EST generation from HERVs; each mixture component is a hidden Markov model (HMM) for ESTs from a particular HERV (See Fig. 1). The component HMM resembles the profile HMM [7], with the exception that it is possible to jump from the start state to any of the match states and from any match state either to the end or to a special EEMIT state that is used to emit the low quality end of an EST. The match states, one for each

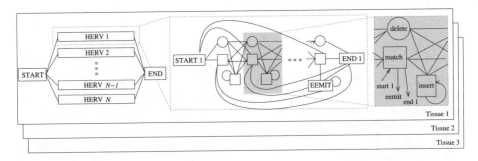

Fig. 1. The structure of the HMM mixture. The model is constrained by sharing parameters. The shaded box is the basic block of the sub-HMM and is repeated length-2 times. It is identical in all sub-HMMs; all other parameters are shared except the emission distribution of the match state which varies between blocks, according to the HERV sequence each sub-HMM corresponds to. EEMIT-state emits the low-quality end part. The plates illustrate that the same model is learned separately for each tissue.

position of the HERV sequence, can either emit the nucleotide in that position of the HERV sequence (with probability p_t) or one of the other nucleotides (with probabilities $(1-p_t)/3$). The parameter p_t is shared between all match states in the mixture model. The EEMIT states and all the insert states share parameters: they emit nucleotides using the same distribution. The transition parameters are also shared throughout the mixture (see Fig. 1). In summary, the component HMM generates data that roughly matches a subsequence of the source HERV, but with mismatches, insertions, deletions, and a low-quality end part.

The mixture model corresponds to one large HMM where the first transition chooses one of the N HERV-specific sub-HMMs (see Fig. 1). The Baum-Welch algorithm is used to learn the whole mixture. The learned probabilities of the first transition (the mixture weights) are estimates of the HERV activities. We use heuristics to reduce HMM training time to reasonable limits [11].

The hidden Markov mixture model can be extended to estimation of expression profiles. We can simply learn a separate model for each tissue and then combine the results meaningfully. In practice, we need to collect several sets of EST sequences, one set for each tissue. Then we learn the model for each EST set. This results in the relative activity distributions of the HERVs for each tissue.

The relative activity distributions of HERVs from different tissues can be combined in two ways to form the HERV expression profiles. 1) The relative activities of a HERV in different tissues are used directly as the expression profile. In this setting it is assumed that each EST set, irrespective of its size, is a sample of all HERV derived mRNAs in the tissue. 2) The relative activities of a HERV in different tissues are first scaled according to the number of ESTs available from the tissues. This way the expression profile of a HERV is more clearly related to the number ESTs available from the HERV and the activity value of the HERV can be seen as a *probabilistic EST count*. In this setting it is assumed that the size of the EST set is relevant. In this work we will use this second approach.

3 Data

We study the expression profiles of HERVs of the HML2 group. This HERV group is the youngest one and thus has the largest proportion of relatively intact elements. It contains sixty members, some of which are full-length, i.e. have retained the typical retrovirus structure LTR-gag-pro-pol-env-LTR. A few of these elements even have open reading frames for the env gene, i.e. they could produce retroviral env proteins.

The HML2 group is the most difficult one to study because the sequences within the young HML2 group are more similar to each other than sequences in other groups. It is impossible to match ESTs to individual HML2 HERVs unambiguously. Our statistical approach is able to alleviate this problem to some extent. But, even with our method, the activities of nearly identical HERVs will be correlated.

We studied the expression of HML2 HERVs in five tissue types: brain, lung, breast, placenta and male reproductive tissues (RT). This selection was mainly due to the availability of the ESTs, but some of these tissues are also interesting *per se*: HERV transcripts have been detected in brain related diseases, HERVs active in reproductive tissues could produce new HERV integrations and some HERVs are known to have beneficial functions in placenta. In addition, we know from earlier studies that HERV-K elements are active at least in testis and brain tissues [9].

The HERVs were automatically detected from the human genome by the program RetroTector[1]. Sixty of the HERVs were similar to HML2 reference sequence and were included into the HML2 HERV set.

ESTs matching the HML2 HERVs were searched from the dbEST database [20] with BLAST [1]. The ESTs were divided into tissue-specific sets using eVoc Ontologies [19]. We used a match threshold of E-value 10^{-40} in BLAST.

In addition to HML2 HERVs, some elements from other HERV groups were included in the HERV set. This was done to ensure reliable activity estimates for the HML2 HERVs: If the extra HERVs would not be included, then EST originating from them would be distributed over HML2 HERVs, falsely increasing their activity estimates. In other words, adding the extra HERVs reduces the error due to ESTs that match a non-HML2 HERV better than any of the HML2 HERVs.

The set of extra HERVs was selected based on a heuristic BLAST activity. The BLAST activity of a HERV is the number of EST matching that HERV better than any other HERV. ESTs that match several HERVs equally well are divided to all those HERVs. In our earlier work [11] BLAST activity was shown to correlate with activity estimates from the HMM model. We included all highly BLAST active HERVs (those with more than 2.5 ESTs) and then the most active

---

[1] RetroTector is a program used for detecting retroviral sequences in genomes. It searches for conserved retroviral motifs and then combines the motifs into chains fulfilling distance constraints. It was developed by Jonas Blomberg and Göran Sperber at Uppsala University [15].

Table 1. Data set sizes for different tissues. "HERV-EST pairs" is the number of EST to HERV matches returned by BLAST.

Tissue	HERVs	ESTs	HERV-EST pairs
Brain	94	471	7076
Lung	86	279	4661
Placenta	85	219	2770
Breast	73	164	2987
Male reproductive tissue	89	249	4157

HERV (still required to have at least one EST match) from each HERV group into the analysis. The set of extra HERVs was different for each tissue. Table 1 list the data sets sizes for all tissues.

4   Results

The method is able to estimate the relative activities of the HERVs. The activity profiles for HML2 HERVs are shown in Fig. 2A. Many of the HERVs exhibit tissue specific expression. There are also some HERVs that are active in all tissues as well as HERVs that are not active in any of them. The activities of most HML2 HERVs were previously unknown. A portion of the full-length HML2 HERVs have been studied before in [16] using a heuristic BLAST approach. Some individual HERVs are analyzed more closely in Section 4.1.

The results show that adding the extra HERVs was necessary to get reliable estimates for the HML2 HERVs. In each case the probability mass allotted to the HML2 HERVs was less than half of the total (ranging from 37% in the placenta to 48% in the lungs). If the extra HERVs would not have been included, then the probability mass now belonging to them would have been distributed over the HML2 HERVs, falsely increasing their activity estimates. Furthermore, some of the non-HML2 HERVs were very active in comparison to the mean activity level of the HML2 HERVs (see Fig. 2B). The high activity of the non-HML2 HERVs indicates that there is a lot of cross-talk between the HERV groups (the ESTs retrieved using the HML2 sequences as queries also match HERVs from the other groups). Some of the cross-talk might be due to portions of the HERVs that resemble other types of retrotransposons (see section 4.1).

We estimated the reliability of the results with a bootstrap-like method. The EST data was resampled with replacement 1000 times, and the activities were reoptimized for each replicate while other parameters were kept fixed (see [11] for more details). Fig. 3 shows the means and standard deviations of these replicates for the HERV activities in the lung tissue. The behavior in the other tissues is very similar. The standard deviations are small compared to the differences in HERV activities and the means are very close to the activities learned from all data. The standard deviations of the clearly active HERVs (probabilistic EST count above 5) and almost inactive HERVs (probabilistic EST count below 1) do not overlap. Thus we can trust the active-looking ones to be truly active.

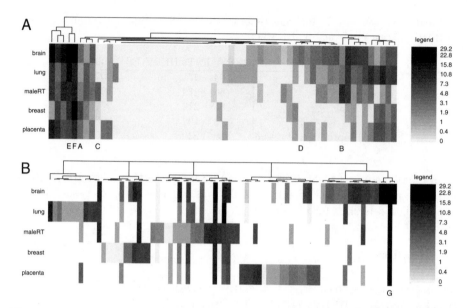

Fig. 2. The activities of the HML2 (panel **A**) and non-HML2 (panel **B**) HERVs. In both panels the rows depict the HERV activity distributions in different tissues and the columns the expression profiles of individual HERV sequences. Letters below the columns are labels for the HERVs analyzed in Section 4.1. The activity values are shown on a logarithmic scale, as can be seen from the legends on the right. The scale is the same in both panels. The numbers next to the legend are the probabilistic EST counts for each gray shade. The highest activity for a HML2 HERV is 24.8 (HERV F in the brain tissue). The columns have been ordered according to a hierarchical clustering based on the (unlogarithmic) Euclidean distances between the HERV expression profiles.

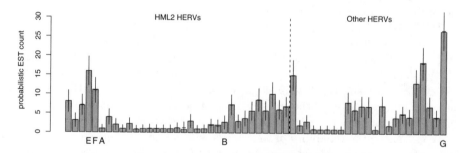

Fig. 3. The activities (probabilistic EST counts) of the HML2 and non-HML2 HERVs in the lung tissue. The crosses are the means and standard deviations from the bootstrap resamples (see text for details) and the bars the activities learned from complete data. The HERVs are in the same order as in Figs. 2A and 2B, but inactive HERVs (with probabilistic EST count below 10^{-7}) have been left out of the visualization to save space. The letters below the columns are the labels for the HERVs analyzed in Section 4.1.

4.1   Closer Look on Individual Active HERVs

Here we take a closer look on some of the individual HERVs. These have been selected as examples of the typical expression patterns of the active HERVs observed in the data. The HERVs analyzed in this subsection are summarized in Table 2. The labels used to denote the HERVs are letters with no special meaning.

HERV A is full-length element with an open reading frame for the env gene. This HML2 HERV is known as HERV-K102. It is somewhat active in all tissues — its highest activity is observed in the breast tissue. The activity is due to ESTs that match the LTRs and the env gene area of the HERV. This HERV is a potential retrovirally active HERV that could produce env protein. HERV A is also mentioned in [16], but no exact details are given. UCSC Genome browser shows a new hypothetical human protein overlapping the LTRs of this HERV. This supports our finding that this HERV is retrovirally active.

HERV B is an almost full-length HERV with no open reading frames and a missing end-LTR. The HERV is active in the brain, lung and male reproductive tissues. Its activity is concentrated on gag and pol genes. This HERV has been studied earlier in [16], where it was found to be expressed in the brain, placenta, testis and prostate tissues. It had low activity in the lung and breast tissues. These results agree with our observations except for placenta and lung, for which our results are just the opposite.

HERV C is active only in the male reproductive tissues. The ESTs match this full-length HERV near the end of pol and at the end-LTR. The ESTs might be coming from the end of a pol gene transcript, however, ESTs from the beginning of the transcript are not observed. UCSC Genome browser shows a short gene sequence, annotated as a retroviral rec gene, between and partly overlapping the sequence segments detected as active by our method. This further supports the observation that this relatively intact HERV locus is active.

HERV D exhibits a clear tissue-specific expression: it is active only in the brain tissues. This non-full-length HERV is active in the gag gene area. However, there is no open reading frame for a gag protein. The observed expression does not resemble that of a retrovirally active HERV [4]. Hence, it seems that this HERV might have been used as a building block for something else than retroviral proteins.

The data set contains some HERVs that are very active in all studied tissues; for example, the HERV sequence E. ESTs match this HERV in the end of the pol gene and parts of env. However, when we look at this genome area at the UCSC Genome Browser, the pol gene area is annotated there as an L1 repeat. Thus, it may be that the (probabilistic) EST count of this HERV is actually measuring L1 derived ESTs. Similar situation applies to the highly active HERV F, where the expression also seems to be L1 derived. These HERVs are examples of broken down sequences that are harder to detect automatically. For these HERVs the Retro-Tector program may have misinterpreted some portion of the L1 structure, which as a retrotransposon is similar to that of a retrovirus, as retrovirus-derived DNA.

HERV G measures expression of SVA elements that are composite retrotransposons consisting of an Alu like portion, a tandem repeat portion and a portion originating from the HML2 LTR sequence [12]. The end portion of the SVA

Table 2. Details about the HERVs analyzed in Section 4.1. "Label" is the label of the HERV used in the text and figures. "Chr", "strand", "start" and "end" tell the chromosome, the strand, and the sequence start and end positions for the HERV, respectively (in the July 2003 version (hg16) of the human genome). "Subgenes" describes the structure and "group" the group of the HERV. The last column in the upper part of the table gives the name used for the HERV in [16]. The "orf" columns describe how intact the retrovirus protein reading frame is: 0 is intact, i.e. the HERV has a open reading frame for the protein. "Age" is the estimated age of the element measured in percentage of LTR unsimilarity. The two LTRs of a retrovirus are identical on integration and mutate afterwards. The "gene context" column gives the gene nearest to or overlapping with the HERV locus.

label	chr	start	end	strand	subgenes	group	name in [16]
A	1	152822428	152813249	-	LTRgagpropolenvLTR	HML2	K102
B	22	22203232	22213324	+	LTRgagpropolenv	HML2	22q11
C	11	101103511	101112976	+	LTRgagpropolenvLTR	HML2	11q22.1
D	7	140863179	140859365	-	gagpropol	HML2	
E	16	35307416	35314276	+	polenv	HML2	
F	1	75265364	75273509	+	LTRgagpro	HML2	
G	19	21682582	21697392	+	LTRLTR	unknown	

label	gagorf	proorf	polorf	envorf	age	gene context
A	3	0	1	0	0.21	3' LTR the last exon of a hypothetical gene
B	1	5	12	9	-	gene IGLL1 2.5 Kb away downs. (antisense)
C	3	0	1	1	0.41	part annotated as retroviral rec gene
D	8	1	16	-	-	gene SSBP1 1 Kb away downs. (antisense)
E	-	-	15	2	-	nearest gene 20 Kb downstream (sense)
F	2	0	-	-	-	HERV in a long intron of an antisense gene
G	-	-	-	-	10.59	gene ZNF100 100b downstream (antisense)

repeat is about 95% similar to the HML2 LTR. For this reason, some of the ESTs retrieved using BLAST may actually come from a SVA element. As a consequence, it is necessary to include a SVA like sequence into the HERV set so that possible SVA derived ESTs will not confuse the activity estimates of LTR-containing HML2 HERVs. The SVA-ESTs will match the SVA like "HERV" better and thus have low probability on matches to HERVs. It turns out that one sequence in the HERV collection (marked with G in the figures) obtained by RetroTector is very similar to a SVA element and actually portions of this sequence are annotated as SVA in the UCSC Genome Browser. It was included into the HERV set to serve as the SVA like element. The results show that this "HERV" is very active in all tissues and the activity is in the SVA repeat areas. This indicates SVA activity in all the analyzed tissues.

5   Discussion and Conclusions

We have used a generative model-based method to estimate the expression profiles of individual HERVs rather than those of HERV groups. Such detailed analysis is vital for understanding the functions and control mechanisms of HERVs. Our method allows the exploration of expression patterns within a HERV group and will reveal interesting potentially active HERVs. These can then be studied further and their activity levels in different tissues can be verified with laboratory methods. By contrast, exhaustive search of active HERVs in the laboratory would be too expensive and/or difficult.

The advantage of our method over a simple "find the best matching HERV for each EST" approach (such as the BLAST activity method described in section 3) is the ability to take uncertainties into account. Our model is able to learn the underlying activities from data where the error rate (noise) in the ESTs is larger than differences between two HML2 HERV sequences. In our earlier work [11] we showed with experiments on simulated data that the HMM model outperforms the simple BLAST activity estimation method. The difference was most notable in the case of HML2 HERVs.

The number of ESTs available from each tissue was not as high as we would have hoped: The EST sets were small with only about three ESTs per HERV. As a result, the activity estimates are not as accurate as they would have been with a larger data set. Still, our results were reliable according to bootstrap resampling and as such can give valuable pointers to HERVs that should be studied more closely.

There are few examples of active and potentially protein-coding HERVs. Most of the active HERVs (such as HERVs B and D discussed in section 4.1) display fragmented expression that could be explained by RNA mediated activity or by function as exons, beginnings or ends of nearby human genes.

Some of the observed expression may be due to active non-retroviral repeat sequences. In this study we wanted to study fragmented HERVs in addition to the full-length elements. The fragmented HERVs are harder to detect and in the process of ensuring that the more mutated HERVs are not missed some elements that are combinations of retrovirus and retrotransposon sequences may be included into the RetroTector produced HERV set. Actually, some of the most active HERVs were found to contain sequence portions which the RepeatMasker[2] [14] had annotated as L1, L2 or SVA repeats. The fact that we observe expression similar to L1 or SVA elements is interesting as these elements have been shown to be active recently: The comparison of human chimpanzee genomes revealed thousands of species specific integrations for both L1 and SVA elements [18]. Our results indicate both L1 and SVA elements are still actively expressed in the human genome.

The hidden Markov mixture model can also be applied to other kinds of mRNA data sources or to other types of repetitive elements. For example our method could be used as a post-processing step in a RT-PCR reaction [9] where

---

[2] RepeatMasker is a widely used program for detecting repeats. It relies on a database, the RepBase [5], of consensus sequences for various kinds or repeats. The repeat annotations in the UCSC Genome Browser come from RepeatMasker predictions. A discussion on the differences between RetroTector and RepeatMasker can be found from [15].

a broadly targeting primer (all members of a HERV group are amplified) has been used. When the PCR products are sequenced, they can be compared to the members of the targeted HERV group using our hidden Markov mixture model. This way it can be determined which elements within the group of very similar sequences are active. This can be done in one or several tissues.

Acknowledgments

We would like to acknowledge the Microbes and Man (MICMAN) research programme of the Academy of Finland (decision 202502). The author belongs to the Adaptive Informatics Research Centre, which is a Centre of Excellence of the Academy of Finland. We would like to thank Göran Sperber and Jonas Blomberg, Uppsala University, for RetroTector data. We are also grateful to Panu Somervuo, University of Helsinki, for his help with the HMM code. Finally, we would like to thank Samuel Kaski and Jaakko Peltonen, Helsinki University of Technology, and Jonas Blomberg for their valuable comments during the research and writing processes.

References

1. Altschul, S.F., Gish, W., Miller, W., Myers, E.W., Lipman, D.J.: Basic local alignment search tool. Journal of Molecular Biology 215, 403–410 (1990)
2. Blomberg, J., Uschameckis, D., Jern, P.: Evolutionary aspects of human endogenous retroviral sequences and disease. In: Sverdlov, E. (ed.) Retroviruses and Primate Evolution, pp. 208–243. Eurekah Bioscience (2005)
3. Lander, E., et al.: Initial sequencing and analysis of the human genome. Nature 409, 860–921 (2001)
4. Griffiths, D.J.: Endogenous retroviruses in the human genome sequence. Genome Biology 2(6), 1017.1–1017.5 (2001)
5. Jurka, J.: RepBase update: a database and an electronic journal of repetitive elements. Trends in genetics 16(9), 418–420 (2000)
6. Kim, T.-H., Jeon, Y.-J., Kim, W.-Y., Kim, H.-S.: HESAS: HERVs expression and structure analysis system. Bioinformatics 21(8), 1699–1700 (2005)
7. Krogh, A., Brown, M., Mian, I.S., Sjölander, K., Haussler, D.: Hidden Markov models in computational biology: Applications to protein modeling. Journal of Molecular Biology 235(5), 1501–1531 (1994)
8. Muir, A., Lever, A., Moffett, A.: Expression and functions of human endogenous retroviruses in the placenta: An update. Placenta 25(suppl. 1), S16–S25 (2004)
9. Muradrasoli, S., Forsman, A., Hu, L., Blikstad, V., Blomberg, J.: Development of real-time PCRs for detection and quantitation of human MMTV-like (HML) sequences. HML expression in human tissues and cell lines. J. Virol. Meth. 136, 83–92 (2006)
10. Nelson, P.N., Carnegie, P.R., Martin, J., Ejtehadi, H.D., Hooley, P., Roden, D., Rowland-Jones, S., Warren, P., Astley, J., Murray, P.G.: Demystified human endogenous retroviruses. Molecular Pathology 56, 11–18 (2003)
11. Oja, M., Peltonen, J., Blomberg, J., Kaski, S.: Methods for estimating human endogenous retrovirus activities from EST databases. BMC Bioinformatics 8(suppl. 2), S11 (2007)

12. Ostertag, E.M., Goodier, J.L., Zhang, Y., Kazazian, H.H.J.: SVA elements are nonautonomous retrotransposons that cause disease in humans. The American Journal of Human Genetics 73(6), 1444–1451 (2003)
13. Seifarth, W., Frank, O., Zeifelder, U., Spiess, B., Greenwood, A.D., Hehlmann, R., Leib-Mösch, C.: Comprehensive analysis of human endogenous retrovirus transcriptional activity in human tissues with a retrovirus-specific microarray. Journal of Virology 79(1), 341–352 (2005)
14. Smit, A.F.A., Hubley, R., Green, P.: RepeatMasker open-3.0 (1996-2004), http://www.repeatmasker.org
15. Sperber, G., Jern, P., Airola, T., Blomberg, J.: Automated recognition of retroviral sequences; RetroTector©. Nucleic Acids Research, Accepted with revision (2007)
16. Stauffer, Y., Theiler, G., Sperisen, P., Lebedev, Y., Jongeneel, C.V.: Digital expression profiles of human endogenous retroviral families in normal and cancerous tissues. Cancer Immunity 4(2) (2004)
17. Su, A.I., Wiltshire, T., Batalov, S., Lapp, H., Ching, K.A., Block, D., Zhang, J., Soden, R., Hayakawa, M., Kreiman, G., Cooke, M.P., Walker, J.R., Hogenesch, J.B.: A gene atlas of the mouse and human protein-encoding transcriptomes. PNAS 101(16), 6062–6067 (2004)
18. The Chimpanzee Sequencing and Analysis Consortium. Initial sequence of the chimpanzee genome and comparison with the human genome. Nature, 437, 69–87 (2005)
19. eVOC ontologies, http://www.evocontology.org
20. Expressed sequence tags database (dbEST), http://www.ncbi.nlm.nih.gov/dbEST/

A Framework for Path Analysis in Gene Regulatory Networks

Ramesh Ram and Madhu Chetty

Gippsland School of IT, Monash University,
Churchill, Victoria 3842, Australia
{Ramesh.ram,Madhu.chetty}@infotech.monash.edu.au

Abstract. The inference of a network structure from microarray data providing dynamical information about the underlying Gene regulatory network is an important and still outstanding problem. Recently, a causal modeling approach was presented in our publications to recover the structure of this network. However, issues like spurious arcs and time delay were not dealt with previously. The graph-theoretical measure d-separation provides criteria to recover the network structure edge-by-edge by calculating the partial correlation. However, the estimation of partial correlations from small sample sizes is a practical problem. As our approach attempts to find networks that closely match the observed partial correlation constraints in the data, main aim to this paper is to attempt to maximize the scoring metric used. In this paper, we formulate a framework for path analysis as a post processing step after learning gene regulatory network using causal modeling. The approach is tested with both artificial and real gene regulatory network scenario and the structure recovered after post processing better fits the data.

Keywords: Causal model, d-separation, conditional independence.

1 Introduction

Simultaneous monitoring of genome wide expression (microarray technology) allows us to gain insight into concerted activity of interacting genes well-known as transcriptional gene regulatory network (GRN). The network of regulatory relationships is inferred by various computational approaches directly from the expression profiles. For example, Boolean network, Bayesian network, neural network and genetic algorithm based approaches have been successfully applied to infer networks from expression profiles [1, 2, 3, 4]. Recently, we have developed linear causal model approach to infer gene networks [5, 6], which are based on graphical models [7, 8]. Linear causal models are closely related to Bayesian networks, which a number of researchers have used to model gene regulation [9, 10, 11, 12]. In fact, a linear causal model is a special case of a Bayesian network (BN) that has linear Gaussian conditional densities at each node. Our method infers a simple network structure where the conditional independence between variables is estimated by the partial correlation coefficient, and corresponds to a graph based on the Markov properties. Network components are nodes, V that are genes in a GRN

J.C. Rajapakse, B. Schmidt, and G. Volkert (Eds.): PRIB 2007, LNBI 4774, pp. 264–273, 2007.
© Springer-Verlag Berlin Heidelberg 2007

and an edge set E representing the dependency structure of the nodes. Thus, the graph G = (V, E) the topology of the gene network. In our approach, we defined a set of scoring functions that can be used to measure the quality of this network G. These include fitness of the Markov Blanket (MB) of every node, fitness of the links between nodes that indicate the direction of influence and the + or - indicate positive or negative influence of the nodes. A search algorithm that can help us to find the best score network structure needed. Since the number of BNs is super-exponential in the number of nodes available and an exhaustive comparison of all the structures is impossible. So we implemented a GA which uses local search of Markov Blankets in BN structure space, moving from one BN to the next one by performing simple graphical modifications such as addition and deletion of edges during crossover and mutation. As score-based approach might suffer the local minima a guided genetic algorithm was proposed.

The goal is to obtain a network that is minimal in the number of links, or representation size, necessary to fit the data. However, the search algorithm is only able to guarantee the quality of the returned structure to a certain extent because there exist multiple BN's representing the same dataset due to the nature of the problem being NP-Hard. Not only in the final returned network, but also in intermediate stage networks of the discovery process using genetic algorithm, there exist many equivalent possibilities of combination of edges having the same fitness posing difficulty is making a choice for drawing causal conclusions. It is general statistical knowledge that extra care should be taken when drawing causal conclusions from statistical analysis. This is in particular valid for Causal Bayesian networks. A "wrong" choice at an intermediate stage and final stage may prevent from finding a minimal network for the data.

Theoretically, given a faithful Bayesian network structure of the training data set, those features that are identified by the Markov blanket indeed block all the influence of the other features. The selection of Markov blanket is based on the d-separation rule of the Bayesian network. When given a specific attribute, which is node in the Bayesian network, Markov blanket for the attribute is the set of nodes composed of the attribute's parents, its children, and its children's parents. Direct-dependent separation or d-separation is a graphical procedure that establishes the conditional statistical independence of certain sets of these random variables, i.e. if the set of nodes X is independent of nodes Y, then there exist nodes Z such that they separate the X's and Y's. Further, there exist multiple paths (a path is a series of variables connected by line segments) between two nodes in a directed graph representing causal flows. Directed graphs provide the visual representation of that flow; the set of independence or conditional independence conditions which are implied by that graph are not (necessarily) obvious. Here, we are interested in analysis of independence and conditional independence of variables under alternative causal flows between variables in order to obtain a minimal network. Since time delay is an important characteristic of gene networks, delay propagation should also taken into account while analyzing these alternative causal flows (path delay analysis).

We propose a path analysis framework algorithm exploiting d-separation and time delay to enhance the quality of the discovered network. More precisely, it is divided into four distinct phases. In the first one, we use a method to find alternative causal flows. As the final network is made of Markov blanket of individual nodes, we obtain upward causal flows from parents, downward causal flows through children and sideward causal flows through the spouses. A set of conditional independence is obtained from these paths. Phase 2 analyses the various paths obtained at the end of phase 1 for their consistency in terms of conformance with d-separation property, conflicting alternative explanations of causality and path delay propagation. These are explained in detail in section 3. During the process of analysis, the edges that are non-conforming to the constraints are marked for deletion. Then, in a third phase, this BN is refined into one that better fits data by making choice of either deleting an edge that is marked for deletion or not. This last phase tests and verifies the final network and if the test fails phase 3 is performed again till a better network is obtained. Preliminary experimental results using small artificial network suggest that our algorithm produces better quality network than the one obtained at the end of GA search. The rest of the paper is organized as follows. Section 2 provides some background on causal model and d-separation. Then, Section 3 describes the methodology. Section 4 shows results from artificial network and *S. cerevisiae* yeast network [13]. Finally, Section 5 provides conclusion and mentions future work.

2 Background

To introduce our approach, in this section we briefly review the concept of d-separation which plays an important role in our algorithm.

2.1 Causal Model for GRN

The inference of causal network structures is an important and challenging problem. A causal structure can be represented by a directed graph whose nodes represent the variables of the system and edges between nodes indicate a causal relationship along the direction of the edge. Important contributions in this problem were made by Pearl *et al*, and Sprites *et al* [7, 8] who suggested algorithms to infer a causal structure from experimental data by using partial correlations if the underlying causal structure is a directed, acyclic graph (DAG). A central step in determining the likelihood of the data given the whole network is decomposed into set of score of local models that includes fitness of structure, direction of causality and sign (positive/ negative) of regulation. The task of network reconstruction is cast into a search for candidate gene networks whose scores are high. To implement a heuristic search method, we apply a genetic algorithm (GA), whereby creating and evolving different networks to eventually obtain a network that best fits the microarray data. Due to the stochastic nature of the GA, the GA is repeated and the resulting network structures are combined to reconstruct the final gene network. While evaluating the fitness, the putative network is actually decomposed into Markov Blankets (MB) and conditional independence tests are applied in order to detect whether or not connections are direct or indirect. The direction and sign of regulation are recovered by estimating the time delay and correlation between expression profiles of pairs of genes. Further, this methodology is applied to a toy dataset generated in the same fashion as discussed in our previous work [6] and *Saccharomyces cerevisiae* (yeast) [13] microarray dataset and the results are promising and agreeing with known biological findings.

2.2 D-Separation

D-Separation is defined as: Two nodes X and Y in a directed acyclic graph are d-separated if every path between them is blocked. Consider 3 disjoint sets of variables X, Y, and Z, represented as nodes on a DAG. Definition: A path is a sequence of consecutive edges (of any directionality) in the graph. A path is said to be d-separated, or blocked, by a set of variables Z iff the path (a) contains a chain (b) or a fork (c) contains an inverted fork, or collider, such that the middle variable m is not in Z and such that no descendant of m is in Z. (Fig. 1)

A set Z is said to d-separate X from Y iff Z blocks every path from a variable in X to a variable in Y.

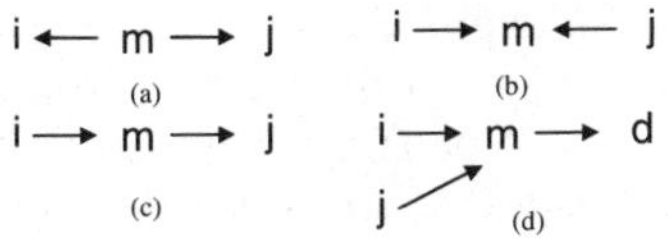

Fig. 1. 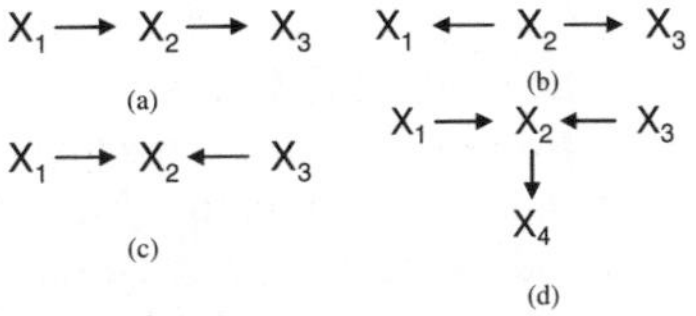   **Fig. 2.**

The *no descendant* example: (Fig. 1 (d))

If d is in Z, then the path from i to j is unblocked even if m is not in Z.

Based on this definition of d-separation, the useful theorem can be stated as follows.

Theorem: If X and Y are d-separated by Z in a DAG, then $X \perp\!\!\!\perp Y \mid Z$. Conversely, if X and Y are not d-separated by Z in the DAG, then X and Y are dependent conditional on Z.

To help understanding this theorem, four basic graphical structures (Fig.2) and the independences implied by each.

1) X_2 is an intermediate variable. The only independence implied by this structure is $X_1 \perp\!\!\!\perp X_3 \mid X_2$. It is NOT true that $X_1 \perp\!\!\!\perp X_3$.

2) X_2 is a common cause. The only independence implied by this structure is $X_1 \perp\!\!\!\perp X_3 \mid X_2$. It is NOT true that $X_1 \perp\!\!\!\perp X_3$.

3) X_2 is a common effect. The only independence implied by this structure is $X_1 \perp\!\!\!\perp X_3$. It is NOT true that $X_1 \perp\!\!\!\perp X_3 \mid X_2$.

4) X_2 is a common effect. X_4 is an effect of a common effect. The independences implied by this structure are $X_1 \perp\!\!\!\perp X_3$, $X_4 \perp\!\!\!\perp X_1 \mid X_2$, and $X_4 \perp\!\!\!\perp X_3 \mid X_2$. It is NOT true that $X_1 \perp\!\!\!\perp X_3 \mid X_2$ or that $X_1 \perp\!\!\!\perp X_3 \mid X_4$. This is the trickiest structure you will find.

2.3 Paths in a Markov Blanket

Path is a sequence of distinct vertices, successive vertices are adjacent. We view a gene network as a network system of information channels, where each node is a valve that is either active or inactive and the valves are connected by noisy information channels. The information flow can pass through an active valve but not an inactive one. When all the valves (nodes) on one undirected path between two nodes are active, we say this path is *open*. If any one valve in the path is inactive, we say the path is *closed*. When all paths between two nodes are *closed* given the status of a set of valves (nodes), we say the two nodes are *d-separated* by the set of nodes. The status of valves can be changed through the instantiation of a set of nodes. The amount of information flow between two nodes can be measured by using mutual information, when no nodes are instantiated, or conditional mutual information, when some other nodes are instantiated.

The two limitations of path analysis algorithms using exclusively partial correlation and d-separation to infer the structure of the underlying graph are: first, for large graphs the search for a set of d-separating paths between two nodes X and Y can be hard, because of the combinatorial explosion of possible sets. Second, the partial correlation does not necessarily vanish for variables not directly connected in the true model. We propose to solve these limitations using our approach.

Since our model [6] works by finding Markov blankets of the main network, one approach to search set of d-separating paths is by separating them as upward, downward and sideway paths (Fig. 3). The upward path (red arrows) is the blocking path to a node from the parents, the downward path (green arrows) is the open path from the node through its children and sideway path (blue arrows) is the path between the node and its spouse node.

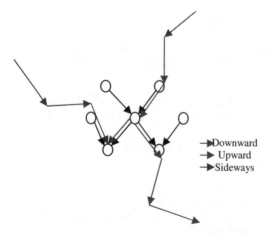

Fig. 3. Paths in Markov Blanket

The explicit goal of the proposed post processing procedure is the inference from a set of causal paths got from the network obtained from the learning algorithm used. The approach is explained in the next section.

3   The Algorithm for Path Analysis Framework

In this section, we propose path analysis approach for incorporating missing d-separation, multiple paths, alternative causal explanation, and effect of path time delay. The d-separation path analysis algorithm is extended to handle signal transition time delays, and to propagate their effects in the circuit using the 'If..then' time functions. This algorithm has four phases: searching, marking for arc deletion, thinning (actual deletion) and finalizing the network. In the first phase, this algorithm finds the sets of paths for each node using its Markov blanket. In phase 2, each of the paths is analyzed for missing d-separation and d-separation for compliance with important property. This is done using the functions described below. Following that, each of the paths are analyzed for variance conformances and the arc that are non conforming are marked for deletion. The result of Phase 2 is a list of arcs marked to be deleted under various conditions. The Phase 3 performs the actual deletion of the arcs those actually affect the fitness of the network. The result of Phase 3 is the final network and phase 4 finalizes the network and if any mistakes identified the network is rolled back and sent back to Phase 3.

3.1 The d-Separation Algorithm

The selection of Markov blanket is based on the d-separation rule of the Bayesian network. When given a specific node in the Bayesian network, Markov blanket for the attribute is the set of nodes composed of the attribute's parents, its children, and its children's parents. Theoretically, given a Bayesian network structure of the training data set, those nodes that are identified by the Markov blanket indeed block all the influence of the other nodes in the network. This helps to identify the d-sep condition set. The problem is to obtain a network that has minimal in the number of links, or representation size, necessary to fit the data. The properties of Markov blanket and d-separation are combined for this reason.

Concept of missing D-Separation: D-separation property in a DAG implies conditional statistical independence, and missing d-separation implies missing conditional independence D-Connection: X and Y are d-connected if and only if either (1) there is a causal path between them or (2) there is evidence that renders the two nodes correlated with each other.

Missing D-Separation
Let G be a complete network (path diagram)

1. Initialize counter k=0 indicating the maximum number of nodes in the steps below.
2. For each pair of nodes X,Y, connected in G by an egde and possessing more than k neighbors in Markov blanket each, check if for any subset of neighbors of X with cardinality (size) exactly k, the variables X,Y are *conditionally independent*. If so, mark the arc (X,Y) for DELETION from G.
3. k=k+1. if more than k neighbors each, go to step 2. Otherwise go to step 4.
4. Repeat for 3 variables X,Y,Z paths, where the end nodes of the path (unconnected edges) are checked for D-separation with their respective neighbors
5. Repeat for each four variables X,Y,Z,T

Similarly, missing D-Connection algorithm is formulated.

3.2   The Framework Algorithm

Phase 1: Search
   Start with a fully connected graph G from the output.
1. Search for paths of the format (x, Z, y) for each pair (x, y) ∈ G from node's Markov blanket.
2. From the search enlist the d-separation rules (condition set) identified in G which is to be tested.
3. Prepare arcs list

Phase 2: Conformance / Compliance
   Check for the following conformances and mark edges for deletion.
   Belief 1: Perform d-separation testing on the condition set. Also perform following two algorithms to obtain arcs that are marked for deletion at later stage.
   1. Missing D-separation algorithm
   2. Missing D-connection algorithm
   Given Model1 constraints and actual model in network, mark belief1 as PASS OR FAIL
   Belief 2: Alternative paths hypothesis is tested here which includes corresponding constraints
   Given Model2 constraints and actual model in real network, mark Belief2 as PASS or FAIL. Mark edges for deletion on the contradictory paths.
   Belief 3:
   Alternative explanation hypothesis is tested. Here the constraints are again brought upto 4 node paths. And *Belief 3* PASS or FAIL is decided.
   Belief 4:
   Paths are converted to IF THEN statements and constraints are laid down. We confine our discussion to 4 node path delay analysis. Short path analysis is a straightforward adaptation which, for the most part, is limited to finding max and min delays. If the constraints match the real network the model is PASS otherwise FAIL. The non conforming edges are marked for deletion.

Phase 3: Thinning (Edge DELETION phase)
   To try to avoid contradicting deletions
While there are edges marked for DELETION, do
   1.   Does the edge satisfy the following criteria:
      (a) Island property
      (b) Sink property
      (c) Acyclic property
   2.   If Model 1 and Model 3 are PASS and the edge for deletion by both models then permanently DELETE the arc and so on. (similar rules are applied)

Phase 4: Finalizing Network
   Check the final DAG
1. Test that Markov blanket of every node (parents and children)
2. Test directionality and delay
3. If both tests are successful return G

4   Experiments and Results

We have carried out experimental evaluation of this framework algorithm. Due to the intractability to test all candidate sets that could d-separate two given nodes discussed earlier, we could not test all possibilities, but have to restrict the complexity of the analysis. Further, it has been suggested by de la Fuente *et al* [9] to calculation the partial correlation only up to order n whose value is practically one or two. In our path analysis, we considered controlling over 4 variables whereas d-separation can be applied when multiple variables are controlled (observed). For our studies, we use a graph called random artificial network generated by connecting possible pairs of nodes having only a few connections per node that approximately matches the observed partial correlation constraints in the artificial data. The method used to generate this data is borrowed from our previous work [6].

The result is shown in Fig.4. Fig.4 (a) is the actual network and Fig.4 (b) is the network obtained after post processing step carried out. It is clear that even under ideal experimental conditions the networks structure can not be inferred perfectly if a method is applied solely based on partial correlations because with at least 30% or more of the network was found wrong.

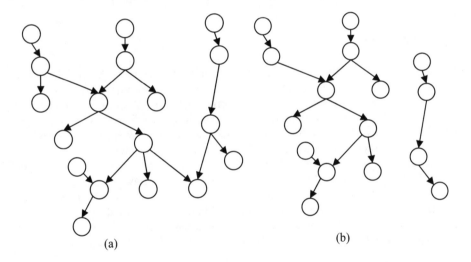

(a)   (b)

Fig. 4. (a) and (b) Results from using artificial data and Network

Fig.5 is the section of the actual yeast network [6, 13] structure and the network after post processing step is carried out where the thick dark lines indicate the barrier and arcs cutting through the barrier were deleted after post processing step. When the fitness measure was re-computed after post processing was carried out, nearly 20% accuracy improvement was noticed in result. This shows that the algorithm delivers more plausible networks close to the actual network.

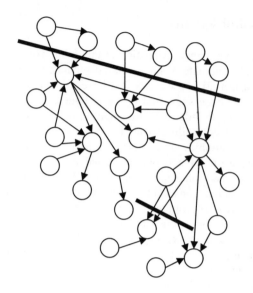

Fig. 5. Results from using Real dataset and network

5 Conclusion

We developed a heuristic approach used to reconstruct gene networks from low-order partial correlations and a GA. In this paper we devised a post processing step for path analysis to improve the accuracy of the inferred network. The framework approach incorporates d-separation, alternative causal hypothesis and time delay as tools. The framework is tested with subsets of the artificial and real yeast networks and has shown overall precision of the inferable network structure improved by upto 20%. Further studies are necessary to demonstrate that these results hold also for different network scenarios, and also whose structure is more close to the structure of biological gene networks. This approach is slightly away from normal methods; as it involves delay analysis and d-separation. Although the method is attractive, for undesirable or impossible cases, such as a situations were time delay is irrelevant, the corresponding beliefs can be turned off in phase 2 of the framework algorithm.

References

[1] Friedman, N., Linial, M., Nachman, I., Pe'er, D.: Using Bayesian networks to analyse expression data. Journal on Computational Biology 7, 601–620 (2000)
[2] Liang, S., Fuhrman, S., Somogyi, R.: REVEAL, a general reverse engineering algorithm for inference of genetic network architecture. In: Pacific Symposium on Biocomputing, vol. 3, pp. 18–29 (1998)
[3] Ando, S., Iba, H.: Inference of gene regulatory model by genetic algorithms. In: Proc. Conference on Evolutionary Computation, pp. 712–719 (2001)

 [4] Wahde, M., Hertz, J.: Modeling genetic regulatory dynamics in neural development. Journal on Computational Biology 8, 429–442 (2001)
 [5] Ram, R., Chetty, M., Dix, T.I.: Causal Modeling of Gene Regulatory Network. In: IEEE Symposium on Computational Intelligence in Bioinformatics and Computational Biology (CIBCB). IEEE Computer Society Press, Los Alamitos (2006)
 [6] Ram, R., Chetty, M., Dix, T.I.: Learning Structure of Gene Regulatory Networks. In: 6th IEEE International Conference on Computer and Information Science (ICIS) (accepted)
 [7] Spirtes, P., Glymour, C., Scheines, R.: Causation, Prediction, and Search. Springer, Heidelberg (2001)
 [8] Pearl, J.: Causality: Models, Reasoning and Inference. Cambridge University Press, Cambridge (2001)
 [9] de la Fuente, A., Bing, N., Hoeschele, I., Mendes, P.: Discovery of meaningful associations in genomic data using partial correlation coefficients. Bioinformatics 20(18), 3565–3574 (2004)
[10] Barabasi, A.L., Oltvai, Z.N.: Network biology: Understanding the cell's functional organization. Nature Reviews 5, 101–113 (2004)
[11] Basso, K., Margolin, A.A., Stolovitzky, G., Klein, U., Dalla-Favera, R., Califano: A Reverse engineering of regulatory networks in human b cells. Nature Genetics 37(4), 382–390 (2005)
[12] van Noort, V., Snel, B., Huymen, M.A.: The yeast coexpression network has a small-world, scale-free architecture and can be explained by a simple model. EMBO reports 5(3), 280–284 (2004)
[13] Spellman, P.T., Sherlock, G., Zhang, M.Q., Iyer, V.R., Anders, K., Eisen, M.B., Brown, P.O., Botstein, D., Futcher, B.: Comprehensive identification of cell cycle-regulated genes of the yeast Saccharomyces cerevisiae by microarray hybridization. Mol. Biol. Cell. 9(12), 3273–3297 (1998)

Transcriptional Gene Regulatory Network Reconstruction Through Cross Platform Gene Network Fusion

Muhammad Shoaib B. Sehgal[1,3], Iqbal Gondal[1,3], Laurence Dooley[1],
Ross Coppel[2,3], and Goh Kiah Mok[4]

[1] Faculty of IT, Monash University, Churchill VIC. 3842, Australia
{Shoaib.Sehgal,Iqbal.Gondal,
Laurence.Dooley}@infotech.monash.edu.au
[2] Department of Microbiology, Monash University Clayton VIC. 3800, Australia
Ross.Coppel@med.monash.edu.au
[3] Victorian Bioinformatics Consortium Wellington Road, Clayton VIC.
3800, Australia
[4] SIMTech Institute of Technology, Nanyang Drive, Singapore
kmgoh@SIMTech.a-star.edu.sg

Abstract. Microarray gene expression data is used to model differential activity in *Gene Regulatory Networks* (GRN) to elucidate complex cellular processes, though network modeling is susceptible to errors due to both noisy nature of gene expression data and platform bias. This intuitively provided the motivation for the development of an innovative technique, which effectively integrates GRN using cross-platform data to minimize the two aforementioned effects. This paper presents a GRN integration (GeNi) framework that fuses cross-platform GRN to remove platform and experimental bias using the *Dempster Shafer Theory of Evidence*. The proposed model estimates gene co-regulation strength by using mutual information and removes spurious co-regulations through data processing inequality. The method automatically adapts to the data distribution using Belief theory, which does not require a preset threshold to accept co-regulated links. [1]GeNi is applied to identify common cancer-related regulatory links in ten different datasets generated by different microarray platforms including *cDNA* and Affymetrix arrays. Experimental results demonstrate that GeNi can be effectively applied for GRN reconstruction and cross-platform gene network fusion for any gene expression data.

1 Introduction

Gene expression analysis has been widely used for different biological studies. Several statistical and computational intelligence modeling methods have been applied for this purpose. While these techniques provide biologists with valuable insights of

---

[1] GeNi Software and Supplementary Material can be downloaded from www.gscit.monash.edu.au/ ~shoaib or can be requested by emailing at {Shoaib.Sehgal AT gmail.com}.

J.C. Rajapakse, B. Schmidt, and G. Volkert (Eds.): PRIB 2007, LNBI 4774, pp. 274–285, 2007.
© Springer-Verlag Berlin Heidelberg 2007

different biological processes, most analyses are based on over/under expressed genes studies despite the fact that differential expression analysis do not fully harness the potential of microarray gene expression data because genes are treated independently and interactions between them are overlooked [1].

Gene Regulatory Networks (GRN) model how genes regulate different metabolism and can map the casual pathways. GRN reconstruction is however error prone due to the noisy nature of microarray data and microarray platform and experimental bias. One possible solution is to integrate networks constructed through different microarray platforms, e.g. *cDNA* or Affymetrix high density oligonucleotide arrays, under either different or similar experimental conditions, though this is a challenging task because data generated by different platforms is not directly comparable. The objectives include being able to construct models capable of inferring knowledge from thousands of genes at a time, assist in understanding complex genetic interactions and to integrate networks constructed from heterogeneous microarray datasets [2, 3].

Previous attempts to integrate cross platform GRN include Zhou *et al.* [3] who proposed cross platform GRN fusion using second-order expression analysis while Choi *et al.* [1] studied different types of cancer links using cross-platform analysis. In both studies the regulatory pathway was only considered if it was present in more than a certain number of experiments T, where the selection of this threshold T was empirically derived with no formal mathematical foundations so that selection of an incorrect T could inevitably lead to erroneous results. Furthermore, most GRN modeling techniques incur several limitations, including exponential time space complexity, unrealistic GRN assumptions such as acyclic network by bayesian networks, overfitting and under constrained regression analysis [4]. This has created a need for suitable techniques that are scaleable and do not impose unrealistic assumptions on the network structure.

This paper proposes a novel *Gene Regulatory Network Integration* (GeNi) Framework to model GRN. The proposed model integrates GRN generated from different platforms using *Dempster Shafer Theory of Evidence* (DSTE) [5]. The GeNi computes gene to gene co-regulation using mutual information. Mutual Information is selected due to its proven improved performance compared to commonly used correlation based methods and Bayesian Networks [4]. The other advantage of using mutual information is that it does not enforce the acyclic assumption as posed by Bayesian networks and is more scaleable than dynamic Bayesian networks, which remove this acyclic restriction. Mutual Information for GRN reconstruction has been used by Baso *et al.* [4] and Zhao *et al.* [6] though these methods can only be used for single data and doesn't reconstruct network through cross platform network integration to remove bias and minimize the impact of noise. Also, GeNi has added advantage over other mutual information based techniques that it does not require threshold to select co-regulated links because it uses belief theory to accept/reject co-regulated links. After mutual information computation, GeNi then prunes the network using data processing inequality to remove the spurious co-regulations. Finally, the fusion of different gene networks is performed by using the belief theory.

The proposed model is tested for its application to find tumor specific links in various cancer datasets generated by different cDNA and Affemtrix microarray platforms. The results corroborate that GeNi can be effectively used to fuse cross-platform GRN.

The rest of the paper is organized as follows: Section 2 presents GeNi model in detail. Analysis of results is presented in Section 3 while conclusions are drawn in Section 4.

2 Gene Regulatory Network Integration (GeNi) Model

The complete GeNi framework is formalized in Fig. 1. Gene expression data is firstly preprocessed to remove noise and outliers followed by gene to gene mutual information computation to measure gene co-regulation strength. The network is then pruned using data processing inequality, before network fusion is undertaken using DSTE. Each of these constituent blocks is now considered in the following sub-sections, with the rationale for the choice of each algorithm being delineated.

2.1 Pre-processing

The data is preprocessed to minimize the affect of noise on subsequent analysis. Negative values in Microarray data are considered as missing and genes with greater than 70% missing values and less than 4 observations are filtered out. Gene expression data is then re-parameterized using rank transformation to convert each gene into equally spaced expressions between the interval [0 1] [4] (Step1 - Fig. 1). Missing values in the data are then imputed by their gene averages. Finally, each clone was mapped to UniGene accession build # 162 to manage heterogeneous data.

Pre Condition: Gene expression matrices Y_N and Y_T for normal and cancerous data.

---

    1. Preprocess (Section 2.1)
    2. Construct GRN using mutual information (Section 2.2).
    3. Remove spurious gene links using data processing inequality (Section 2.3).
    4. Fuse cross platform networks using Dempster Shafer theory (Section 2.4).
    5. Compare Normal and Cancerous fused networks to find out *Conserved*, *Broken* and *Tumor links* (Section 2.5).
    6. Stop

Post Condition: Gene Regulatory Networks N_n, N_t for normal and cancerous data.

Fig. 1. GRN integration algorithm

The pre-processing step is followed by GRN reconstruction for each data set (See Step2 - Fig. 1). Following sub-section explains this step in detail.

2.2 GRN Reconstruction

After pre-processing, GeNi computes pair wise mutual information for all genes to construct gene networks. The mutual information $I(g_1,g_2)$, between two genes g_1 and g_2 is computed using *Gaussian Kernel Estimator* as:

$$I(g_1, g_2) = \frac{1}{m} \sum_{i=1}^{m} log \left[\frac{p(g_{1i}, g_{2i})}{p(g_{1i}) p(g_{2i})} \right] \qquad (1)$$

where

$$p(g_{1i}) = \frac{1}{\sqrt{2\pi} N \alpha_1} \sum_j e^{\frac{(g_{1i} - g_{1j})}{2\alpha_1^2}}, \qquad (2)$$

$$p(g_{2i}) = \frac{1}{\sqrt{2\pi} N \alpha_1} \sum_j e^{\frac{(g_{2i} - g_{2j})}{2\alpha_1^2}}, \qquad (3)$$

$$p(g_{1i}, g_{2i}) = \frac{1}{\sqrt{2\pi} N \alpha_2} \sum_j e^{\frac{(g_{1i} - g_{1j}) + (g_{2i} - g_{2j})}{2\alpha_2^2}}. \qquad (4)$$

where α_1 and α_2 are tunable parameter and computed by *Monte Carlo Simulations* [7] using bi-variate normal probability densities [4].

Mutual information computation step is followed by network pruning using Data Processing Inequality which is explained in the next sub-section.

2.3 Network Pruning

When two genes g_1 and g_2 are interacting through a third gene G_3 and $I(G_1, G_2| G_3)$ is zero then these genes are directly interacting with each other if:

$$I(g_1, g_3) \leq I(g_1, g_2) \text{ and } I(g_1, g_3) \leq I(g_2, g_3). \qquad (5)$$

As this property is asymmetric it has the possibility of rejecting some of the loops or interactions between three genes whose information may not be fully modeled by pair wise mutual information. The introduction of a tolerance threshold addresses this problem as well as provides the advantage of avoiding rejection of some of thetriangular links and loops [4].

2.4 Cross Platform GRN Fusion

GeNi fuses cross platform networks using DSTE (Step 4 - Fig. 1.) [5], as alluded to in Section 1. The theory extends Bayesian theory to evaluate beliefs from different evidences. The DSTE allows beliefs to be represented by upper and lower probability intervals normally referred to as belief and plausibility respectively [5].

The DSTE assumes that the information sources are independent of each other. This assumption makes it further feasible to use in GeNi as it first constructs GRN independent of each other using data generated by heterogeneous platforms under independent studies.

The application of this theory for cross platform GRN fusion requires a definition of the degree of belief (mass functions) to assign masses, normally referred to as probability value. The DSTE doesn't mandate the method of computing these masses (probabilities) [8], which makes it a more generalized approach than Bayesian theory [9-11]. It adds flexibility to GeNi, as belief masses can be calculated using any GRN reconstruction method (Correlation, Probability value or Mutual Information) to compute the gene co-regulation.

For GRN fusion, the $\Omega = \{R, NR\}$ represents mutually exclusive event space for *Co-Regulated* (R) and *Non Co-Regulated* (NR) links, called frame of discernment or universe of discourse. The $2^{\Omega} = \{ \phi, \{R\}, \{NR\}, \{R, NR\}\}$ represents the set of all subsets of Ω, and classes in Ω are considered mutually exclusive. Let A be a non-zero degree of belief in 2^{Ω}, called the focal element where:

$$\sum_{A \subseteq \Omega} m(A) = 1 \ and \ m(\phi) = 0 \tag{6}$$

Focal elements and their masses construct an evidence structure, which can be expressed as:

$$\{(A, m(A)) | A \subseteq \Omega, m(A) > 0\} \tag{7}$$

The value $m(A)$ represents the weight of evidence in favor of complete set A. The belief function, which is a sum of masses of all subsets of hypothesis for R and NR, can be computed as:

$$Bel(R) = \sum_{A \subseteq R} m(R) \ and \ Bel(NR) = \sum_{A \subseteq NR} m(NR) \tag{8}$$

The same information can be computed by calculating plausibility or upper probability value, which is the sum of the masses of all sets whose intersection with the hypothesis is empty [12]. The plausibility of R can be defined as:

$$Pl(R) = \sum_{A \cap R = \phi} m(R) \ and \ Pl(NR) = \sum_{A \cap NR = \phi} m(NR) \tag{9}$$

Similarly plausibility for ϕ is

$$Pl(\phi) = 0 \tag{10}$$

The relation between plausibility and belief masses can be expressed as: $Bel(R) \leq Pl(R) \ and \ Pl(R) = 1 - Bel(\overline{R})$ where $\overline{R} = \Omega - R$.

The belief masses are the gene co-regulation probability or correlation values computed in STEP 2 (Fig. 1). Figure 2 represents a schematic diagram for the fusion of belief masses where g_1, g_2, g_3 and g_4 are the genes sets. These genes are triggered by different regulation weights (P_1, P_2 ...P_n) that represent belief masses $m(R)$) and $m(NR)$ for co-regulated and not co-regulated weights respectively, in experiments $\{E_1, E_2 ... E_n\}$. The combined belief F_k for the gene co-regulation is computed by:

$$Bel(F_k) = \frac{\sum_{R_i \cap NR_j = F_k ; F_k \neq \phi} m(R_i) \oplus m(NR_j)}{1 - \sum_{R_i \cap NR_j = \phi} m(R_i) \oplus m(NR_j)} \qquad (11)$$

where $\oplus$ represents the orthogonal sum and can be computed for n experiments as:

$$\oplus_{i=1}^{n} m_i(A) = \sum_{A_1 \cap \ldots \cap A_n = A} \prod_{i=1} m_i(A_i) \qquad (12)$$

The combination can be normalized by introducing normalization factor N_b such that:

$$\oplus_{i=1}^{n} m_i(A) = N_b \sum_{A_1 \cap \ldots \cap A_n = A} \prod_{i=1} m_i(A_i) \qquad (13)$$

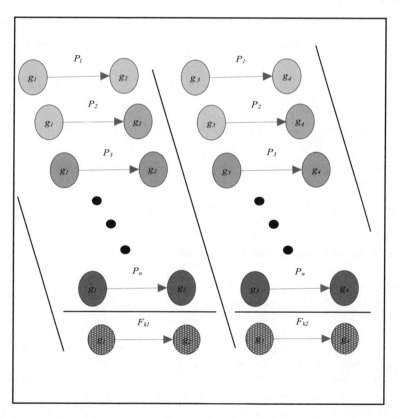

Fig. 2. A schematic diagram presenting fusion of belief masses in GeNi. The probabilities $P_1 \ldots P_n$ are the co-regulation weights of $g_1 \rightarrow g_2$ links computed from heterogeneous datasets. The F_k is the final co-regulation weight calculated using DSTE. The link is accepted or rejected based on the value of F_k.

The value of N_b was set to 0.5; however, GeNi doesn't restrict the use of any normalization value. The denominator in (11) normalizes the output to the safer belief function and serves to distribute any mass associated with ϕ intersections of beliefs to the non empty intersections [13]. The genes co-regulation link is added to the fused network if the combined belief of R is higher than NR where F_k represents the link weight.

After the integrated network is constructed, the network is pruned to remove the platform bias and the links that occur by chance. Only those regulatory links that are present in more than E_n experiments where $E_n > n/4$ and $E_n > 8$ [3] are added to the final fused network. It should be noted, however, that network pruning is an optional step in GeNi and this step is different from threshold-based integration based methods, as they don't consider the link co-regulation weight-age while selecting the links for the final integration, as mentioned earlier, GeNi adds/removes links, primarily based on belief masses $m(R)$ and $m(NR)$.

2.5 Network Comparison

Once the fused networks have been constructed they are compared for Broken, Conserved and Tumor links (See Step 5 - Fig. 1). The precise definition for each of these links is now given:

Definition 1. A link is a *Conserved Link* if it is present in both normal and tumor networks.

Definition 2. A link is a *Broken Link* if it is present in normal network and is missing in tumor network.

Definition 3. A link is a *Tumor Link* if it is not present in normal network but exists in tumor networks.

The next section provides analysis of GRN constructed using GeNi.

3 Analysis of Results and Discussion

For cross platform GRN fusion, 10 different datasets under 11 different experimental conditions (Table 1), designed for the comparison of primary cancer and non cancer counterpart, were used. These datasets were generated using different microarray platforms including cDNA and Affmetrix GeneChip. The datasets were collected from breast, pancreas, colon, brain, bladder, ovary, uterus, kidney, liver, lung, lymphoma, stomach and prostate tissues and had 5603, 17660, 3697, 3732, 5575, 13171, 12065, 5983, 4615, 24822, 6593 genes respectively (Table 1). The total number of genes in all experiments were 103,516 (Choi *et al.* [1] for further details). To construct the fused network, we selected 61 commonly present, regulated genes from the above datasets. The gene networks were first individually constructed using (Steps 1-3 - Fig. 1) and then these network were integrated using belief theory to form fused

Table 1. Datasets

Tissues	Platforms	Normal Samples	Tumor Samples
Breast [14]	cDNA	13	13(72)
Colon [15]	Hu6800	22	22
Kidney [16]	cDNA	81	81
Liver [17]	cDNA	76	76(104)
Lung [18]	U95A	17	17(127)
Lymphoma [19]	cDNA	31	31(77)
Pancreas [20]	cDNA	14	22
Prostate [21]	U95A	50	52
Stomach [22]	cDNA	29	29(103)
Brain,	Hu6800	8	20
Bladder,	Hu35KSubA	7	11
Ovary	Hu35KSubA	3	11
Uterus [23]	Hu35KSubA	6	10

networks for both normal and tumor data (Figs. 3-5). These fused normal and tumor networks were then compared to search for *Broken*, *Conserved* and *Tumor* links.

Table 2 shows selected Broken, Conserved and Tumor links. The results demonstrate that 52% of the links were broken links in tumor tissue samples which were present in normal tissues while only 2% links were newly created in tumor cells compared to normal tissues. Only 45% of the links were conserved between normal and tumor tissues. These links can be used to monitor patient's response to certain treatment. For instance, if the response of patient to the treatment is positive then the number of conserved links should increase while concomitantly decreasing the broken and tumor links.

Figure 3 plots a selected section of normal and tumor networks for comparison where complete normal and tumor networks for commonly selected genes are shown in Figs. 4 and 5 (Individual networks can be downloaded from www.gscit.monash. edu.au/~shoaib/GeNi.html). It is evident from Figs. 3-4 that normal data has high percentage of connected nodes compared to tumor network. Figure 3 shows several inserting observations for instance, a link from *nuclear factor of activated T-cells, cytoplasmic, calcineurin-dependent 3* (HS.172674) to *protein phosphatase 2 (formerly 2A), regulatory subunit A (PR 65), beta isoform* (HS.431156) is present in normal network but is broken in the tumor network. A new link is created between *protein phosphatase 2 (formerly 2A), regulatory subunit A (PR 65), beta isoform* (HS.431156) and *Sulfotransferase family, cytosolic, 1A, phenol-preferring, member* (HS.368950) in tumor network, which was not present in the normal network. Figure 3 also shows the conserved link between *nuclear factor of activated T-cells, cytoplasmic, calcineurin-dependent 3* (HS.172674) and *Sulfotransferase family, cytosolic, 1A, phenol-preferring, member 1* (HS.368950), which is present in both datasets.

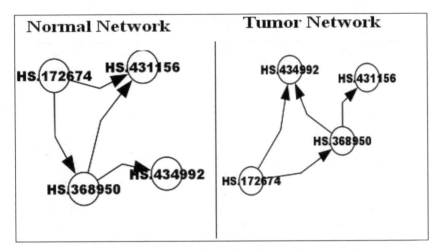

Fig. 3. Cross-section of normal and tumor tissue networks

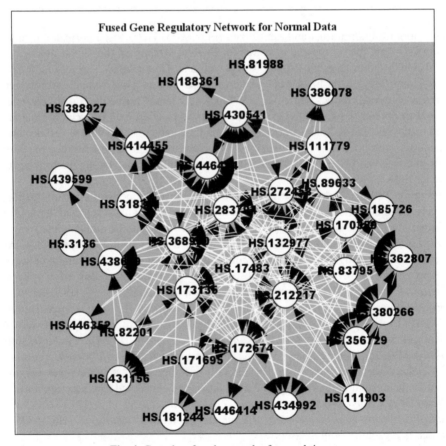

Fig. 4. Complete fused network of normal tissues

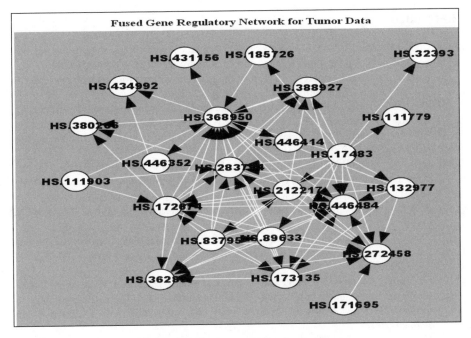

Fig. 5. Complete fused network of tumor tissues

Table 2. Number of genes involved in GRN links

GRN Links	Broken Links	Conserved Links	Tumor Links
% Links	52%	45%	2%

These results above all demonstrate that GeNi can indeed be used for cross-platform network fusion however; further wet laboratory results are required in order to completely verify the model.

4 Conclusions

The paper has presented GRN integration (GeNi) framework to fuse cross-platform GRN in order to remove platform and experimental bias. The proposed model estimates gene co-regulation strength by using mutual information and removes spurious co-regulations by using data processing inequality. The method automatically adapts to the data distribution using Belief theory and hence does not require preset threshold to accept the co-regulated links which makes method more robust for GRN reconstruction. The GeNi was used to find common cancer related regulatory links in ten different datasets generated by different microarray platforms including cDNA and Affymetrix arrays. The experimental results demonstrated that GeNi can be applied successfully for GRN reconstruction and cross-platform gene network fusion for various types of genetic data.

References

[1] Choi, J.K., Yu, U., Yoo, O.J., Kim, S.: Differential coexpression analysis using microarray data and its application to human cancer. Bioinformatics 21, 4348–4355 (2005)

[2] Fort, G., Lambert-Lacroix, S.: Classification using partial least squares with penalized logistic regression. Bioinformatics 21, 1104–1111 (2005)

[3] Zhou, X.J., Ming-Chih, Kao, J., Huang, H., Wong, A., Nunez-Iglesias, J., Primig, M., Aparicio, O.M., Finch, C.E., Morgan, T.E., Wong, W.H.: Functional annotation and network reconstruction through cross-platform integration of microarray data. Nature Biotechnology 23, 238–243 (2005)

[4] Basso, K., Margolin, A.A., Stolovitzky, G., Klein, U., Dalla-Favera, R., Califano, A.: Reverse engineering of regulatory networks in human B cells. Nature Genetics 37, 382–390 (2005)

[5] Shafer, G.: Mathematical Theory of Evidence. Princeton Univ. Press, Princeton, NJ (1976)

[6] Zhao, W., Serpedin, E., Dougherty, E.R.: Inferring gene regulatory networks from time series data using the minimum description length principle. Bioinformatics 22(17), 2129–2135 (2006)

[7] Casella, G., Robert, C.P.: Monte Carlo Statistical Methods. Springer, Heidelberg (2005)

[8] Malpicaa, J.A., Alonsoa, M.C., Sanz, M.A.: Dempster–Shafer Theory in geographic information systems: A survey, Expert Systems with Applications, vol. 32. Elsevier, Amsterdam (2007)

[9] Hegarat-Mascle, S.L., Bloch, I., Vidal-Madjar, D.: Application of Dempster-Shafer evidence theory to unsupervised classification in multisource remote sensing. IEEE Trans. Geosci. Remote Sensing 35, 1018–1031 (1997)

[10] Bloch, I.: Some aspects of Dempster-Shafer evidence theory for classification of multimodality medical images taking partial volume effect into account. Pattern Recognition Letters 17, 905–919 (1996)

[11] Rombaut, M., Zhu, Y.M.: Study of Dempster–Shafer for image segmentation applications. Image Vision Comput. 20, 15–23 (2002)

[12] Barnett, J.A.: Calculating Dempster-Shafer plausibility. IEEE Transactions on Pattern Analysis and Machine Intelligence 13, 599–602 (1991)

[13] Murphy, R.R.: Dempster-Shafer Theory for Sensor Fusion in Autonomous Mobile Robots. IEEE Transactions on Robotics and Automation 14, 197–206 (1998)

[14] Sorlie, T., Perou, C., Tibshirani, R., Aas, T., Geisler, S., Johnsen, H., Hastie, T., Eisen, M., Rijn, M.v.d., Jeffrey, S., Thorsen, T., Quist, H., Matese, J., Brown, P., Botstein, D., Lonning, P.E., Borresen-Dale, A.: Gene expression patterns of breast carcinomas distinguish tumor subclasses with clinical implications. Proc. Natl. Acad. Sci. 11, 98(19), 10869–10874 (2001)

[15] Notterman, D.A., Alon, U., Sierk, A.J., Levine, A.J.: Transcriptional Gene Expression Profiles of Colorectal Adenoma, Adenocarcinoma, and Normal Tissue Examined by Oligonucleotide Arrays. Cancer Res. 61, 3124–3130 (2001)

[16] Boer, J.M., et al.: Identification and classification of differentially expressed genes in renal cell carcinoma by expression profiling on a global human 31,500-element cDNA array. Genome Research 11, 1861–1870 (2001)

[17] Chen, X., Cheung, S.T., So, S., Fan, S.T., et al.: Gene Expression Patterns in Human Liver Cancers. Mol. Biol. Cell 13, 1929–1939 (2002)

[18] Bhattacharjee, A., Richards, W.G., Staunton, J., Li, C., Monti, S., Vasa, P., Ladd, C., Beheshti, J., Bueno, R., Gillette, M., Loda, M., Weber, G., Mark, E.F., Lander, E.S., Wong, W., Johnson, B.E., Golub, T.R., Sugarbaker, D.J., Meyerson, M.: Classification of human lung carcinomas by mRNA expression profiling reveals distinct adenocarcinoma subclasses. Proc. Natl. Acad. Sci. 13790–13795 (2001)

[19] Alizadeh, A.A., et al.: Distinct types of diffuse large B-cell lymphoma identified by gene expression profiling. Nature 403, 503–511 (2000)

[20] Lacobuzio-Donahue, C.A., et al.: Exploration of Global Gene Expression Patterns in Pancreatic Adenocarcinoma Using cDNA Microarrays. Am. J. Pathol. 162, 1151–1162 (2003)

[21] Singh, D., et al.: Gene expression correlates of clinical prostate cancer behavior. Cancer Cell 1 (2002)

[22] Chen, X., et al.: Variation in gene expression patterns in human gastric cancers. Mol. Biol. Cell 14, 3208–3215 (2003)

[23] Ramaswamy, S., et al.: Multiclass cancer diagnosis using tumour gene expression signatures. Proc. Natl. Acad. Sci. 98(26), 15149–15154 (2001)

Reconstruction of Protein-Protein Interaction Pathways by Mining Subject-Verb-Objects Intermediates

Maurice HT Ling[1,2], Christophe Lefevre[3],
Kevin R. Nicholas[2], and Feng Lin[1]

[1] BioInformatics Research Centre, Nanyang Technological University, Singapore
[2] CRC for Innovative Dairy Products, Department of Zoology,
The University of Melbourne, Australia
[3] Victorian Bioinformatics Consortium, Monash University, Australia
mauriceling@acm.org, k.nicholas@zoology.unimelb.edu.au,
Chris.Lefevre@med.monash.edu.au, ASFLIN@ntu.edu.sg

Abstract. The exponential increase in publication rate of new articles is limiting access of researchers to relevant literature. This has prompted the use of text mining tools to extract key biological information. Previous studies have reported extensive modification of existing generic text processors to process biological text. However, this requirement for modification had not been examined. In this study, we have constructed Muscorian, using MontyLingua, a generic text processor. It uses a two-layered generalization-specialization paradigm previously proposed where text was generically processed to a suitable intermediate format before domain-specific data extraction techniques are applied at the specialization layer. Evaluation using a corpus and experts indicated 86-90% precision and approximately 30% recall in extracting protein-protein interactions, which was comparable to previous studies using either specialized biological text processing tools or modified existing tools. Our study had also demonstrated the flexibility of the two-layered generalization-specialization paradigm by using the same generalization layer for two specialized information extraction tasks.

Keywords: biomedical literature analysis, protein-protein interaction, monty lingua.

1 Introduction

PubMed currently indexes more than 16 million papers with about one million papers and 1.2 million added in the years 2005 and 2006 respectively. A simple keyword search in PubMed showed that nearly 900 thousand papers on mouse and more than 1.3 million papers on rat research had been indexed in PubMed to date, and in the last four years, more than 150 thousand papers have been published on each of mouse and rat research. This trend of increased volume of research papers indexed in PubMed over the last 10 years makes it difficult for researchers to maintain an active and productive assessment of relevant literature. Information extraction (IE) has been used as a tool to analyze biological text to derive assertions on specific biological domains [30], such as protein phosphorylation [19] or entity interactions [1].

J.C. Rajapakse, B. Schmidt, and G. Volkert (Eds.): PRIB 2007, LNBI 4774, pp. 286–299, 2007.
© Springer-Verlag Berlin Heidelberg 2007

A number of IE tools used for mining information from biological text can be classified according to their capacity for general application or tools that considers biological text as specialized text requiring domain-specific tools to process them. This has led to the development of specialized part-of-speech (POS) tag sets (such as SPECIALIST [28]), POS taggers (such as MedPost [33]), ontologies [11], text processors (such as MedLEE [15]), and full IE systems, such as GENIES [16], MedScan [29], MeKE [4], Arizona Relation Parser [10], and GIS [5]. On the other hand, an alternative approach assumes that biological text are not specialized enough to warrant re-development of tools but adaptation of existing or generic tools will suffice. To this end, BioRAT [12] had modified GATE [8], MedTAKMI [36] had modified TAKMI [27], originally used in call centres, Santos [31] had used Link grammar parser [32].

Although both systems demonstrated similar performance, either developing these systems or modifying existing systems were time consuming [20]. Although work by Grover [17] suggested that native generic tools may be used for biological text, a recent review had highlighted successful uses of a generic text processing system, MontyLingua [14, 23], for a number of purposes [22]. For example, MontyLingua has been used to process published economics papers for concept extraction [35]. The need to modify generic text processors had not been formally examined and the question of whether an un-modified, generic text processor can be used in biological text analysis with comparable performance, remains to be assessed.

In this study, we evaluated a native, generic text processing system, MontyLingua [23], in a two-layered generalization-specialization architecture [29] where the generalization layer processes biological text into an intermediate knowledge representation for the specialization layer to extract genic or entity-entity interactions. This system demonstrated 86.1% precision using Learning Logic in Languages 2005 evaluation data [9], 88.1% and 90.7% precisions in extracting protein-protein binding and activation interactions respectively. Our results were comparable to previous work which modified generic text processing systems which reported precision ranging from 53% [24] to 84% [5], suggesting this modification may not improve the efficiency of information retrieval.

2 System Description

We have developed a biological text mining system, known as Muscorian, for mining protein-protein inter-relationships in the form of subject-relation-object (for example, protein X bind protein Y) assertions. Muscorian is implemented as a 3-module sequential system of entity normalization, text analysis, and protein-protein binding finding, as shown in Figure 1. It is available for academic and non-profit users through http://ib-dwb.sf.net/Muscorian.html.

2.1 Entity Normalization

Entity normalization is the substitution of the long form of either a biological or chemical term with its abbreviated form. This is essential to correct part-of-speech

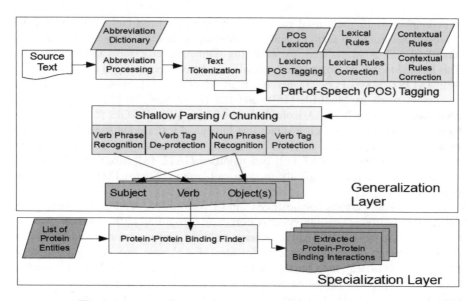

Fig. 1. Schematic Diagram Illustrating the Operations of Muscorian

tagging errors which are common in biological text due to multi-worded nouns. For example, the protein name "phosphatase and tensin homolog deleted on chromosome 10" has to be recognized as a single noun and not a phrase. In this study, we attempt to mine protein-protein interactions and consolidate this knowledge to produce a map. Therefore, the naming convention of the protein entities must be standardized to allow for matching. However, this is not the case for biological text and synonymous protein names exist for virtually every protein. For example, "MAP kinase kinase", "MAPKK", "MEK" and "MAPK/Erk kinase" referred to the same protein. Both of these problems could be either resolved or minimized by reducing multi-worded nouns into their abbreviated forms.

A dictionary-based approach was used for entity normalization to a high level of accuracy and consistency. The dictionary was assembled as follows: firstly, a set of 25000 abstracts from PubMed was used to interrogate Stanford University's BioNLP server [3] to obtain a list of long forms with its abbreviations and a calculated score. Secondly, only results with the score of more than 0.88 were retained as it is an inflection point of ROC graph [3], which is a good balance between obtaining the most information while reducing curation efforts. Lastly, the set of long form and its abbreviations was manually curated with the help of domain experts.

The domain experts curated dictionary of long forms and its abbreviated term was used to construct a regular expression engine for the process of recognition of the long form of a biological or chemical term and substituting it with its corresponding abbreviated form.

2.2 Text Analysis

Entity normalized abstracts were then analyzed textually by an un-modified text processing engine, MontyLingua [14], where they were tokenized, part-of-speech tagged, chunked, stemmed and processed into a set of assertions in the form of 3-element subject-verb-object(s) (SVO) tuple, or more generally, subject-relation-object(s) tuple. Therefore, a sequential pattern of words which formed an abstract was transformed through a series of pattern recognition into a set of structurally-definable assertions.

Before part-of-speech tagging is possible, an abstract made up of one or more sentences had to be separated into individual sentences. This is done by regular expression recognition of sentence delimiters, such as full-stop, ellipse, exclamation mark and question mark, at the end of a word (regular expression: ([?!]+|[.][.]+)$) with an exception of acronyms. Acronyms, which are commonly represented with a full-stop, for example "Dr.", are not denoted as the end of a sentence and were generally prevented by an enumeration of common acronyms.

Individual sentences were then separated into constituent words and punctuations by a process known as tokenization. Tokenization, which is essential to atomize a sentence into atomic syntactic building blocks, is generally a simple process of splitting of an English sentence in words using whitespaces in the sentence, resulting in a list of tokens (words). However, there were three problems which were corrected by examining each token. Firstly, punctuations are crucial in understand a written English sentence, but typographically a punctuation is usually joined to the presiding word. Hence, punctuation separation from the presiding word is necessary. However, it resulted in incorrect tokenization with respect to acronyms and decimal numbers. For example, "... an appt. for ..." will be tokenized to "... an appt . for ..." and "$4.20'" will be "$ 4 . 20". This problem was prevented by pre-defining acronyms and using regular expressions, such as "^[$][0-9]{1,3}[.][0-9][0-9](?[.]?)$". Lastly, common abbreviated words, such as "don't", were expanded into two tokens of "do" and "n't". Despite the above error correction measures, certain text such as mathematical equations, which might be used to describe enzyme kinetics in biological text, will not be tokenized correctly. In spite of this limitation, the described tokenization scheme is still appropriate as extraction of enzyme kinetics or mathematical representations are not the aims of this study.

Each of the tokens (words and punctuations) in a tokenized sentence is then tagged using Penn TreeBank Tag Set [25] by a Brill Tagger, trained on Wall Street Journal and Brown corpora, which operates in two phases. Using a lexicon, containing the likely tag for each word, each word is tagged. This is followed by a phase of correction using lexical and contextual rules, which were learnt using training with a tagged corpora, in this case, Wall Street Journal and Brown corpora. Lexical rules uses a combination of preceding tag and prefix or suffix of the token (word) in question. For example, the rule "NN ing fhassuf 3 VBG" defines that if the current token is tagged as a noun (NN) and has a 3-character suffix of "ing", then the tag should be a verb (VBG). On the other hand, contextual rules uses only the preceding or proceeding tags and hence, must be applied after lexical rules for effectiveness. The contextual rule "RB JJ NEXTTAG NN" defines that an abverbial tag (RB) should be changed to an adjective (JJ) if the next token was tagged as a noun (NN). A table of Penn Treebank Tag Set [25] without punctuation tags is given in Table 1.

Table 1. Penn Treebank Tag Set without Punctuation Tags (Adapted from [25])

Tag	Description	Tag	Description
CC	Coordinating conjunction	PRP$	Possessive pronoun
CD	Cardinal number	RB	Adverb
DT	Determinant	RBR	Adverb, comparative
EX	Existential *there*	RBS	Adverb, superlative
FW	Foreign word	RP	Particle
IN	Preposition or subordinating conjunction	SYM	Symbol
JJ	Adjective	TO	to
JJR	Adjective, comparative	UH	Interjection
JJS	Adjective, superlative	VB	Verb, base form
LS	List item marker	VBD	Verb, past tense
MD	Modal	VBN	Verb, past participle
NN	Noun, singular or mass	VBG	Verb, gerund or present participle
NNS	Noun, plural	VBP	Verb, non-3rd person singular present
NNP	Proper noun, singular	VBZ	Verb, 3rd person singular present
NNPS	Proper noun, plural	WDT	Wh-determiner
PDT	Predeterminer	WP	Wh-pronoun
POS	Possessive ending	WP$	Possessive wh-pronoun
PRP	Personal pronoun	WRB	Wh-adverb

By tagging, the complexity of an English sentence (ie, the number of ways an English sentence can be grammatically constructed with virtually unlimited words and unlimited ideas) was collapsed into a sequence of part-of-speech tags, in this case, Penn TreeBank Tag Set [25], with only about 40 tags. Therefore, tagging reduced the large number of English words to about 40 "words" or tags.

Generally, an English sentence is composed of a noun phrase, a verb, and a verb phase, where the verb phrase may be reduced into more noun phrases, verbs, and verb phrases. More precisely, the English language is an example of subject-verb-object typology structure, which accounts for 75% of all languages in the world [7]. This concept of English sentence structure is used to process a tagged sentence into higher-order structures of phrases by a process of chunking, which is a precursor to the extraction of semantic relationships of nouns into SVO structure. Using only the sequence of tags, chunking was performed as a recursive 4-step process: protecting verbs, recognition of noun phrases, unprotecting verbs and recognition of verb phrases. Firstly, verb tags (VBD, VBG and VBN) were protected by suffixing the tags. The main purpose was to prevent interference in recognizing noun phrases. Secondly, noun phrases were recognized by the following regular expression pattern of tags:

```
((((PDT )?(DT  |PRP[$]  |WDT |WP[$] )(VBG |VBD |VBN |JJ
|JJR |JJS |, |CC |NN |NNS |NNP |NNPS |CD )*(NN |NNS
NNP |NNPS |CD )+)|((PDT )?(JJ |JJR |JJS |, |CC |NN
NNS |NNP |NNPS |CD )*(NN |NNS |NNP |NNPS |CD )+)|EX
|PRP |WP |WDT )POS )?(((PDT )?(DT |PRP[$] |WDT |WP[$]
)(VBG |VBD |VBN |JJ |JJR |JJS |, |CC |NN |NNS |NNP
|NNPS |CD )*(NN |NNS |NNP |NNPS |CD )+)|((PDT )?(JJ
JJR |JJS |, |CC |NN |NNS |NNP |NNPS |CD )*(NN |NNS
|NNP |NNPS |CD )+)|EX |PRP |WP |WDT )
```

Thirdly, the protected verb tags in the first step were de-protected by removing the suffix appended onto the tags. Lastly, verb phrases were recognized by the following regular expression:

```
(RB |RBR |RBS |WRB )*(MD )?(RB |RBR |RBS |WRB )*(VB
|VBD |VBG |VBN |VBP |VBZ )(VB |VBD |VBG |VBN |VBP |VBZ
|RB |RBR |RBS |WRB )*(RP )?(TO (RB )*(VB |VBN )(RP )?)?
```

After chunking, each word (token) was stemmed into its root or infinite form. Firstly, each word was matched against a set of rules for specific stemming. For example, the rule "dehydrogenised verb dehydrogenate" defines that if the word "dehydrogenised" was tagged as a verb (VBD, VBG and VBN tags), it would be stemmed into "dehydrogenate". Similarly, the words "binds", "binding" and "bounded" were stemmed to "bind". Secondly, irregular words which could not be stemmed by removal of prefixes and suffixes, such as "calves" and "cervices", were stemmed by a pre-defined dictionary. Lastly, stemming was done by simple removal of prefixes or suffixes from the word based on a list of common prefixes or suffixes. For example, "regards" and "regarding" were both stemmed into "regard".

Given the general nature of an English sentence is an aggregation of noun phrase, a verb, and a verb phase, where the verb phrase may be reduced into more noun phrases, verbs, and verb phrases, each verb phrase may be taken as a sentence by itself. This allowed for recursive processing of a chunked-stemmed sentence into SVO(s) by a 3-step process. Firstly, the first terminal noun phrase, delimited by "(NX" and "NX)" was taken as the subject noun. Secondly, proceeding from the first terminal noun phrase, the first terminal verb would be taken as the verb in the SVO. Lastly, the rest of the phrase was scanned for terminal noun phrases and would be taken as the object(s). The recursive nature of SVO extraction also meant that the subject, verb, and object(s) will be contiguous, which had been demonstrated to have better precision than non-contiguous SVOs [26].

2.3 Protein-Protein Binding Finding

The protein-protein binding finder module is a data miner for protein-protein binding interaction assertions from the entire set of subject-relation-object (SVO) assertions from the text analysis process using apriori knowledge. That is, the set of proteins of interest must be known, in contrast to an attempt to uncover new protein entities, and their binding relationships with other protein entities, that were not known to the researcher.

Protein-protein binding assertions were extracted in a three step process. Firstly, a set of SVOs was isolated by the presence of the term "bind" in the verb clause resulting in a set of "bind-SVOs" assertions. Non-infinite forms of "bind" (such as, "binding" and "binds") were not used as verbs were stemmed into their infinite forms during text processing. Secondly, the set of bind-SVOs were further characterized for the presence of protein entities in both subject and object clauses by comparing with the desired list of protein entities. A pairwise isolation of bind-SVOs for protein entities resulted in a set of bind-SVOs, "entity-bind-SVOs", containing SVOs describing binding relationship between the protein entities. Lastly, entity-bind-SVOs were cleaned so that the subject and object clauses only contains protein entities. For example, "MAPK in the cytoplasm" in the object clause will be reduced to just the entity name "MAPK", the full subject and object clauses could be used in other information extraction tasks, such as determining protein localization, but is not explored in this study. This step is required to allow for the construction of network graphs, such as using Graphviz, without reference to the list of protein names during construction. Given that protein_entities is the list of desired proteins, table SVO contains the SVO output from MontyLingua and table entity_bind_SVO contains the isolated and cleaned SVOs, the pseudocode for Protein-Protein Binding Finding module is given as:

```
for subject_protein in protein_entities_{1 to n}
     for object_protein in protein_entities_{1 to n}
          insert (pmid, subject_protein, object_protein) into entity_bind_SVO
               from select pmid
               from (select * from SVO where verb = 'bind')
               where subject is containing subject_protein
               and object is containing object_protein
```

3 Experimental Results

Four experiments were carried out to evaluate the performance of Muscorian and demonstrate the flexibility of the two-layered generalization-specialization approach in constructing systems that could be readily be adapted to related problems. The results are summarized in Table 2.

3.1 Benchmarking Muscorian Performance

The performance of Muscorian, in terms of precision and recall, could only be evaluated using a defined data set with known results. For such purpose, the data set

Table 2. Summary of the Experimental Results Comparing the Precision and Recall Measures

	LLL05 Directional	*LLL05 Un-directional*	*Protein-Protein Binding*	*Protein-Protein Activation*
Precision	55.8%	86.1%	88.1%	90.7%
Recall	19.8%	30.7%	Not measured	Not measured

for Learning Languages in Logic 2005 (LLL05) [9] was used to benchmark Muscorian on genic interactions, which is a superset of protein-protein binding interactions. LLL05 had defined a genic interaction as an interaction between 2 entities (agent and target) but the nature of interaction was not considered under the challenge task. LLL05 provided a list of protein entities found in the data set, which was used to filter subject-relation-object assertions from text analysis (MontyLingua) output where both subject and object contained protein entities in the given list. The filtered list of assertions was evaluated for precision and recall, which was found to be 55.6% and 19.8% respectively.

LLL05 required that the agent and target (subject and object) to be in the correct direction, making it a vector quality. However, this requirement was not biologically significant to protein-protein binding interactions, which is scalar. For example, "X binds to Y" and "Y binds to X" have no biological difference. Hence, this requirement of directionality was eliminated and the precision and recall was 86.1% and 30.7% respectively.

3.2 Verifying Protein-Protein Binding Interactions

Precision of Muscorian for mining protein-protein binding interactions from published abstracts was evaluated by manual verification of a sample of assertions (n=135) yielded by the protein-protein binding finder module against the original abstracts. Each of the sampled assertions was assumed to be atomic, in the form of "X binds Y". In cases where there were more than one target, such as "X binds Y and Z", they would be reduced to atomic assertions. In this case, "X binds Y and Z" would be reduced to 2 assertions, "X bind Y" and "X bind Z". These were then checked with the original abstract, traceable by the PubMed IDs, and precision was measured as the ratio of the number of correct assertions to the number of sampled atomic assertions (which is 135). A 95% confidence interval was estimated by bootstrapping (re-sampling with replacement) [13] of the manual verification results. Our results suggested a precision of 88.1%, with a 95% confidence interval between 82.4% to 93.7%.

An IE trial was performed using the Protein-Protein Binding Finding module to search for the binding partners of CREB and insulin receptor and a sample network diagram of the results are shown in Figure 2 and 3 respectively.

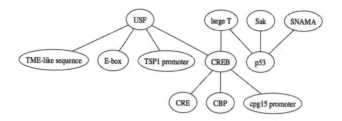

Fig. 2. Preliminary Protein Binding Network of CREB

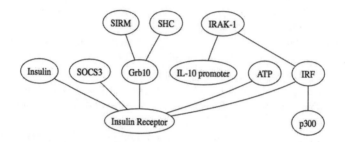

Fig. 3. Preliminary Protein Binding Network of Insulin Receptor

3.3 Large Scale Mining of Protein-Protein Binding Interactions

A large scale mining of protein-protein binding interactions was carried out using all of the PubMed abstracts on mouse (about 860000 abstracts), which were obtained using "mouse" as the keyword for searches, with a predefined set of about 3500 abbreviated protein entities as the list of proteins of interest (available from http://cvs.sourceforge.net/viewcvs.py/ib-dwb/muscorian-data/protein_accession.csv? rev=1.2&view=markup). In this experiment, the primary aim was to apply Muscorian to large data set and the secondary aim was to look for multiple occurrences of the same interactions as multiple occurrences might greatly improve precision confidence.

For example, given our lower confidence estimate that the precision of Muscorian with respect to mining protein-protein binding interactions is 82%, which means that every binding assertion has an 18% likelihood of not having a corresponding representation in the published abstracts. However, if 2 abstracts yielded the same binding assertion, the probability of both being wrong was reduced to 3.2% (0.18^2), and the corresponding probability that at least one of the 2 assertions was correctly represented was 96.8% ($1-0.18^2$). The more times the same assertion was extracted from multiple sources text (abstracts), the higher the possibility that the mined interaction was represented at least once in the set of abstracts. For example, if 5 abstracts yielded the same assertion, the possibility that at least one of the 5 assertions was correctly represented would be 99.98% ($1-0.18^5$).

Our experiment mined a total of 9803 unique protein-protein binding interactions, of which 7049 binding interactions were from one abstract (P=82%), 1297 binding interactions were from two abstracts (P=96.8%), 516 binding interactions were from three abstracts (P=99.4%), 235 binding interactions were from four abstracts (P=99.9%), 164 binding interactions were from five abstracts (P=99.98%), 105 binding interactions were from six abstracts (P=99.997%), 69 binding interactions were from seven abstracts (P=99.9993%), 398 binding interactions were from more than seven abstracts (P>99.9993%).

3.4 Pilot Study - Protein-Protein Activation Interactions

In order to demonstrate the adaptability of our proposed two-layered model, a small pilot study for mining protein-protein activation interactions was carried out. For this study, the protein-protein binding finder module, the data mining module for mining protein-protein binding interaction, was replaced with a protein-protein activation finder module.

The protein-protein activation finder was semantically similar to the original protein-protein binding finder module as described in Section 3.3 previously. The only difference was that raw assertion output from MontyLingua was filtered for activation-related assertions, instead of binding-related assertions, before analysis for the presence of protein names in both subject and object nouns from a pre-defined list of proteins of interest. For example, by modifying the Protein-Protein Binding Finding module to look for the verb 'activate' instead of 'bind', it can then be used for mining protein-protein activation interactions. A trial was done for insulin activation and a subgraph is illustrated in Figure 4 below.

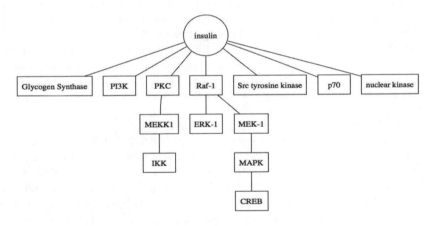

Fig. 4. Preliminary Protein Activation Network of Insulin

The precision measure of Muscorian for mining protein-protein activation interactions was calculated using identical means as described for protein-protein binding interactions. Using a sample of 85 atomic assertions, the precision of Muscorian for mining protein-protein activation interactions was estimated to be 90.7%, with a 95% confidence interval of precision between 84.7% to 96.4% by bootstrapping [13].

4 Discussion

New research articles in gene expression regulation networks, protein-protein interactions and protein docking are emerging at a rate faster than what most biologists can manage to extract the data and generate working pathways. Information extraction technologies have been successfully used to process research text and automate fact extraction [1]. Previous studies in biological text mining have developed specialized text processing tools and adapted generic tools to relatively good performance of more than 80% in precision [5, 11, 20, 31]. However, either specialized tool development or modifying existing tools often require much effort [20]. The need to modify existing tools has not been formally tested and the possibility of using an un-modified generic text processor for biological text for the purpose of extracting protein-protein interaction remains unresolved. Using a two-layered approach [29] of generalizing biological text into a structured intermediate form, followed by specialized data mining, we present Muscorian, which uses MontyLingua natively in the generalized layer, as a tool for extracting either protein-protein or genic interactions from about 860000 published biological abstracts.

Benchmarking Muscorian against LLL05, a tested data set, demonstrated a precision of 55.6%, which is about 5% higher than that reported in the conference and a recall of 19.7% is similar to that reported by other participants of LLL05 [9]. This may be due to the emphasis of LLL05 on F-measure, which is the harmonic mean of precision and recall, rather than putting more emphasis on precision. Nevertheless, this also suggested that Muscorian is able to perform text analysis for the purpose of extracting genic interactions effectively, which is comparable to specialized systems reported in LLL05. In addition, directionality of genic interactions was not a concern for protein-protein binding interactions as binding interaction is scalar rather than vector. By eliminating directionality of genic interactions, the precision and recall of Muscorian was 86.1% and 30.7% respectively. This suggested that Muscorian is a suitable tool for mining quality genic interactions from biological text compared to other tools reported in LLL05 [9].

Our results on protein-protein binding and activation interactions show the insulin receptor binds to IL-10 promoter through IRF and IRAK-1, which is an important insulin receptor signalling pathway. In addition, our data shows insulin activates CREB via Raf-1, MEK-1 and MAPK, which is consistent with the MAP kinase pathway. Combining these data (Figures 2 and 4) indicated that insulin activates CREB via MAP kinase pathway, and CREB binds to cpg15 promoter in the nucleus. A simple keyword search on PubMed, using the term "cpg15 and insulin" (done on 30th of April, 2007), did not yield any results, suggesting that the effects of insulin on cpg15, also known as neuritin [2], had not been studied thoroughly. This might also suggest limited knowledge shared between insulin investigators and cpg15 investigators as suggested by Don Swanson in his classical paper describing the links between fish oil and Raynaud's syndrome [34]. Neuritin is a relatively new research area with less than 20 papers published (as of 30th of April, 2007) and had been implicated as a lead for neural network re-establishment [18], suggesting potential collaborations between endocrinologists and neurologists.

Our experiments in extracting two different forms of relations demonstrated that despite using specialized dictionaries in the generalized layer, it is still general to the extend that specific application (the type of relationships to extract) was not built into the generalized layer.

At the same time, these 2 experiments also illustrated the relative ease in re-targeting the system for extracting another form of relationship by modifying the specialized layer. The Protein-Protein Activation Finder module is a slight modification of the original Protein-Protein Binding Finder module where the original SQL statement that selects 'bind'-related SVOs from total SVOs, *"select * from SVO where verb = 'bind'"*, was changed to *"select * from SVO where verb = 'activate'"* to select for 'activation'-related SVOs from total SVOs. Hence, it is plausible that similar changes may suffice for extracting other relationships, such as 'inhibition'. This relative ease of re-targeting the system for extracting other relationships also demonstrated the robustness of the generalization layer, as implied by Novichkova et. al. [29] – *"the adaptability of the system to related problems other than the problem the system was designed for"*.

Given large numbers of published abstracts, the performance of Muscorian on precision was comparable with published values of BioRAT (58.7%) [12], GIS (84%) [5], Cooper and Kershenbaum (74%) [6] and CONAN (53%) [24] while Muscorian's recall was comparable with published values of Arizona Relations Parser (35%) [10] and Daraselia et. al. (21%) [11]. Poor precision was considered unacceptable because incorrect information is more detrimental than missing information (1 - recall) when protein-protein binding interactions were used to support other biological analyses. Muscorian's mediocre recall of 30% (from LLL05 test set evaluation) could be supplemented by the fact that the same interaction could be mentioned or described by multiple abstracts; thus, the actual recall when tested on a large corpus may be higher. For example, 30% recall essentially means a loss of 70% of the information; however, if the same information (in this case, protein interactions) were mentioned in 3 or more abstracts, there is still a reasonable chance to believe that information from at least 1 of the 3 or more abstracts will be extracted. This is supported by our results indicating that almost 30% (2754 of 9803) of binding interactions were extracted from more than one abstract.

Multiple isolation of 2754 binding interactions enabled a higher confidence that these interactions were correctly extracted with reference to the source literature. Based on this analysis, 2754 binding interactions could be assigned higher confidence based on their occurrences [21], in this case more than 95% chance of being correct based on literature. In addition, the number of multiple interaction occurrence varies inversely with the number of abstracts these interactions were found in is in line with expectation. Although this line of argument is based on the assumption that the appearance of protein names across abstracts were independent, it can be reasonably held as this study uses abstracts rather than full text – abstracts tends to describe what main results of the particular article while the introduction of a full text article tends to be a brief background review of the field. Hence, independence of protein names can be better assumed in abstracts than in full text articles.

An evaluation of a sample of atomic assertions (interactions) of binding and activation interactions between entities was performed by domain experts comparing the assertions with their source abstracts. Both approaches gave similar precision measures and are consistent with the evaluation using LLL05 test set. The ANOVA test demonstrated that there was no significant differences between these three precision measures. Taken together, these evaluations strongly suggested that

Muscorian performed with precisions between 86-90% for genic (gene-protein and protein-protein) interactions, which was similar to that reported by studies either modifying existing tools [31] or developing specialized tools [11]. This suggested that MontyLingua could be used natively (un-modified), with good precision, to process biological text into structured subject-verb-objects tuples which could be mined for protein interactions.

Acknowledgments. We wish to thank Prof. I-Fang Chung, Institute of Biomedical Informatics, National Yang Ming University, Taiwan, for his comments on improving the initial drafts. This work is sponsored by the CRC for Innovative Dairy Products, Australia, and Postgraduate Overseas Research Experience Scholarship, The University of Melbourne, Australia.

References

1. Abulaish, M., Dey, L.: Biological relation extraction and query answering from MEDLINE abstracts using ontology-based text mining. Data & Knowledge Engineering 61, 228 (2007)
2. Cappelletti, G., Galbiati, M., Ronchi, C., Maggioni, M.G., Onesto, E., Poletti, A.: Neuritin (cpg15) enhances the differentiating effect of NGF on neuronal PC12 cells. Journal of Neuroscience Research (2007)
3. Chang, J.T., Schutze, H., Altman, R.B.: Creating an online dictionary of abbreviations from MEDLINE. Journal of the American Medical Informatics Association 9, 612–620 (2002)
4. Chiang, J.H., Yu, H.C.: MeKE: discovering the functions of gene products from biomedical literature via sentence alignment. Bioinformatics 19, 1417–1422 (2003)
5. Chiang, J.H., Yu, H.C., Hsu, H.J.: GIS: a biomedical text-mining system for gene information discovery. Bioinformatics 20(1), 120 (2004)
6. Cooper, J.W., Kershenbaum, A.: Discovery of protein-protein interactions using a combination of linguistic, statistical and graphical information. BMC Bioinformatics 6, 143 (2005)
7. Crystal, D.: The Cambridge Encyclopedia of Language, 2nd edn. Cambridge University Press, Cambridge (1997)
8. Cunningham, H.: Software Architecture for Language Engineering. PhD Thesis. Department of Computer Science: University of Sheffield (2000)
9. Cussens, J. (ed.): Proceedings of the Learning Languages in Logic Workshop 2005 (2005)
10. Daniel, M.M., Hsinchun, C., Hua, S., Byron, B.M.: Extracting gene pathway relations using a hybrid grammar: the Arizona Relation Parser. Bioinformatics 20, 3370 (2004)
11. Daraselia, D., Yuryev, A., Egorov, S., Novichkova, S., Nikitin, A., Mazo, I.: Extracting human protein interactions from MEDLINE using a full-sentence parser. Bioinformatics 20, 604–611 (2004)
12. David, P.A.C., Bernard, F.B., William, B.L., David, T.J.: BioRAT: extracting biological information from full-length papers. Bioinformatics 20, 3206 (2004)
13. Efron, B., Tibshirani, R.: Bootstrap Methods for Standard Errors, Confidence Intervals, and Other Measures of Statistical Accuracy. Statistical Science 1, 54–75 (1986)
14. Eslick, I., Liu, H.: Langutils – A natural language toolkit for Common Lisp. In: Proceedings of the International Conference on Lisp 2005 (2005)
15. Friedman, C., Alderson, P.O., Austin, J.H., Cimino, J.J., Johnson, S.B.: A general natural-language text processor for clinical radiology. Journal of the American Medical Informatics Association 1, 161–174 (1994)

16. Friedman, C., Kra, P., Yu, H., Krauthammer, M., Rzhetsky, A.: GENIES: a natural-language processing system for the extraction of molecular pathways from journal articles. Bioinformatics 17, S74–S82 (2001)
17. Grover, C., Klein, E., Lascarides, A., Lapata, M.: XML-based NLP Tools for Analysing and Annotating Medical Language. In: Proc. of the 2nd Int. Workshop on NLP and XML (NLPXML-2002), Taipei (2002)
18. Han, Y., Chen, X., Shi, F., Li, S., Huang, J., Xie, M., Hu, L., Hoidal, J.R., Xu, P.: CPG15, A New Factor Upregulated after Ischemic Brain Injury, Contributes to Neuronal Network Re-Establishment after Glutamate-Induced Injury. Journal of Neurotrauma 24, 722–731 (2007)
19. Hu, Z., Narayanaswamy, M., Ravikumar, K., Vijay-Shanker, K., Wu, C.: Literature mining and database annotation of protein phosphorylation using a rule-based system. Bioinformatics 21, 2759–2765 (2005)
20. Jensen, L.J., Saric, J., Bork, P.: Literature mining for the biologist: from information retrieval to biological discovery. Nature Review Genetics 7, 119–129 (2006)
21. Jenssen, T.K., Laegreid, A., Komorowski, J., Hovig, E.: A literature network of human genes for high-throughput analysis of gene expression. Nature Genetics 28, 21–28 (2001)
22. Ling, M.H.T.: An Anthological Review of Research Utilizing MontyLingua, a Python-Based End-to-End Text Processor. The Python Papers 1, 5–12 (2006)
23. Liu, H., Singh, P.: ConceptNet: A Practical Commonsense Reasoning Toolkit. BT Technology Journal 22, 211–226 (2004)
24. Malik, R., Franke, L., Siebes, A.: Combination of text-mining algorithms increases the performance. Bioinformatics 22, 2151–2157 (2006)
25. Marcus, M.P., Santorini, B., Marcinkiewicz, M.A.: Building a Large Annotated Corpus of English: The Penn Treebank. Computational Linguistics 19, 313–330 (1993)
26. Masseroli, M., Kilicoglu, H., Lang, F.M., Rindflesch, T.: Argument-predicate distance as a filter for enhancing precision in extracting predications on the genetic etiology of disease. BMC Bioinformatics 7, 291 (2006)
27. Nasukawa, T., Nagono, T.: Text analysis and knowledge mining system. IBM System Journal 40, 967–984 (2001)
28. National Library of Medicine, UMLS Knowledge Sources, 14th edn. (2003)
29. Novichkova, S., Egorov, S., Daraselia, N.: MedScan, a natural language processing engine for MEDLINE abstracts. Bioinformatics 19, 1699–1706 (2003)
30. Rebholz-Schuhmann, D., Kirsch, H., Couto, F.: Facts from Text - Is Text Mining Ready to Deliver? PLoS Biology 3, e65 (2005)
31. Santos, C., Eggle, D., States, D.J.: Wnt pathway curation using automated natural language processing: combining statistical methods with partial and full parse for knowledge extraction. Bioinformatics 21, 1653–1658 (2005)
32. Sleator, D., Temperley, D.: Parsing English with a Link Grammar. In: Proceedings of the 3rd International Workshop on Parsing Technologies (1991)
33. Smith, L., Rindflesch, T., Wilbur, W.J.: MedPost: a part-of-speech tagger for bioMedical text. Bioinformatics 20, 2320–2321 (2004)
34. Swanson, D.R.: Fish oil, Raynaud's syndrome, and undiscovered public knowledge. Perspectives in Biology and Medicine 30, 7–18 (1986)
35. van Eck, N.J., van den Berg, J.: A novel algorithm for visualizing concept associations. In: Andersen, K.V., Debenham, J., Wagner, R. (eds.) DEXA 2005. LNCS, vol. 3588, Springer, Heidelberg (2005)
36. Uramoto, N., Matsuzawa, H., Nagano, T., Murakami, A., Takeuchi, H., Takeda, K.: A text-mining system for knowledge discovery from biomedical documents. IBM System Journal 43, 516–533 (2004)

Validation of Gene Regulatory Networks from Protein-Protein Interaction Data: Application to Cell-Cycle Regulation

Iti Chaturvedi[1,2], Meena Kishore Sakharkar[2], and Jagath C. Rajapakse[1]

[1] Bioinformatics Research Center, Nanyang Technological University, Singapore
[2] Adams Lab, MAE, Nanyang Technological University, Singapore
asjagath@ntu.edu.sg

Abstract. We develop a technique to validate large-scale gene regulatory networks (GRN) by comparing with corresponding protein-protein interaction (PPI) networks. The GRN are obtained with Bayesian networks while PPI networks are obtained from database of known PPI interactions. We look for exact matches and then reduced networks by skipping one or more genes in GRN. We demonstrate our technique on expression profiles of differentially expressed genes in the *S. cerevisiae* cell cycle. We validate GRNs against a merged database of 53235 genes. The precisions of GRN obtained over all genes were from 0.82 to 0.95 in all the phases. In particular we realized that one-skip and two-skip model significantly improved accuracy of the GRN of different phases of cell cycle.

Keywords: Dynamic Bayesian networks, gene regulatory networks, genetic algorithms, protein-protein interactions.

1 Introduction

A *protein-protein interaction network* (PPIN) has protein as nodes and the edges can be signaling, regulatory and biochemical interactions of the proteome. However, a *Gene Regulatory Network* (GRN) shows interaction of DNA segments of the genome with other substances of the cell, which results in regulating rates at which genes are transcribed to mRNA. This high throughput data has a large scope for organization in context of disease and biological function [1]. There is a need to explain the cellular machinery of a GRN in a systems biology perspective as seen by a PPIN. A common representation of GRN is a '*pathway model*', a graph where vertices represent genes (or larger chromosomal regions) and arcs represent casual pathways. A vertex can either be off/normal or on/abnormal. Bayesian networks (BN) have recently become popular in deriving and deciphering GRN [2] and PPIN [3]. BN is a directed acyclic graph representing casual relations among interacting variables at the nodes. Pathway models have natural representations as BN.

GRN is a model based on mRNA abundance, measured usually by microarrays, rendering an effective network of gene to gene interactions. DNA hybridization arrays simultaneously measure the expression levels of thousands of genes. Clustering-based

J.C. Rajapakse, B. Schmidt, and G. Volkert (Eds.): PRIB 2007, LNBI 4774, pp. 300–310, 2007.

visual tools, such as hierarchical clustering [7] and SOM [8] assume that each gene belongs to only one cluster. Such algorithms attempt to locate groups of genes having similar expression patterns over a set of experiments and hence possibly co-regulated or having similar functions. This assumption fails where genes belong to two or more independent expression patterns. Traditional statistical methods for computing low-dimensional or hidden representations of these data sets, such as principal component analysis (PCA)[9] and independent component analysis (ICA)[10], ignore the underlying interactions and provide a decomposition based purely on *a priori* statistical constraints on the computed component signals.

Here our knowledge about a biological system is not directly expressed by a parameter vector of state variables, but instead is about the statistical dependencies (or independencies) called casual relationships among the variables. The casual dependencies among variables are represented by BN in terms of conditional probabilities, so they infer 'cause and effect' relationships. The nodes of BN mimicking GRN represent gene expressions, either by analog or discrete variables, and interactions by discrete and continuous multidimensional distributions [4]. Further, dynamic Bayesian networks (DBN) can model the stochastic evolution of a set of random genes over time and therefore temporal information of interactions efficiently [5]. DBN have advantages over hidden Markov models (HMM) whose parameterization grows exponentially with the number of state variables and over Kalman filters which is capable of handling only unimodal posterior distributions. BN and DBN are defined by a graphical structure and a set of parameters, which together specify a joint distribution over the variables it represents. The nodes in Bayesian network could represent either binary or continuous variables. One advantage of representing state variables as continuous Gaussian rather than discrete is that the posterior can be marginalized efficiently over time [6]. A special class of regulatory network models is one of linear time continuous models [11]. Analysis of gene expression reveals a considerable amount of time delayed interactions, suggesting that time delay is ubiquitous in gene regulation. State-space models with time delays of gene regulatory networks use Boolean variables to capture the existence of discrete time delays of the regulatory relationships among the internal variables [12].

Various tools are now available to generate GRN from Microarray data using above models. *Gene Networks* [13] offers four models including the linear model, and 3 genetic algorithm based models, S-system, Boolean networks, and Bayesian networks. BN uses a genetic algorithm adapted from REVEAL[14] to optimize the cost function which is a NP-hard problem. Linear differential model assumes that the change of each component over time is given by a weighted sum of all other components. In this model, the expression state at one time point determines the expression state observed at the next point However assumption of linear gene-regulation relationship in unrealistic, complex systems, such as gene expression networks and metabolic pathways, are comprised of numerous richly interacting components. By representing states as binary variables and then connections by multinomial distributions, non-linear interactions among nodes can be represented in Bayesian networks.

The GRN derived from gene expression data are often over-fitted. And some of the genes are masked by the activation of highly expressed similar genes. Here we try to enhance and validate GRN derived using Bayesian networks with corresponding

PPIN discovered from PPI databases. Validation of GRN is of vital importance for making inference on large scale pathways. Here we assume skipping of one or more genes in predicted gene interaction networks and, when mapping to a protein-protein interaction, allow for prodigies of genes. As seen later, this enhances the accuracy of GRN derived from gene expression data and increases true prediction of interactions without altering biological pathways.

We demonstrate our technique with the yeast cell-cycle data, which contain differentially expressed genes in different phases of cell-cycle. Our results show that the sensitivity of BN in detecting genes of a common pathway can be improved with the validation using PPI. This paper is organized as follows: in Section 2, we explain how GRN are derived using BN Section 3 describes how GRN and PPIN are mapped. Experiments and results with yeast cell-cycle data are given in Section 4. Lastly, we draw conclusions from our findings.

2 Gene Regulatory Networks

2.1 Dynamic Bayesian Networks

A BN is a graphical model representing joint multivariate probability distributions to capture the properties of conditional independencies among variables and consists of two components: a directed acyclic graph (DAG) structure, S, and a set of conditional distributions with parameters θ, of each variable, given its parents [15]. BN are unable to model stochastic systems evolving over time. Furthermore, they are unable to construct cyclic regulations (positive and negative feedback loop mechanisms) to regulate the activities of state variables at nodes typical of biological processes. Hence, we use dynamic Bayesian networks (DBN) to generate GRN. DBN makes the following assumptions: (1) the genetic regulation process is first-order Markovian, i.e., the expression state of one gene at one time point is dependent only on the expression state of other genes observed at the previous time point; (2) the dynamic casual relationships between genes are invariable over all the time slices, that is, the set of variables and probability definitions of a DBN are the same for each time points (i.e., stationarity).

The dynamics of the DBN are hence defined in a *transition* network over two time slices, taken at time t and time $t+1$ as illustrated in Figure 1: The parameters are the probabilities of each variable, conditioned on the other variables at the previous one time point. Given the transition network over two time slices, the DBN is obtained by unrolling static transition BN over all time instances to determine the dynamics of stochastic variables over entire experiment.

In a GRN, the nodes of the BN are represented by the expressions of genes and the edges by the causal effects. Let us consider a Bayesian network representing a set of gene expressions $X = \{X_1, X_2, ... X_n\}$ in a GRN consisting of n genes. The joint probability of the expression of the genes is then be represented by $P(X) = \prod_{i=1}^{n} P(X_i \mid \prod_i)$ where $\prod_i$ denotes the set of gene expressions of parent nodes of gene i with expression X. We see that this metric is NP-hard but decomposable.

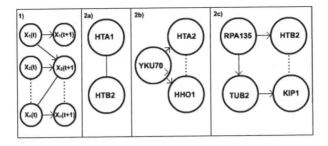

Fig. 1. Illustration of the transition network defining a dynamic Bayesian network consisting of n nodes. **Fig. 2.** Dotted line show the predicted interaction by a GRN (a) HTA1 interaction with HTB2 predicted by PPI which is same as by GRN (0-skip model) (b) HTA2 interaction with HHO1 predicted by PPI (1-skip model), GRN skips the gene YKU70 (c) HTB2 interaction with KIP1 predicted by PPI (2-skip model), GRN skipped 2 genes RPA135 and TUB2.

Finding a Bayesian network that fits the gene expressions best requires a search over the model space of both structure S and the interactions. Hence, a proper scoring function is needed to rank possible solutions and find the optimal solution. The posterior probability of a GRN, S, given gene expression data X, is given by, $P(S \mid X) \propto P(X \mid S)P(S)$ where $P(S)$ gives the prior probability of the network structure and $P(X \mid S)$ the likelihood. We have taken 6 important assumptions. Firstly we assume a multinomial sample, given domain U and database X, let X_1 denote the first $1-1$ cases in the database. In addition, let X_{il} and $\prod_{il}$ denote the variable x_i and the parent set $\prod_i$ in the 1th case, respectively. Then for all network structures B_s in U, there exist positive parameters Θ_{B_s} such that, for $i = 1,...,n$ and for all $k, k_1,...., k_{i-1}$,

$p(x_{il} = k \mid x_{1l} = k_1,...., x_{(i-1)l} = k_{i-1}, X_1, \Theta_{B_s}, B_s^h, \xi) = \theta_{ijk}$. Where ξ is the current state of information. Second assumption is of parameter independence, given network structure B_s if $p(B_s^h \mid \xi) > 0$ then $\rho(\Theta_{B_s} \mid B_s^h, \xi) = \prod_{i=1}^{n} \rho(\Theta_i \mid B_s^h, \xi)$,

for $i = 1,....,n : \rho(\Theta_i \mid B_s^h, \xi) = \prod_{j=1}^{q_i} \rho(\Theta_{ij} \mid B_s^h, \xi)$. Third assumption is that of parameter modularity which says that given two network structures B_{s1} and B_{s2} such that $p(B_{s1}^h \mid \xi) > 0$ and $p(B_{s2}^h \mid \xi) > 0$, and x_i has the same parameters in B_{s1} and B_{s2}, then $\rho(\Theta_{ij} \mid B_{s1}^h, \xi) = \rho(\Theta_{ij} \mid B_{s2}^h, \xi)$ $j = 1,....,q_i$. Fourth is the assumption that the distribution is Dirichlet. Given the network structure B_s such that

$p(B_s^h \mid \xi) > 0$. $\rho(\Theta_{ij} \mid B_s^h, \xi)$ is Dirichlet for all $\Theta_{ij} \subseteq \Theta_{B_s}$. That is, there exists exponents N_{ijk}' which depend on B_s^h and ξ, that satisfy $\rho(\Theta_{ij} \mid B_s^h, \xi) = c.\prod_k \theta_{ijk}^{N_{ijk}-1}$ where c is a normalization constant. The fifth assumption is that the database is complete. That is there are no missing data. The final assumption is of likelihood equivalence that given two network structures B_{s1} and B_{s2} such that $p(B_{s1}^h \mid \xi) > 0$ and $p(B_{s2}^h \mid \xi) > 0$, if B_{s1} and B_{s2} are equivalent, then $\rho(\Theta_U \mid B_{s1}^h, \xi) = \rho(\Theta_U \mid B_{s2}^h, \xi)$. The assumption of likelihood equivalence when combined with the previous assumptions introduces constraints on the Dirichlet exponents N_{ijk}'. The result is a likelihood-equivalent specialization of the BD metric , which we call the BDe metric. The marginal likelihood can be represented by the BDe metric [16].

$$BDe = \prod_{i=1}^{n}\prod_{j=1}^{q_i} \frac{\Gamma(N_{ij}')}{\Gamma(N_{ij}' + N_{ij})} \prod_{k=1}^{r_i} \frac{\Gamma(N_{ijk}' + N_{ijk})}{\Gamma(N_{ijk}')}$$

where $\Gamma(x)$ is a Gamma function Dirichlet distribution. Each gene i can take a finite number of distinct states r such that $X_i = \{x_1, x_2, ... x_{r_i}\}$ and is assumed to have a finite number of distinct state combinations of the parents, q_i such that $\prod_i = \{a_1, a_2, ... a_{q_i}\}$. N_{ijk}' represents the Dirichlet prior parameters and N_{ijk} the counts of interactions.

2.2 Derivation of GRN Using a Genetic Algorithm

A Genetic Algorithm (GA) is applied to effectively search the large solution space and to learn the network structure optimizing the BDe metric. We only consider binary interactions and therefore a solution individual is represented as a binary matrix which indicates the interaction states between genes and their parent genes (the genes that regulate them) where 1 denotes a regulation and 0 means no interaction. The solution $C = \{c_{i,j}\}_{n \times n}$, where $c_{ij} \in \{0,1\}$ is the interaction between genes i and j. Using the solution C we can calculate the terms N_{ijk}', the parameters of prior [17], and N_{ijk}, the number of observations (for the state defined by i , j and k) respectively where $X_i = x_k$, hence k is state of gene i , also $\prod_i = a_j$, j is the state combination of parents of i. Further, $N_{ij}' = \sum_{k=1}^{r_j} N_{ijk}'$ and $N_{ij} = \sum_{k=1}^{r_i} N_{ijk}$. Then, using the equation above, we can get the BDe metric of the solution C .

The inputs to the genetic algorithm is a time-series data of expression of all genes. Genes in consecutive time points having similar expression levels can be said to have an interaction. The algorithm is as follows:

Procedure for DBN-GA
Begin
> **Initialize:** Randomly create P initial individuals that can be represented as a binary interaction matrix.
>> **While**(until G generations)
>>> Evaluate the fitness function of each individual using BDe metric
>>> **Select** the elite individual to be passed on to next generation
>> **Generate** new individuals by selection, crossover and mutation. With the exception of the elite individual, the design code of each child (new individual) is created based on the design codes of two parents (old individual). Two parents are selected from the P individuals according to the probability proportional to their order of fitness (ranking or roulette strategy).
> **End of While**
>> **Build** the gene regulation matrix based on the individual that has the largest fitness.
End

2.3 Missing Data

A key problem for all models is a shortage of data. The raw gene expression data, usually in the form of large matrix, may contain missing values. This is a result of insufficient resolution, image corruption, or simply due to dust or scratches on the slide. *KNNimpute* (K Nearest Neighbors) method [18] is used to predict missing Microarray expression levels.

3   Mapping of GRN and PPI

3.1   Protein-Protein Interaction Networks (PPIN)

Proteins frequently bind together in pairs or larger complexes to take part in biological processes. Most biological phenomena is due to a protein-protein interaction. There are several experimental techniques for determining protein-protein interaction data. Synthetic lethality[19], Affinity Capture-MS[20], and Yeast-2-Hybrid [21]being the top few in our biogrid dataset.

3.2   Motivation

The derivation of BN, using the GA, is very sensitive to the population set of structures. Since we are trying to achieve a final maximum fitness, it is at the expenses of finding the set of solutions that are together most likely to be correct, which means individual correct solutions are left out because of this evolutionary population model. The networks or the solutions on the other hand aim to connect genes which have similar expression profiles. Since a child gene follow the expression pattern of a parent gene which is regulating it. This results in skipping or missing genes in GRN, especially those with highly expressed genes. Therefore, often the GRNs derived using BN are often underestimated in the number of genes. In order to overcome this, we propose a technique that incorporates the knowledge from corresponding PPIN to infer the missing interactions in GRN.

3.3 K-Skip Validation

In order to account for the missing genes and interactions in GRN, we employ k-skip models of GRN which assumes that k-genes are skipped in estimating GRN between two parent genes. The simplest is called the *one-skip* model where one gene is skipped in GRN due to an interaction between two genes. One reason for this could be that mRNA from gene1 might not be directly interacting with mRNA from gene2. Rather the protein product from gene1 may alter the level of mRNA from gene2. An example could be a transcription factor, which may not occur by making more of it, but just by phosphorylation (post-translational modification) [22]. Also we are interested in finding genes which lie in the same pathway. Hence these one-skip and two-skip predictions are also of high importance to us.

These models are defined as follows:

0-skip Model: Indicates a direct interaction between proteins A and B
1-skip Model: There exists a protein C such that both A and B interact with C according to 0-skip Model
2-skip Model: There exists a protein D such that D interacts with A by 0-skip model and B by 1-skip model or vice versa.
3-skip Model: There exists a protein D such that D interacts with A by 0-skip model and B by 2-skip model or vice versa.

We illustrate the above different models in the Figure 2. Figure 2 (a) shows a Gene Interaction predicted : HTA1-HTB2, which has a corresponding interaction in PPI db. This will lie in the 0-skip model. (b) Shows an interaction HHT1-HTB2 which is not found in the PPI db, however a missing gene HTA1, shows they lie in the same pathway. This is called the 1-skip model. Similarly, (d) is an example of 2-skip model. We run BN on each of the 4 sets of genes under different values of two parameters namely, the number of generations and number of individuals in each generation (i.e. population size) at the genetic algorithm step. It is possible that the interaction incorrectly bypassed a single or multiple genes. The Gene Network software provides us with the Regulatory Matrix of the final optimal solution C.

4   Cell-Cycle Regulation

4.1   Data

We illustrate our method using an application to cell-cycle regulation in yeast. Yeast has 40% genes have orthologus to human. Also it is non-pathogenic and hence can be tested for different interactions safely. We model GRN of the genes involved in the cell-cycle from an extended Spellman yeast dataset, which consists of mRNA measurement of 6,178 genes of yeast *S. cerevisiae* [24]. Here we use the cdc15 experimental data where cdc15 yeast strain is given a cdc-15 arrest (to the cell-cycle) by moving into an incubator at 37°C. The arrest is then removed by moving back to 23°C. Cells are then monitored together at different time points for presence of new buds. 24 such time points are available from 10 to 290 mins. Cell-cycle control of transcription seems to be a universal feature of proliferating cells. Three main transcriptional waves which roughly coincide with three main cell-cycle transitions: initiation of DNA replication, entry into mitosis, and exit from mitosis. Proliferation of all cells is mediated though cell-division cycle which consists of four main phases: genome duplication (S phase) and nuclear division (mitosis or M phase), separated by two gap phases (GI and G2). Transcription of a number of genes peaks at specific cell-cycle phases. At the end of G1 phase, cells decide whether to commit to cell division in a process called start in yeast or restriction point in mammalian cells [23]. In this paper, we attempt to demonstrate our method by modeling GRNs involved in different phases of yeast cell cycle and then validating with the use of PPI data.

We downloaded the list of phase specific genes from [24]. Our dataset consists of 118 genes in G1, 36 genes in S phase, 34 genes in G2 and 60 genes in M phase respectively. Figure 3 shows the expression patterns of the 4 sets. We can see that G1 genes peak in time points 10 to 70 mins, then the S phase genes peak from 30 to 90mins, next is the G2 phase peaking 70 to 100 mins and lastly the M phase genes from time points 90 to 130 mins. Hence we can say that they are all differentially expressed.

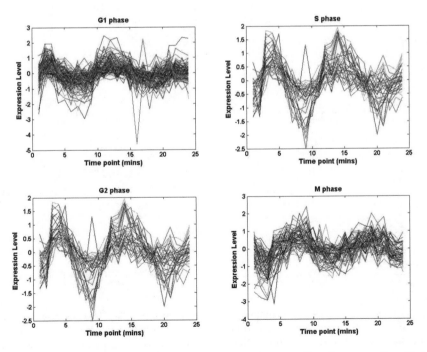

Fig. 3. Expression levels of genes in different phases of yeast cell-cycle measured at 24 time points in the cdc-15 experiment

4.2 Experiments and Results

Sensitivity of the model is very low. One of the methodologies proposed by us to overcome this is the K-skip model. The GRN software allows us to choose the number of Generations and population size of each generation, allowing for choice of combinations for tuning the correct number of predictions. Bayesian nets on our 4 gene sets of cell-cycle under five experimental settings is presented (Figure 4). These predictions were then validated against the PPI data for inferring the correct predictions under the k-skip model. As seen the accuracy of DBN first increase and then decreases with the increase in complexity of searches.

The cumulative curves for the correct number of predictions for four datasets is shown in Figure 4, G1 phase, S phase, G2 phase and M phase. We downloaded yeast data from BIOGRID [25] and got a non-redundant validations dataset of 53,235 protein interactions. It is observed that in all the graphs, there is a steep increase in the number of predictions by the one-skip model. Further increase is seen with the two-skip model. However the three-skip model shows 0 interactions in all datasets. Hence while reading Bayesian nets one must take into account that the predictions might be bypassing one or two genes in the pathways.

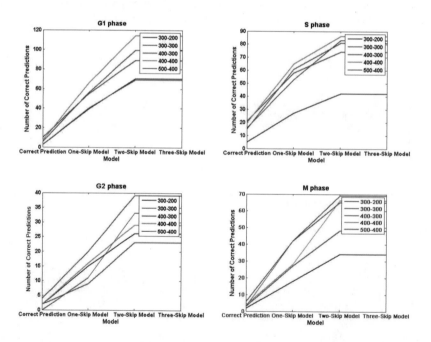

Fig. 4. Cumulative number of correct predictions for correct, one-skip, two-skip and three-skip model under 5 parameter settings (i) 300 Generations, 200 Individuals, (ii) 300 Generations, 300 Individuals, (iii) 400 Generations, 300 Individuals, (iv) 400 Generations, 400 Individuals, (v) 500 Generations, 400 Individuals

Table 1. Precision of Bayesian Networks for different cell-cycle phases. Average Precision is calculated over 5 runs with different parameter settings.

	Number of Genes	Average Precision	Maximum Precision
G1	118	0.76	0.82
S	36	0.89	0.93
G2	34	0.89	0.95
M	60	0.80	0.84

Precision for each GRN for different phases of cell-cycle which is defined as True Positive / Total Linkages was calculated. (Table 1). We notice a very high precision of over 80% in most trials. Which indicates that the Bayesian network is indeed picking up most interactions, however the accuracy is constrained by the one-skip/two-skip model.

Thus the advantages of Dynamic Bayesian Network include the ability to model stochasticity, to incorporate prior knowledge, and to handle hidden variables and missing data in a principled way. However, the discretization of gene expression by Bayesian network can lead to information loss. Also determining optimal structure of Bayesian networks is an NP-hard problem. Domain experts like the readability of trees in Bayesian networks however this is at the cost of accuracy.

5 Conclusion

We see a similar trend in all the 4 phases, confirming that a one-skip or two-skip bias exists in the model. This seems like a limitation of the model, as it looks for the best possible pathway. The proposed method may have diverse applications in understanding pathways involved in diseases.

However we must realize the constraints of the model. Some genes are redundant in different stages of the cell-cycle. This can alter the graph. Also we know that protein interactions can be stable or transient . Transient interactions are on/off and require a set of conditions that promote them. Finally we are testing the accuracy of a GN against a PPI database which is mostly generated from scientific literature and is not completely experimentally verified. Future work would involve accuracy testing against previous methods and other databases.

References

1. Nikolsky, Y.: Biological networks and analysis of experimental data in drug discovery. Drug discovery today 10(9), 653 (2005)
2. Wyrick, J.J., Young, R.A.: Deciphering gene expression regulatory networks. Curr. Opin. Genet. Dev. 12(2), 130–136 (2002)
3. Fromont-Racine, M., Rain, J.C., Legrain, P.: Toward a functional analysis of the yeast genome through exhaustive two-hybrid screens. Nature Genetics 16, 277–282 (1997)
4. Duda, R.O.: Pattern classification [Book]
5. Friedman, N., Linial, M., Nachman, I., Pe'er, D.: Using bayesian networks to analyze expression data. J. Computational Biology 7(3), 601–620 (2000)

6. Lauritzen, S.: Graphical Models. Oxford University Press, Oxford (1996)
7. Luo, F., Tang, K., Khan, L.: Hierarchical clustering of gene expression data. Bioinformatics and Bioengineering, 2003. In: Proceedings. Third IEEE Symposium, 10-12 March, pp. 328–335 (2003)
8. Wei, W., Xin, L., Min, X., Jinrong, P., Setiono, R.: A hybrid SOM-SVM method for analyzing zebra fish gene expression. Pattern Recognition, 2004. In: ICPR 2004. Proceedings of the 17th International Conference, vol. 2, pp. 323–326 (2004)
9. Alter, O., Brown, P.O., Botstein, D.: Singular value decomposition for genome-wide expression data processing and modeling. Proc. Natl. Acad. Sci. USA 97, 10101–10106 (2000)
10. Lee, S., Batzoglou, S.: Application of Independent component analysis to microarrays. Genome Biology 4, R76 (2003)
11. Bar-Joseph, Z.: Analyzing Time Series Gene Expression Data. Bioinformatics 20(16), 2493–2503 (2004)
12. Fang-Xiang, et al.: A Genetic Algorithm for Inferring Time Delays in Gene Regulatory Networks, CSB (2004)
13. Wu, C.C., Huang, H., Juan, H., Chen, S.: Gene Networks: an interactive tool for reconstruction of genetic networks from microarray data. Bioinformatics advanced access (2004)
14. Liang, S., Fuhrman, S., Somogyi, R.: REVEAL: a general reverse engineering algorithm for inference of genetic network architectures. In: Pacific Symposium on Biocomputing, vol. 3, pp. 18–29 (1998)
15. Friedmann, N., Linial, M., Nachman, I., Pe'er, D.: Using Bayesian Networks to Analyze expression data. Journal of Computational Biology 7, 601–620 (2000)
16. Heckerman, D., Geiger, D., Chickering, D.M.: Learning Bayesian networks: The combination of knowledge and statistical data. Machine Learning 9, 309–347 (1999)
17. Friedman, N., Murphy, K., Russell, S.: Learning the structure of dynamic probabilistic networks. In: Proc. Fourteenth Conference on Uncertainty in Artificial Intelligence (UAI '98), pp. 139–147 (1998)
18. Troyanskaya, O., et al.: Missing value estimation methods for DNA microarray. Bioinformatics 17(6), 520–525 (2001)
19. Davierwala, A.P., et al.: Synthetic genetic interaction spectrum of essential genes. Nature Genetics 37(10), 1147–1152 (2005)
20. Krogan, N.J., et al.: High-definition macromolecular composition of yeast RNA-processing complexes. Mol. Cell 13(2), 225–239 (2004)
21. Young, K.H.: Yeast Two-Hybrid: So Many Interactions (in) So Little Time. Biology of reproduction 58, 302–311 (1998)
22. Page, D., Ong, I.M.: Experimental design of time series data for learning from dynamic Bayesian networks. PSB 11, 267–278 (2006)
23. Bahler, J.: Cell-Cycle Control of Gene Expression in Budding and Fission Yeast. Annu. Rev. Genet. 39, 69–94 (2005)
24. Spellman, et al.: Comprehensive Identification of Cell Cycle–regulated Genes of the Yeast Saccharomyces cerevisiae by Microarray Hybridization. Molecular Biology of the Cell 9, 3273–3297 (1998)
25. Stark, C., Breitkreutz, B.J., Reguly, T., Boucher, L., Breitkreutz, A., Tyers, M.: BioGRID: a general repository for interaction datasets. Nucleic Acids Res. 34, D535–D539 (2006)

Rough Sets and Fuzzy Sets Theory Applied to the Sequential Medical Diagnosis

Andrzej Zolnierek and Marek Kurzynski

Wroclaw University of Technology, Faculty of Electronics, Chair of Systems and
Computer Networks, Wyb. Wyspianskiego 27, 50-370 Wroclaw, Poland
{andrzej.zolnierek,marek.kurzynski}@pwr.wroc.pl

Abstract. Sequential classification task is typical in medical diagnosis, when the investigations of the patient's state are repeated several times. Such situation always takes place in the controlling of the drug therapy efficacy. A specific feature of this diagnosis task is the dependence between patient's states at particular instants, which should be taken into account in sequential diagnosis algorithms. In this paper methods for performing sequential diagnosis using fuzzy sets and rough sets theory are developed and evaluated. For both soft methodologies several algorithms are proposed which differ in kind of input data and in details of classification procedures for particular instants of decision process. Proposed algorithms were practically applied to the computer-aided medical problem of recognition of patient's acid-base equilibrium states. Results of comparative experimental analysis of investigated algorithms in respect of classification accuracy are also presented and discussed.

1 Introduction

In many pattern recognition tasks there exist dependencies among the patterns to be recognized. Such a task, henceforth called the sequential classification (SC) task, involves dealing with a complex decision problem in which the sequences of patterns should be recognized. For instance, in the medical diagnosis we have to deal with such problems in which the patient's state at a given time depends on the preceding states. Although there remains no doubt about existence of this dependence, it may be of a different nature and range; its simplest instance can be a one-instant-backwards dependence to so complex arrangements as those in which the current state depends on the whole former course of the disease including the sequence of applied treatment as well. From the theoretical point of view, during construction of an appropriate decision algorithm we must not limit our approach only to the current feature vector but we have to consider all the available measurement and observed data instead, as they may contain important information about the recognized patient's state at a given instant. The measurement data can comprise all the features vectors and applied treatments observed so far, thus the amount of data is very large and grows over the time from one instant to another. In such a situation performing of SC task various simplifications and compromises must be made. The dependence can be

J.C. Rajapakse, B. Schmidt, and G. Volkert (Eds.): PRIB 2007, LNBI 4774, pp. 311–322, 2007.

included at an as early stage as that of formulating a mathematical model for the SC task, or as late as at the stage of selecting the appropriate input data set in the decision algorithm which otherwise does not differ from the classical recognition task. In this paper the second approach is presented. It can be called data-oriented, while it uses methods developed in the field of computational intelligence such as fuzzy logic, rough sets theory and genetic algorithms. These methods are recently becoming increasingly popular in the pattern recognition as an attractive alternative to statistical approach. They can perform classification from both labeled and unlabeled training sets as well as acquire and explore the human expert knowledge. They have been successfully applied in classical pattern recognition tasks, i.e. without taking into account the context [6], and in the sequential classification task [7], [17]. Although the information about former applied treatment can be useful for physicians but including such information in model of recognition requires simplifying assumptions. In consequence in such case, in respective pattern recognition algorithms with learning the number of unknown parameters is growing. It requires more medical data in learning process, but the number of medical data is as usual limited. Then in this chapter comparative analysis of such methodologies, but without taking into account the treatment process, to the problem of medical sequential diagnosis (classification of states of acid-base balance) is described. Let us stress that there are many possible methods for performing sequential diagnosis, but the aim of this work was to compare approaches using fuzzy sets and rough set theory, because both are data-oriented and do not require any additional assumptions. After preliminaries and problem statement, we present several algorithms of SC, which differ from each other using different kind of input data and using either fuzzy logic or rough sets methodology. All presented algorithms were practically applied to the problem of sequential medical diagnosis and results can be found at the end of this chapter.

2   Preliminaries and the Problem Statement

We will treat the sequential classification (SC) task as a discrete controlled dynamical process. The pattern (patient) is at the n-th instant in the state $j_n \in \mathcal{M}$, where $\mathcal{M}$ is an m-element set of possible states numbered with the successive natural numbers, thus

$$j_n \in \mathcal{M} = \{1, 2, \ldots, m\}. \tag{1}$$

Obviously, the notion of instant has no specific temporal meaning here, as its interpretation depends on the character of the medical case under consideration. The actual used measure may be minutes, hours, days or even weeks. The patient's state j_n is unknown and does not undergo our direct observation. What we can only observe are the symptoms by which a state manifests itself. We will denote an d-dimensional symptom vector measured at the n-th instant by $x_n \in \mathcal{X}$ (thus $\mathcal{X}$ is the observation space). As already mentioned, the patient's current state depends on the history and thus in the general case the decision algorithm must take into account the whole sequence of the preceding symptom vectors $\bar{x}_n = \{x_1, x_2, \ldots, x_n\}$ and the course of treatment process. It must be underlined here that sometimes it may be difficult to include all the available data, especially for bigger n. In such cases we have to allow some simplifications (e.g. do not take into account the treatments or take into account only several recent values in the sequence of symptom vectors $\bar{x}_n$.

In order to classify such sequences of patient's states we need some more general information to take a valid decision, namely the *a priori* knowledge concerning the general associations that hold between decisions on the one hand, and the sequence of feature on the other. This knowledge may have multifarious forms and various origins. From now on we assume that it has the form of so called *training set*, which in the investigated diagnostic task consists of N training sequences (patient's records):

$$S = \{S_1, S_2, \ldots, S_N\}, \tag{2}$$

A single patient's record:

$$S_k = ((x_{1,k}, j_{1,k}), (x_{2,k}, j_{2,k}), \ldots, (x_{L,k}, j_{L,k})) \tag{3}$$

denotes a single dynamic process course that comprises L feature observations instatnts and the patient's states. Analysis of the SC task implies that, when considered in its most general form, the explored decision algorithm should use in the n-th instant the whole available observed data i.e. the sequences of all feature vectors $\bar{x}_n$ as well as the knowledge included in the training set S. In consequence, the algorithm is of the following form:

$$i_n = \Psi_n(S, \bar{x}_n), \ n = 1, 2, \ldots, i_n, \ i_n \in \mathcal{M}. \tag{4}$$

The next chapters describe in depth the construction of the sequential diagnosis algorithms (4) using various concepts based on fuzzy and rough sets theory.

3   Algorithms of SC Based on Fuzzy Sets Theory

In this section we will apply the fuzzy sets theory to the construction of SC algorithm (4). Two approaches will be considered:

- Mamdani inference system for fuzzy rules with procedure of generating fuzzy rules from learning set developed for sequential classification,
- application of fuzzy relation defined on Cartesian product of input data and class number set obtained from the learning set as a solution of appropriate optimization problem.

For both concepts corresponding algorithms will be proposed which differ in kind of input data and details of classification procedures for particular instants of decision process.

3.1 Fuzzy Method with Mamdani Inference System

This concept consists in applying the inference engine from a fuzzy rule system to construction decision algorithms for SC task. For all the algorithms presented below we assume the following general form of the k-th rule in the system ($k = 1, 2, ..., K$), which associate an observation vector $a = (a^{(1)}, a^{(2)}, ..., a^{(d_a)})$ with class numbers:

$$IF\ a^{(1)}\ isA_{1,k}\ AND\ \cdots\ AND\ a^{(L)}\ isA_{L,k}\ THEN\ B_k. \qquad (5)$$

$A_{i,k}$ are fuzzy sets (which membership functions are designated by $\mu_{A_{i,k}}$) that correspond to the nature of particular input observations, whereas B_k is a discrete fuzzy set defined on the class number set $\mathcal{M}$, with the μ_{B_k} membership function.

The only difference between the algorithms is the form of observation vector a and its relation with features of pattern to be recognized and, what follows, the procedure for rule system (5) derivation from the learning set (2).

As recognition algorithm the Mamdani fuzzy inference system has been applied [1], [16]. In this system we use the minimum t-norm as AND connection in premises, product operation as conjunctive implication interpretation in rules, the maximum t-conorm as aggregation operation, and finally the maximum defuzzification method.

Two methods can be used to obtain collection of fuzzy if-then rules (5) in the construction of fuzzy system:

- from human expert or based on domain knowledge,
- extraction of rules using numerical input-output data.

One of the best known method of rules generating from the given training patterns set (2), is the method proposed by Wang and Mendel [15], which developed for the SC will be applied in the further algorithms.

Algorithm Without Context (Mamdani-0). In this case the SC is considered as a sequence of single independent tasks without taking into account the associations that may occur between them. Such approach leads to the classical concept of recognition algorithm, which assigns a pattern at the n-th instant to a class on the base of its features only, namely:

$$i_n = \Psi(\mathcal{S}, x_n), n = 1, 2, ..., i_n \in \mathcal{M}. \qquad (6)$$

Thus it will be obtained assuming $a = x_n$ for the n-th instant. Now, rule derivation is performed based on the whole training set S for which neither the division into sequences S_k nor element succession in the sequence is pertinent. Resulting procedure is following:

1. Cover the space $\mathcal{X}^{(l)}$ of the individual feature $x^{(l)}$ ($l = 1, 2, ..., d$) by overlapping fuzzy sets corresponding to the linguistic "values" of this feature (e.g. small, medium, big, etc.). For each fuzzy set define its membership function.

Obtained fuzzy sets state premises $A_{i,k}$ in fuzzy rules (5). For example, in the further practical medical diagnosis task, we used triangular fuzzy numbers with 3 regular partitions [6].

2. For each example generate fuzzy rule with premises corresponding to fuzzy regions with the highest membership grade of appropriate feature.
3. Find the rules with the same premises and aggregate them into one rule.
4. Determine the discrete fuzzy conclusion of the rule (fuzzy set), for which

$$\mu_{B_k}(j) = \frac{n_k(j)}{\sum_j n_k(j)}, \; j \in \mathcal{M}, \tag{7}$$

where $n_k(j)$ denotes the number of learning patterns from j-th class fulfilling the k-th rule.

Algorithm with k-th Order Context (Mamdani-k). This algorithm makes allowance for the k-instant-backwards dependence using full bulk of the measurement data. In effect, we have now $a = (x_n, x_{n-1}, \cdots, x_{n-k})$ and rule derivation is achieved based on the whole training set S, taking into account the succession of particular k elements in sequences S_i .

3.2 Algorithms Using Fuzzy Relations

Algorithm Without Context (Relation-0). This algorithm, as algorithm *Mamdani-0*, includes no inter-state dependences on a state but it utilizes only the current feature values instead. Application of fuzzy relation to the construction of classifier (6) from the learning set (2) containing $N \times L$ patterns (now the order of patterns in the sequences (3) is irrelevant) is well known in literature [10], [12], [13] and resulting procedure comprises the following items:

1. Cover the space $\mathcal{X}^{(l)}$ of the individual feature $x^{(l)}$ ($l = 1, 2, ..., d$) by overlapping fuzzy sets corresponding to the linguistic "values" of this feature (e.g. *small, medium, big*, etc.). For each fuzzy set define its membership function. Obtained fuzzy sets state fuzzified feature space $\mathcal{X}_F^{(l)}$ of individual features. Create fuzzified feature space as a product $\mathcal{X}_F = \mathcal{X}_F^{(1)} \times \mathcal{X}_F^{(2)} \times ... \times \mathcal{X}_F^{(d)}$. Let its cardinality be equal to d_F - this value depends on number of partitions and the size of feature vector. For example, in the further practical medical diagnosis task, $d = 3$ and we used triangular fuzzy numbers with 3 regular partitions, which gave $d_F = 27$.
2. Determine observation matrix $O(\mathcal{S})$ of learning set $\mathcal{S}$, i.e. fuzzy relation defined on product of fuzzified feature space $\mathcal{X}_F$ and learning set $\mathcal{S}$. The ith row of $O(\mathcal{S})$ ($i = 1, 2, ..., N \times L$) contains membership degrees of features of ith learning pattern to fuzzy sets of space $\mathcal{X}_F$. The number of columns of $O(\mathcal{S})$ is equal to d_F.
3. Determine decision matrix $D(\mathcal{S})$, i.e. relation defined on product of learning set $\mathcal{S}$ and the set of decisions (classes) $\mathcal{M}$. For the training data, where the classification is exactly known, the ith row is a fuzzy singleton set, i.e. a

vector of all zeros except for a one at the place corresponding to the class number of ith learning pattern.

4. Find matrix $E(\mathcal{S})$ as a solution of so-called *fuzzy relational equation* ([11], [13]):

$$O(\mathcal{S}) \circ E(\mathcal{S}) = D(\mathcal{S}), \tag{8}$$

or - in approximate way - as a solution of the following optimization problem:

$$\rho(O(\mathcal{S}) \circ E(\mathcal{S}), D(\mathcal{S})) = min_E \, \rho(O(\mathcal{S}) \circ E, D(\mathcal{S})), \tag{9}$$

where criterion $\rho(A, B)$ evaluates difference between matrices A and B, i.e. $\rho(A, B) \geq 0$ and $\rho(A, B) = 0$ iff $A = B$. Operator $\circ$ denotes here max-min-norm composition of relations, i.e. multiplication of matrices O and E with $\times$ and $+$ operators replaced by min and max operators (more general by t-norm and s-norm operators)([1]). In the further practical example we decided to select the method of determination of matrix E, adopting

$$\rho(A, B) = \sum_{i,j} (a_{ij} - b_{ij})^2 \tag{10}$$

and applying as an optimization procedure real-coded genetic algorithm.

Matrix $E(\mathcal{S})$ is a fuzzy relation defined on product of decision set $\mathcal{M}$ and feature space $\mathcal{X}_F$, in which reflects knowledge contained in the learning set. To classify a new pattern x, first the row-matrix of fuzzy observation $O(x)$ is calculated from known vector of its features $[x^{(1)}, x^{(2)}, ..., x^{(d)}]$. Then matrix $E(\mathcal{S})$ is applied to compute an output row-matrix called *target vector* ([14]):

$$O(x) \circ E(\mathcal{S}) = T(x) = [t_1(x), t_2(x), ..., t_M(x)], \tag{11}$$

which gives a fuzzy classification in terms of membership degrees $t_i(x)$ of the pattern x to the given classes $i = 1, 2, ..., m$. When a crisp decision is required, defuzzification has to be applied, typically according to the maximum rule.

Algorithm with k-th Order Context (Relation-k). This algorithm includes k-instant-backwards-dependence ($k < L$) with full measurement data, i.e. the decision at the n-th instant is made on the base of vector of features:

$$\bar{x}_n^{(k)} = (x_{n-k}, x_{n-k+1}, ..., x_{n-1}, x_n). \tag{12}$$

Although, the main concept of the proposed methods of SC is the same as for independent patterns, there are many differences concerning details in procedure of construction of matrix E and course of recognition process.

Before we will describe these algorithms let us first introduce set $\mathcal{S}^{(k)}$ containing sequences of $(k+1)$ learning patterns from $\mathcal{S}$ and set $\mathcal{S}_{\bar{j}^{(k)}}^{(k)}$ - as previously but in which at the first k position additionally the sequence of classes $\bar{j}^{(k)} \in \mathcal{M}^k$ appears. Consequently, the algorithm with k-th order dependence (*Relation k*) and full measurement features can be presented according to the following points:

1. Create the fuzzified feature space $\mathcal{X}_F$ as in the procedure for independent patterns
2. Determine observation matrix $O^{(k)}$, i.e. fuzzy relation in the space $\mathcal{X}_F^k = \mathcal{X}_F \times \mathcal{X}_F \times \cdots \times \mathcal{X}_F$ (k times) and learning subset $\mathcal{S}^{(k)}$. The ith row of observation matrix contains memberships degrees of features $\bar{x}^{(k)}$ of ith learning sequence from $\mathcal{S}^{(k)}$ to the fuzzy sets of space $\mathcal{X}_F^k$.
3. Determine decision matrix $D^{(k)}$, i.e. relation defined on product of learning sequences $\mathcal{S}^{(k)}$ and the set of decisions (classes) $\mathcal{M}$. The ith row of $D^{(k)}$ is a vector of all zeros except for a one at the place corresponding to the last class number of ith sequence in the set $\mathcal{S}^{(k)}$.
4. Find matrix $E^{(k)}$, so as to minimize criterion

$$\rho(O^{(k)} \circ E^{(k)}, D^{(k)}). \tag{13}$$

Next, at the nth step of sequential recognition first the row-matrix of fuzzy observation $O(\bar{x}_n^{(k)})$ is calculated from known sequence of feature observations (12). Then matrix $E^{(k)}$ is applied to compute a target vector of soft decisions:

$$O(\bar{x}_n^{(k)}) \circ E^{(k)} = T(\bar{x}_n^{(k)}), \tag{14}$$

and final crisp decision is obtained after defuzzification step.

4   Algorithms of SC Based on Rough Sets Theory

In this section we will apply the rough sets theory [9] to the construction of SC algorithm (4).

Now, the training set (2) is considered as an *information system* $S = (U, A)$, where U and A, are finite sets called *universe* and the set of *attributes*, respectively. For every attribute $a \in A$ we determine its set of possible values V_a, called *domain* of a. Such information system can be represented as a table, in which every row represents a single sequence (3). In successive column of k-th row of this table we have values of the following attributes:

$$x_{1,k}^{(1)}, x_{1,k}^{(2)}, ..., x_{1,k}^{(d)}, j_{1,k}, x_{2,k}^{(1)}, x_{2,k}^{(2)}, ..., x_{2,k}^{(d)}, j_{2,k}, ..., x_{L,k}^{(1)}, x_{L,k}^{(2)}, ..., x_{L,k}^{(d)}, j_{L,k}. \tag{15}$$

In such an information system we can define in different way the subset $C \subseteq A$ of *condition attributes* and the single-element set $M \subseteq A$ which will be the *decision attribute*. Consequently, we obtain the *decision system* $S = (U, C, M)$ in which, knowing the values of condition attributes, our task is to find the value of decision attribute, i.e. to find appropriate pattern recognition algorithm of sequential classification. Of course, as in algorithms based on fuzzy sets theory, we can choose the subset of condition attributes in different way. Taking into account the set of condition attributes C, let us denote by X_j the subset of U for which the decision attribute is equal to j, $j = 1, ..., m$. Then, for every j we can defined respectively the *C-lower approximation* $C_*(X_j)$ and the *C-upper approximation* $C^*(X_j)$ of set X_j [9], [17]. Hence, the lower approximation of set

X_j is the set of objects $x \in U$, for which knowing values of condition attributes C, for sure we can say that they are belonging to the set X_j. Moreover, the upper approximation of set X_j is the set of objects $x \in U$, for which knowing values of condition attributes C, for sure we can not say that they are not belonging to the set X_j. Consequently, we can define *C-boundary region* of X_j as follows:

$$CN_B(X_j) = C^*(X_j) - C_*(X_j). \tag{16}$$

For every decision system we can formulate its equivalent description in the form of set of decision formulas $For(C)$. Each row of the decision table will be represented by single if-then formula, where on the left side of this implication we have logical product (*and*) of all expressions from C such that every attribute is equal to its value. On its right side we have expression that decision attribute is equal to the one number of class from (1). These formulas are necessary for constructing different pattern recognition algorithms for sequential classification.

4.1   Algorithm Without Context (Rough-0)

As usual, we start with the algorithm without the context which is well known in literature ([3], [5], [9], [10]). In this case our decision table contains $N \times L$ patterns, each having d condition attributes (features) and one decision attribute (the class to which the pattern belongs).

Application of rough set theory to the construction of classifier (6) from the learning set (2) can be presented according to the following items:

1. If the attributes are the real numbers then the discretization preprocessing is needed first. After this step, the value of each attribute is represented by the number of interval in which this attribute is included. Of course for different attributes we can choose the different numbers of intervals in order to obtain their proper covering and let us denote for l-th attribute ($l = 1, ..., d$) by $\nu_p^l l$ its p_l-th value or interval.
2. The next step consists in finding the set $For(C)$ of all decision formulas from (2), which have the following form:

$$IF\ (x^{(1)} = \nu_{p1}^l)\ AND\ ...AND\ (x^{(d)} = \nu_{pd}^d)\ THEN\ \Psi(\mathcal{S}, x) = j. \tag{17}$$

   Of course, it can happen that from the learning set (2) we obtain more than one rule for particular case. Then for such a formula (17) we determine its *strength* factor [5], which is the number of correctly classified patterns during learning procedure. If any case in (2) is single then the strength factor of corresponding rule is equal to one.
3. For the set of formulas $For(C)$, for every $j = 1, ..., m$ we calculate their C-lower approximation $C_*(X_j)$ and their boundary regions $CN_B(X_j)$.
4. In order to classify x_n (after discretization its attributes if it is necessary) we look for matching rules in the set $For(C)$, i.e. we take into account such rules in which the left condition is fulfilled by the attributes of recognized pattern.

5. If there is only one matching rule, then we classify this pattern to the class which is indicated by its decision attribute j, because for sure such rule is belonging to the lower approximation of all rules indicating j, i.e. this rule is *certain*.

6. If there is more then one matching rule in the set $For(C)$, it means that the recognized pattern should be classified by the rules from the boundary regions $CN_B(X_j)$, $j = 1, ..., m$ and in this case as a decision we take the index of boundary region for which the strength of corresponding rule is maximal. In such a case we take into account the rules which are *possible*.

4.2   Algorithm with k-th Order Context (Rough-k)

Although as in [17], we could take into account at n-th instant whole available information about the state of recognized sequential process,for the same reason as previously let us choose the following decision atributes in our decision table:

$$x_{n-k}^{(1)}, ..., x_{n-k}^{(d)}, x_{n-k+1}^{(1)}, ..., x_{n-k+1}^{(d)}, ..., x_{n-1}^{(1)}, ..., x_{n-1}^{(d)}, ..., x_n^{(1)}, ..., x_n^{(d)}. \qquad (18)$$

This means that algorithm includes k-instant-backwards-dependence $(k < L)$ with full measurement data. Let us denote by D the total number of decision attributes (former was $D = d$ and now $D = (k + 1) \times d + k$). Next, from the (2), we can create the decision table which will have $D + 1$ column (the last one is the true classification of n-th recognized pattern) and consequently the number of rows will be equal to $N \times (L - k)$, because from each sequence (3) we can obtain $L - k$ subsequences of the length $k + 1$. The main idea of the proposed methods of SC is the same as for independent patterns but there are differences concerning details in procedure of construction of the set of decision formulas $For(C)$. Now, the decision formulas are of the following form:

$$IF (x^{(1)} = \nu_{p1}^l) AND ...AND (x^{(D)} = \nu_{pD}^D) THEN \Psi(\mathcal{S}, \bar{x}_n^{(k)}) = j_n \qquad (19)$$

The next steps of SC are the same as previously, i.e. we calculate $C_*(X_{j_n})$ and $CN_B(X_{j_n})$ and finally, the decision is made according to same procedure.

All the decision algorithms that are depicted in the previous sections have been experimentally tested in respect of the decision quality (frequency of correct classifications) for real data that concern recognition of human acid-base equilibrium states (ABE).Results of experimental investigations are presented in the next section.

5   Medical Example: Sequential Diagnosis of Acid-Base State Balance

In the course of many pathological states, there occur anomalies in patient's organism as far as both hydrogen ion and carbon dioxide production and elimination are concerned, which leads to disorders in the acid-base equilibrium (ABE).

Thus we can distinguish acidosis and alkalosis disorders here. Each of them can be of metabolic or respiratory origin, which leads to the following ABE states classification: metabolic acidosis, respiratory acidosis, metabolic alkalosis, respiratory alkalosis, correct state.

In the process of treatment, correct recognition of these anomalies is indispensable, because the maintenance of the acid-base equilibrium, e.g. the pH stability of the fluids is the essential condition for correct organism functioning. Moreover, the correction of acid-base anomalies is indispensable for obtaining the desired treatment effects.

In medical practice, only the gasometric examination results are made to establish fast diagnosis, although the symptom set needed for correct ABE estimation is quite large. The utilized results are: the pH of blood, the pressure of carbon dioxide, the current dioxide concentration.

An anomalous acid-base equilibrium has a dynamic character and its changes depend on the previous state, and in consequence they require frequent examinations in order to estimate the current ABE state. It is clear now that the SC methodology presented above suits well the needs of computer aided ABE diagnosing. The current formalization of the medical problem leads to the task of the ABE series recognition, in which the classification basis in the n-th moment constitutes the quality feature consisting of three gasometric examinations. And the set of diagnostic results $\mathcal{M}$ is represented by 5 mentioned acid-base equilibrium states.

The diagnostic algorithms applied to the ABE which state sequential diagnosis task have been worked out on the basis of evidence material that was collected in Neurosurgery Clinic of Medical Academy of Wroclaw and constitutes the set of training sequences (2). The material comprises 78 patients (78 sequences) with ABE disorders caused by intracranial pathological states for whom the gasometric examination results and the correct ABE state diagnosis were regularly put down on the 12-hour basis. There were around 20 examination cycles for each patient, yielding the total of 1416 single examination instances.

To compare the classification accuracy of proposed concepts of SC algorithms and the performance of RGA, ten independent runs of RGA were carried out for each diagnostic algorithm with different random initial populations. The results are shown in Table 1. The values depicted in the table are those of the best solution obtained at the end of a RGA trial. Table 1 contains also the best result, the mean value and standard deviation for each SC algorithm. In testing of Mamdani inference system and algorithms based on rough sets theory, the cross validation method was used, i.e. for every trial ten testing sequences were chosen randomly.

These results imply the following conclusions:

1. Algorithms **Mamdani-0, Relation-0** and **Rough-0** that do not include the inter-state dependencies and treat the sequence of states as independent objects are worse than those that have been purposefully designed for the sequential medical diagnosis task, even for the least effective selection of input data. This confirms the effectiveness and usefulness of the conceptions and

Table 1. Frequency of correct diagnosis for various diagnostic algorithms (in per cent)

Trial	Ma-0	Ma-1	Ma-2	Re-0	Re-1	Re-2	Ro-0	Ro-1	Ro-2
1	66.9	72.1	67.3	80.6	89.6	91.8	83.5	85.7	90.1
2	68.1	69.2	70.2	82.2	86.5	91.9	85.1	86.1	89.6
3	67.5	70.1	70.1	79.4	87.2	88.9	83.8	88.0	91.1
4	68.0	71.2	71.6	78.5	85.9	92.6	84.0	88.5	92.0
5	67.8	69.1	71.2	80.9	90.3	91.9	83.6	87.6	89.8
6	67.9	70.0	69.4	82.1	89.7	91.6	85.7	86.4	89.9
7	67.2	71.2	70.9	81.9	88.1	89.4	83.1	85.1	90.2
8	68.7	69.8	71.2	78.3	87.2	89.0	85.9	86.9	91.4
9	67.1	68.6	70.4	78.5	90.7	92.9	83.4	88.9	91.3
10	67.0	68.2	67.6	81.1	88.7	92.8	86.9	87.8	91.2
Best	68.7	72.1	71.6	82.2	90.7	92.9	84.0	88.9	92.0
Mean	67.6	70.0	70.0	80.3	88.4	91.3	84.5	87.1	90.7

algorithm construction principles presented above for the needs of sequential diagnosis.

2. There occurs a common effect within each algorithm group: the model of the second order dependency (**Mamdani-2, Relation-2, Rough-2**) turns out to be more effective than the first order dependence approach (**Mamdani-1, Relation-1, Rough-1**).

3. There is no essential difference among the algorithms using the same input data which are based either on fuzzy relation method or on rough sets theory.

4. The RGA method is capable of solving the problem of learning of SC algorithm for practical computer-aided medical diagnostic system. Results of RGA performances turn out to be quite repeatable and insensitive to initial conditions.

It must be emphasized that proposed procedures leads to the very flexible sequential recognition algorithm due to optional value of k. In particular the value of k need not be constant but it may dynamically change from step to step. So, the choice $k = n - 1$ for n-th instant of sequential classification denotes the utilization of the whole available information according to the general form of decision rule (4). On the other hand however, such a concept - especially for bigger n - is rather difficult for practical realization.

6   Conclusions

The aim of this work was comparative analysis of soft computing methods applied to the sequential classification tasks in which there exist dependencies among the patterns to be recognized. Two approaches to SC task were considered: using fuzzy sets theory and using rough sets theory. For both of them corresponding algorithms were proposed, without taking into account the treatment directly (of course results of treatment are observed indirectly in feature vectors). The empirical results show that in such case the accuracy of classification can be improved taking into account such dependencies however, more empirical studies are required. Moreover, in presented practical example there is no essential difference among the results of algorithms using the same input data which are based either on fuzzy sets or on rough sets theory. It means, that such soft computing methods in SC task can be considered as complementary.

References

1. Czogala, E., Leski, J.: Fuzzy and neuro-fuzzy intelligent systems. Springer, New York (2000)
2. Dinola, A., Pedrycz, W., Sessa, S.: Fuzzy relation equations theory as a basis of fuzzy modelling: an overview. Fuzzy Sets and Systems 40, 415–429 (1991)
3. Fang, J., Grzymala-Busse, J.: Leukemia Prediction from Gene Expression Data-a Rough Set Approach. In: Rutkowski, L., Tadeusiewicz, R., Zadeh, L., Zurada, J. (eds.) Artificial Intelligence and Soft Computing, pp. 899–908. Springer, New York (2006)
4. Goldberg, D.: Genetic algorithms in search, optimization and machine learning. Addison-Wesley, New York (1989)
5. Grzymala-Busse, J.: A System for Learning from Examples Based on Rough Sets. In: Slowinski, R. (ed.) Intelligent Decision Support: Handbook of Applications and Advances of the Rough Sets Theory, pp. 3–18. Kluwer Academic Publishers, Dordrecht (1992)
6. Kurzynski, M.: Multistage diagnosis of myocardial infraction using a fuzzy relation. LNCS (LNAI), pp. 1014–1019. Springer, New York (2004)
7. Kurzynski, M., Zolnierek, A.: Sequential Classification via Fuzzy Relations. In: Rutkowski, L., Tadeusiewicz, R., Zadeh, L., Zurada, J. (eds.) Artificial Intelligence and Soft Computing, pp. 623–632. Springer, New York (2006)
8. Michalewicz, Z.: Genetic Algorithms + Data Structure = Evolution Programs. Springer, New York (1996)
9. Pawlak, Z.: Rough Sets - Theoretical Aspect of Reasoning About Data. Kluwer Academic Publishers, Dordrecht (1991)
10. Pawlak, Z.: Rough Sets, Decision Algorithms and Bayes' Theorem. European Journal of Operational Research 136, 181–189 (2002)
11. Pedrycz, W.: Fuzzy Sets in Pattern Recognition: Methodology and Methods. Pattern Recognition 23, 121–146 (1990)
12. Pedrycz, W.: Genetic Algorithms for Learning in Fuzzy Relation Structures. Fuzzy Sets and Systems 69, 37–45 (1995)
13. Ray, K., Dinda, T.: Pattern classification using fuzzy relational calculus. IEEE Transactions on Systems, Man and Cybernetics 33(1), 1–16 (2003)
14. Setnes, M., Babuska, R.: Fuzzy relational classifier trained by fuzzy clustering. IEEE Transactions on Systems, Man and Cybernetics 29, 619–625 (1999)
15. Wang, L.X., Mendel, J.M.: Generating fuzzy rules by learning from examples. IEEE Trans. on Systems, Man and Cybernetics 22, 1414–1427 (1992)
16. Wang, L.X.: A course in fuzzy systems and control. Prentice-Hall, New York (1998)
17. Zolnierek, A.: Application of rough sets theory to the sequential diagnosis. In: Maglaveras, N., Chouvarda, I., Koutkias, V., Brause, R. (eds.) ISBMDA 2006. LNCS (LNBI), vol. 4345, pp. 413–422. Springer, Heidelberg (2006)

In silico Identification of Putative Drug Targets in *Pseudomonas aeruginosa* Through Metabolic Pathway Analysis

Deepak Perumal[1,2], Chu Sing Lim[2], and Meena K. Sakharkar[1,*]

[1] School of Mechanical and Aerospace Engineering, Nanyang Technological University, Singapore
[2] BioMedical Engineering Research Centre, Nanyang Technological University, Singapore
mmeena@ntu.edu.sg

Abstract. Comparative genomic analysis between pathogens and the host *Homo sapiens* has led to identification of novel drug targets. Microbial drug target identification and validation has been the latest trend in pharmacoinformatics. In order to identify a suitable drug target for the pathogen *Pseudomonas aeruginosa* an *in silico* comparative analysis of the metabolic pathways between the pathogen and the host *Homo sapiens* was performed. Detection of bacterial genes that are non-homologous to human genes, and are essential for the survival of the pathogen represents a promising means of identifying novel drug targets. Metabolic pathways for the pathogen and *H.sapiens* were obtained from the metabolic pathway database KEGG and were compared to identify unique pathways present only in the pathogen and absent in the host. We identified 361 enzymes from both unique and common pathways between the pathogen and the host of which 50 belong to the 12 unique pathways. Enzymes from both genomes were subject to a BLASTp search and sequences homologous to human were removed as non essential. *P.aeruginosa* targets without human homologs were identified when the e-value threshold was set as 10^{-2}. Of the 214 targets that had no hits only 30 targets belong to unique pathways. These 30 targets were then compared with the list of candidate essential genes identified by mutagenesis. Only 8 targets matched with the essential genes list and these were considered as potential drug targets. We have built homology model for the four target genes lpxC, kdsA, kdsB and waaG using MODELLER software. This approach enables rapid potential drug target identification, thereby greatly facilitating the search for new antibiotics.

Keywords: *Pseudomonas aeruginosa*, *Homo sapiens*, Comparative microbial genomics, KEGG, Homology, MODELLER, kdsA, kdsB, waaG, lpxC, Potential drug targets.

1 Introduction

Pseudomonas aeruginosa is a Gram-negative bacterium and an opportunistic human pathogen as well as an opportunistic pathogen for plants. It mainly target

---

* Corresponding author. Meena Kishore Sakharkar (Ph.D.). Assistant Professor. N3-2C-113B, MAE, Nanyang Technological University, Singapore.

J.C. Rajapakse, B. Schmidt, and G. Volkert (Eds.): PRIB 2007, LNBI 4774, pp. 323–336, 2007.

immunocompromised patients and typically infects the pulmonary tract, urinary tract and even causes blood infections. *P. aeruginosa* is highly resistant to a wide range of antibiotics and disinfectants [23]. The pathogen has been reported to have lower outer membrane permeability to small molecules [10]. There is also the presence of several multidrug efflux pumps from the major facilitator superfamily (MFS), multidrug and toxic compound extrusion (MATE) families, ATP-binding cassette (ABC) and small multi-drug resistance (SMR) that have increased its intrinsic resistance to many efficient antibiotics. Thus, developing new antibacterial drugs against this pathogen has been a challenging problem over these years.

Over the last decade, complete genome sequences of several pathogenic bacteria have been sequenced and many more such projects are currently under investigation. This global effort has focused primarily on pathogens which encompass the majority of all genome projects, and has generated a large amount of raw material for *in silico* analysis. These data pose a major challenge in the post-genomic era, i. e. to fully exploit this treasure trove for the identification and characterization of virulent factors in these pathogens, and to identify novel putative targets for therapeutic intervention [16].

Genomics can be applied to evaluate the suitability of potential targets using two criteria, i. e. "essentiality" and "selectivity" [19]. The target must be essential for the growth, replication, viability or survival of the microorganism, i. e. encoded by genes critical for pathogenic life-stages. The microbial target for treatment should not have any well-conserved homolog in the host, in order to address cytotoxicity issues. This can help to avoid expensive dead-ends when a lead target is identified and investigated in great detail only to discover at a later stage that all its inhibitors are invariably toxic to the host. Genes that are conserved in different genomes often turn out to be essential [7] [25] [12] [11]. A gene is deemed to be essential if the cell cannot tolerate its inactivation by mutation, and its status is confirmed using conditional lethal mutants.

The complete genome sequence of the pathogen *Pseudomonas aeruginosa* [23] and the host *Homo sapiens* [The Genome International Consortium, 2001] is available. *Pseudomonas aeruginosa* PA01 strain is the largest bacterial genome sequenced with 6.3 million base pairs and with 5,570 predicted open reading frames (ORFs). Comparative analysis between the two genomes has led to know about the pathogenicity of the bacterium and offers to identify new novel antimicrobial drug targets. Galperin and Koonin, 1999 suggested that targets that serve as inhibitors of certain bacterial enzymes and specific to bacteria can be developed as potential drug targets. Comparative metabolic pathway analysis results in the identification of unique pathways and enzymes that are present in the pathogen but absent in the host. Our approach by differential genome analysis identified bacterial genes that are non-homologous to human and thus making them attractive targets for new frontline antibiotics. Our *in silico* approach enabled us to identify suitable targets from the pathogen resulting in homology modeling of these targets and further analysis using molecular docking studies.

As a proof of concept, many of the genes identified by our approach are also reported as essential by experimental methods. Of the 30 distinct targets belonging to the unique pathways of *P.aeruginosa* the experimentally determined candidate essential genes generated by Jacobs *et al.,* 2003 listed out only 8 targets as the most essential ones. By further analyses of these genes with PDB structures only 3 were selected as the most suitable antibacterial drug targets. Using homology modeling, a target sequence can be modeled with reasonable accuracy with the template sequence based on the sequence similarity between them. Our approach was successful in modeling 4 potential drug targets enabling us further validation and characterization in the laboratory in near future.

2 Materials and Methods

2.1 Identification of Unique Enzymes as Drug Targets

Metabolic pathway information was obtained from the pathway database Kyoto Encyclopedia of Genes and Genomes [9]. Enzyme commission numbers (EC) of the pathogen *P.aeruginosa* and the host *H.sapiens* were extracted from the KEGG database. Pathways unique to *P. aeruginosa* were filtered out. Twelve unique pathways were observed [Table 1]. These are the pathways that do not appear in the host (*H. sapiens*) but are present in the pathogen. We further identified unique enzymes among shared pathways under carbohydrate metabolism, energy metabolism, lipid metabolism, nucleotide metabolism, amino acid metabolism, glycan biosynthesis and metabolism and metabolism of cofactors and vitamins were obtained from the KEGG database. A total of 361 enzymes that are present in *P. aeruginosa* but absent in *H. sapiens* were obtained and their corresponding protein sequences were retrieved from the KEGG database.

The protein sequences for these 361 unique enzymes were retrieved and were subject to BLAST [1] search against human protein sequences database at an expectation E-value cutoff of 10^{-2} to identify non-homologous genes in *P. aeruginosa*. Removing enzymes from the pathogen that share a similarity with the host protein ensures that the targets have nothing in common with the host proteins, thereby, eliminating undesired host protein-drug interactions. The above search resulted in 214 enzymes that had "no hits" in BLAST search. Thirty of these 214 "no hits" belonged to the unique pathways set and the remaining 184 belong to unique enzymes in shared pathways.

Table 1. Pathways unique to *Pseudomonas aeruginosa*

S.No	Pathways and their enzymes	Gene	EC #
1	**Polyketide sugar unit biosynthesis**		
	Glucose 1-phosphate thymidylyltransfease	rmlA	2.7.7.24
	dTDP-D-Glucose 4,6 dehydratase	rmlB	4.2.1.46
	dTDP-4-dehydrorhamnose 3,5 epimerease	rmlC	5.1.3.13
	dTDP-4-dehydrorhamnose reductase	rmlD	1.1.1.133
2	**Biosynthesis of siderophore group nonribosomal peptides**		
	Isochorismate synthase	pchA	5.4.4.2
	Isochorismate pyruvate lyase	pchB	4.1.99.-
3	**Toluene and xylene degradation**		
	catechol 1,2-dioxygenase	catA	1.13.11.1

Table 1. (*continued*)

4	**1,2 Dichloroethane degradation**		
	Quinoprotein alcohol dehydrogenase	exaA	1.1.99.8
	Probable aldehyde dehydrogenase		1.2.1.3
5	**Type II secretion system**		
	Two-component sensor PilS	pilS	2.7.3.-
	Leader peptidase (prepilin peptidase) / N-methyltransferase	pilD	3.4.23.43
	Methyltransferase PilK	pilK	2.1.1.80
6	**Type III secretion system**		
	Flagellum-specific ATP synthase FliI	fliI	3.6.3.14
7	**Phosphotransfease system (PTS)**		
	phosphotransferase system, fructose-specific IIBC component	fruA	2.7.1.69
	probable phosphotransferase system enzyme I		2.7.3.9
8	**Bacterial Chemotaxis**		
	Methyltransferase PilK	pilK	2.1.1.80
	Two-component sensor PilS	pilS	2.7.3.-
	probable methylesterase		**3.1.1.61
9	**Flagellar Assembly**		
	ATP synthase in type III secretion system		3.6.3.14
10	**D-Alanine metabolism**		
	D-alanine-D-alanine ligase A	ddlA	6.3.2.4
	biosynthetic alanine racemase	alr	5.1.1.1
11	**Lipopolysaccharide Biosynthesis**		
	Probable glucosyltransferases		2.4.-
	3-deoxy-manno-octulosonate cytidylyltransferase	kdsB	**2.7.7.38
	Putative 3-deoxy-D-manno-octulosonate 8-phosphate phosphatase		3.1.3.45
	Tetraacyldisaccharide 4'-kinase	lpxK	**2.7.1.130
	Lipid A-disaccharide synthase	lpxB	**2.4.1.182
	Lipopolysaccharide core biosynthesis protein WaaP	waaP	**2.7.-.-
	Poly(3-hydroxyalkanoic acid) synthase 1	phaC1	2.3.1.-
	UDP-glucose:(heptosyl) LPS alpha 1,3-glucosyltransferase WaaG	waaG	**2.4.1.-
	UDP-2,3-diacylglucosamine hydrolase		**3.6.1.-
	UDP-3-O-acyl-N-acetylglucosamine deacetylase	lpxC	3.5.1.-
	UDP-N-acetylglucosamine acyltransferase	lpxA	2.3.1.129
	ADP-L-glycero-D-mannoheptose 6-epimerase	rfaD	5.1.3.20
	2-dehydro-3-deoxyphosphooctonate aldolase (KDO 8-P synthase)	kdsA	**2.5.1.55
12	**Two component system**		
	Two-component sensor PilS	pilS	2.7.3.-
	Probable 2-(5"-triphosphoribosyl)-3'-dephospho coenzyme-A synthase		2.7.8.25
	Serine protease MucD precursor	mucD	3.4.21.-
	Probable acyl-CoA thiolase		2.3.1.9
	Glutamine synthetase	glnA	6.3.1.2
	Citrate lyase beta chain		4.1.3.6
	Protein-PII uridylyltransferase	glnD	2.7.7.59
	Beta-lactamase precursor	ampC	3.5.2.6
	Anthranilate synthase component II	trpG	4.1.3.27
	Anthranilate phosphoribosyltransferase	trpD	2.4.2.18
	Indole-3-glycerol-phosphate synthase	trpC	4.1.1.48
	Tryptophan synthase alpha chain	tr	4.2.1.20
	Potassium-transporting ATPase	kd	3.6.3.12
	Probable methylesterase		3.1.1.61
	Alkaline phosphatase	phoA	3.1.3.1
	Respiratory nitrate reductase alpha chain	narG	1.7.99.4

** Enzymes that matched with list of candidate essential genes and were considered as potential drug targets.

2.2 Comparison of Unique Enzymes to Essential Gene Data

We further compared the 214 unique enzymes to the list of candidate essential genes of *P.aeruginosa* obtained from transposon mutagenesis studies [8]. It is observed that 83 enzymes in total (8 enzymes from unique pathways and 75 enzymes from shared pathways) are reported as essential [8]. It is noteworthy that 7 of the 8 enzymes from the unique pathways map to a single pathway that of lipopolysaccharide biosynthesis [Table 2]. Literature search revealed that LpxC (UDP-3-O-acyl-N-acetylglucosamine deacetylase) is another enzyme in lipopolysaccharide biosynthesis that is essential but is absent in the transposon mutagenesis data. Our selection of LpxC for further analyses was based on the concept that molecular validation of this enzyme could act as a target for novel antibacterial drugs in *Pseudomonas aeruginosa* [10].

2.3 Comparative Homology Modeling

Annotation screen for the 8 enzymes in unique pathways revealed that one of the enzymes is reported as a conserved hypothetical protein and another one as probable methylesterase. We removed these two proteins for homology modeling. The remaining 6 enzymes from the unique pathways were subject to BLASTp search against PDB. We further removed 3 enzymes that had "no hits" in PDB or had short template sequences and thus modeling is not possible. The final potential drug targets are kdsA (2-dehydro-3-deoxyphosphooctonate aldolase), kdsB (3-deoxy-manno-octulosonate cytidylyl

Table 2. Potential eight drug targets obtained from unique pathways after comparison with the list of candidate essential genes for *P.aeruginosa* [8].The targets which were considered for homology modeling are shaded grey in colour.

EC no	Protein name	Gene
3.6.1.-	conserved hypothetical protein	ybbF
2.7.7.38	3-deoxy-manno-octulosonate cytidylyltransferase	kdsB
2.7.1.130	tetraacyldisaccharide 4*-kinase	lpxK
2.5.1.55	2-dehydro-3-deoxyphosphooctonate aldolase	kdsA
2.4.1.182	lipid A-disaccharide synthase	lpxB
3.1.1.61**	probable methylesterase	
2.7.-.-	LPS biosynthesis protein RfaE	rfaE
2.4.1.-	"UDP-glucose:(heptosyl) LPS alpha 1,3-glucosyltransferase WaaG"	waaG

** Enzyme belonging to bacterial chemotaxis pathway. All other remaining enzymes belong to the lipopolysaccharide biosynthesis pathway.

transferase), waaG ("UDP-glucose: (heptosyl) LPS alpha 1,3- glucosyltransferase WaaG") and lpxC (UDP-3-O-acyl-N-acetylglucosamine deacetylase).

A homology 3D model was built for the four potential drug targets kdsA, kdsB, waaG and lpxC [Figure 1] using MODELLER program [20]. The structural homologues from PDB were used as templates for building the 3D models for the four potential targets. All the selected templates had identity of more than 35% with the target protein and had a resolution of <3.0Å. The structural homologue used as template for kdsA is 2-dehydro-3-deoxyphosphooctonate aldolase (68% identity) from *Escherichia coli* with PDB identifier 1G7U [2], kdsB is 3-deoxy-manno-octulosonate cytidylyltransferase (53% identity) from *Haemophilus influenzae* with PDB identifier 1VIC [3], waaG is a synthetic construct with PDB identifier 2CMU[18] with 41% identity and lpxC is UDP-3-O-acyl-N-acetylglucosamine deacetylase from *Aquifex aeolicus* with PDB identifier 1P42 [26] with 36% sequence identity. The stereo chemical quality of the modeled protein structures was assessed

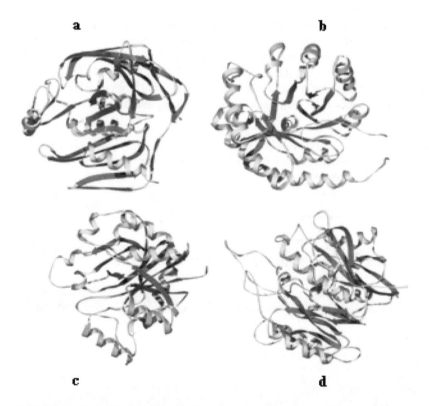

Fig. 1. Models generated by DeepView. Ribbon representation of the following 3D models a) LpxC, b) KdsA, c) KdsB and d) WaaG.

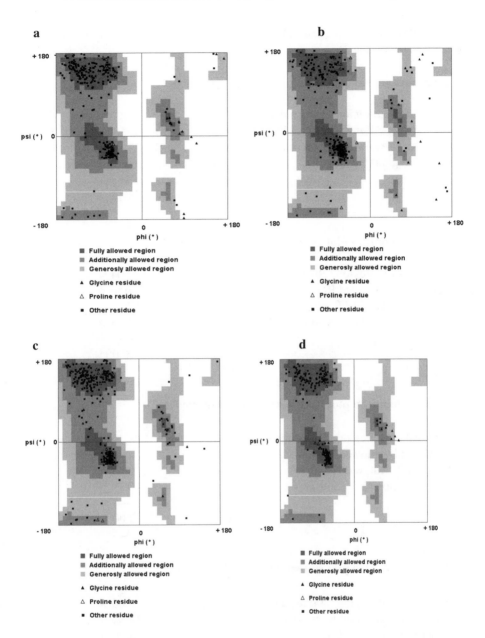

Fig. 2. Ramachandran plot for the following models a) LpxC, b) KdsA, c) KdsB and d) WaaG

by Ramachandran plot (φ vs ψ) for all the 4 models generated using the Ramachandran plot server [22] [Figure 2]. The quality of the model was validated with the PROCHECK program [13] and the main chain parameters for the model were tabulated [Table 3].

a) Ramachandran plot statistics for the LpxC model

Fully Allowed Region (244 residues)	:	81.06 %
Additionally Allowed Region (42 residues)	:	13.95 %
Generously Allowed Region (6 residues)	:	1.99 %
Outside region (9 residues)	:	2.99 %

	Total	100.00 %
		----------

b) Ramachandran plot statistics for the KdsA model

Fully Allowed Region (201 residues)	:	72.04 %
Additionally Allowed Region (49 residues)	:	17.56 %
Generously Allowed Region (20 residues)	:	7.17 %
Outside region (9 residues)	:	3.23 %

	Total	100.00 %
		----------

c) Ramachandran plot statistics for the KdsB model

Fully Allowed Region (222 residues)	:	88.10 %
Additionally Allowed Region (22 residues)	:	8.73 %
Generously Allowed Region (6 residues)	:	2.38 %
Outside region (2 residues)	:	0.79 %

	Total	100.00 %
		----------

d) Ramachandran plot statistics for the WaaG model

Fully Allowed Region (300 residues)	:	80.06 %
Additionally Allowed Region (55 residues)	:	14.82 %
Generously Allowed Region (10 residues)	:	2.70 %
Outside region (6 residues)	:	1.62 %

	Total	100.00 %

Table 3. Main Chain Parameters

a. LpxC

No	Stereochemical parameter	No. of data pts	Parameter value	Comparison Typical value	Values Band width	No. of band widths from mean	
a.	%-tage residues in A,B,L	268	86.2	88.2	10	-0.2	Inside
b.	Omega angle st dev	302	3.5	6	3	-0.8	Inside
c.	Bad contacts/ 100 residues	9	3	1	10	-0.2	Inside
d.	Zeta angle st dev	280	1.2	3.1	1.6	-1.2	Better
e.	H-bond energy st dev	173	0.8	0.7	0.2	0.3	Inside
f.	Overall G-factor	303	-0.1	-0.2	0.3	0.2	Inside

b. KdsA

No	Stereochemical parameter	No. of data pts	Parameter value	Comparison Typical value	Values Band width	No. of band widths from mean	
a.	%-tage residues in A,B,L	238	84.9	88.2	10	-0.3	Inside
b.	Omega angle st dev	269	4.9	6	3	-0.4	Inside
c.	Bad contacts/ 100 residues	6	2.2	1	10	0.1	Inside
d.	Zeta angle st dev	255	1.6	3.1	1.6	-1	Inside
e.	H-bond energy st dev	164	0.8	0.7	0.2	0.5	Inside
f.	Overall G-factor	274	-0.3	-0.2	0.3	-0.3	Inside

c. KdsB

No	Stereochemical parameter	No. of data pts	Parameter value	Comparison Typical value	Values Band width	No. of band widths from mean	
a.	%-tage residues in A,B,L	220	93.6	88.2	10	0.5	Inside
b.	Omega angle st dev	252	3.8	6	3	-0.7	Inside
c.	Bad contacts/ 100 residues	6	2.4	1	10	0.1	Inside
d.	Zeta angle st dev	238	1.4	3.1	1.6	-1	Better
e.	H-bond energy st dev	150	0.7	0.7	0.2	0.2	Inside
f.	Overall G-factor	254	0	-0.2	0.3	0.5	Inside

d. WaaG

No	Stereochemical parameter	No. of data pts	Parameter value	Comparison Typical value	Values Band width	No. of band widths from mean	
a.	%-tage residues in A,B,L	325	83.4	88.2	10	-0.5	Inside
b.	Omega angle st dev	372	4	6	3	-0.7	Inside
c.	Bad contacts/ 100 residues	17	4.6	1	10	0.4	Inside
d.	Zeta angle st dev	348	1.7	3.1	1.6	-0.9	Inside
e.	H-bond energy st dev	186	0.8	0.7	0.2	0.6	Inside
f.	Overall G-factor	373	-0.2	-0.2	0.3	-0.1	Inside

Main- chain parameters for four models generated by PROCHECK. A- Fully Allowed Region, B- Additionally Allowed Region, L-Generously Allowed Region.

3   Results and Discussion

The impact of microbial genomics on drug discovery has led to the identification of new novel antibacterial drugs. Enzymes mediate the synthesis of many complex molecules from simpler ones in a series of chemical reactions. Targeting enzymes present in the pathogen but absent in the host will make sure the elimination of pseudo drug targets in the pathways. It is therefore essential to identify unique pathways and target only those unique enzymes and thus narrowing down to few potential drug targets.

3.1   Pathways and Enzymes Unique to *P.aeruginosa* When Compared to *H.sapiens*

Metabolic pathways belonging to the pathogen and the host were compared and pathways that are present in the pathogen but not in the host are considered to be unique pathways whose enzymes are suitable antibacterial drug targets. Comparative metabolic pathway analysis resulted in 12 unique pathways: Polyketide sugar unit biosynthesis, biosynthesis of siderophore group non-ribosomal peptides, toluene and xylene degradation, D-alanine metabolism, type II secretion system, type III secretion system, phosphotransferase system, bacterial chemotaxis, flagellar assembly, lipopolysaccharide biosynthesis, 2-component system and 1,2 dichloroethane degradation [Table 1]. Enzymes of these pathways are specific and hence they can be explored by finding suitable inhibitors against them.

Among these pathways, lipopolysaccharide biosynthesis, polyketide sugar unit biosynthesis, biosynthesis of siderophore group non-ribosomal peptides, phosphotransferase system (membrane transporter) and D- alanine metabolism are considered to be essential pathways whose enzymes can be targeted for novel antimicrobial drugs [10] [6]. We therefore elaborate on the unique enzymes in these pathways and investigate the lipopolysaccharide pathway in detail.

Polyketides are secondary metabolites playing an important role in defence and intercellular communication. Polymerization of acetyl and propionyl subunits results in the formation of polyketides. Polyketides have important biological activities and pharmacological properties and hence targeting enzymes of these pathways would be an ideal one for drug discovery. Enzymes of this pathway RmlA (EC 2.7.7.24), RmlB (EC 4.2.1.46), RmlC (EC 5.1.3.13) and RmlD (EC 1.1.1.133) synthesize deoxy-thymidine di-phosphate (dTDP)-L-rhamnose from dTTP and glucose-1-phosphate and are important targets for the development of new antimicrobial drugs [14]. Cell wall is necessary for viability and hence they are attractive targets with new drugs being developed for inhibition of cell wall synthesis.

Iron chelating compounds secreted by microorganisms are called Siderophores. These are nonribosomal peptides and they help in dissolving $Fe3+$ ions as soluble $Fe3+$ complexes that can be taken up by active transport mechanism. *Pseudomonas* B 10 produces a siderophore namely pseudobactin. Nonribosomal peptides are synthesized by specialized nonribosomal peptide-synthetase (NRPS) enzymes and this biosynthesis is in similar with that of polyketide and fatty acid biosynthesis. Enzymes pchB (Isochorismate pyruvate-lyase EC 4.1.99.-) and pchA (Isochorismate synthase EC 5.4.4.2) are the two target enzymes of this pathway catalyzing the salicylate biosynthesis [4] [21].

Recent studies reveal that pathogens depend on their hosts for nutrients and hence transport of substrates and other products becomes essential thus making bacterial transport proteins as potential drug targets [6]. PTS or Phosphotransferase system is involved in transporting many sugars into pathogen and involves enzymes of the plasma membrane and the cytoplasm making it a multicomponent system. Protein-N (pi)-phosphohistidine--sugar phosphotransferase (EC 2.7.1.69) belongs to enzyme II of phosphotransferase system with a phosphocarrier protein substrate of low molecular mass (9.5 kDa). The other enzyme that has no human homologue is Phosphoenol-pyruvate—protein phosphotransferase (EC 2.7.3.9) which can also serve as a suitable drug target [5]. The presence of many multidrug efflux pumps increases antibiotic resistance and hence prevents the antibiotic action. It therefore becomes necessary to inhibit these efflux pumps enabling the transport of antibiotic molecules [17].

Enzymes D-alanine-D-alanine ligase A ('ddlA' EC 6.3.2.4) and alanine racemase ('alr' EC 5.1.1.1) catalyze the alanine biosynthesis. Alanine is a non-essential α-amino acid and exists as two enantiomers L-form and D-form. D-alanine mainly exists in bacterial cell walls and serves as a precursor for peptidoglycan biosynthesis. The catalytic action of enzymes alanine racemases makes L-form get racemized to its D-form. These two enzymes are rarely present in eukaryotes and hence they can be developed as suitable drug targets [24].

3.1.1 Lipopolysaccharide Biosynthesis

P.aeruginosa is a gram-negative bacterium producing lipopolysaccharide, a major constituent of the outer cell membrane. Lipopolysaccharide (LPS) serves as selectively permeable membrane for organic molecules and also increases the negative charge of the cell wall and stabilizes the overall membrane structure. LPS consist of a polysaccharide chain covalently linked to a lipid moiety, known as lipid A. Lipopolysaccharide has an important role in the structural integrity of the bacteria and its defense against the host and hence the pathways of these enzymes are attractive drug targets. The enzymes of this pathway had no human homologues and hence they served as potential targets. A total of thirteen enzymes formed this pathway of which seven enzymes matched with the list of candidate essential genes obtained by transposon mutagenesis study and four of them lpxC, kdsA, kdsB and waaG were selected for homology modeling LpxC, KdsA, KdsB and WaaG [Figure 1].

3.2 Targets from Pathways Common to Both *P.aeruginosa* and *H.sapiens*

Unique enzymes in common pathways between *H.sapiens* and *P.aeruginosa* present another source for exploration of drug targets. These targets may be responsible for the pathogenicity and other important biological functions of *P.aeruginosa*. Though our approach concentrated mainly on those enzymes present in the unique pathways we also investigated to find potential targets from the common pathways. There are about 84 targets in carbohydrate metabolism, 30 targets in energy metabolism, 14 targets in lipid metabolism, 31 targets in nucleotide metabolism, 122 targets in amino acid metabolism, 65 targets from metabolism of cofactors and vitamins and 10 targets from secondary metabolite biosynthesis [Figure 3]. It must be noted that several targets were functional in more than one pathway.

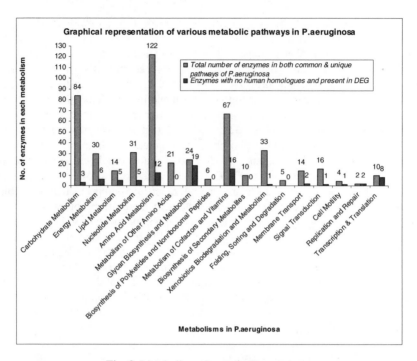

Fig. 3. Metabolic pathways in *P.aeruginosa*

Amino acid metabolism consists of maximum number of targets since amino acids, as precursors of proteins, are essential to all organisms [15]. Many of the enzymes are involved in the biosynthesis of glutamate, lysine, arginine and many other amino acid biosynthesis. Lipid metabolism consists of enzymes that function for lipid biosynthesis as well as for lipid degradation. Many virulence factors including phospholipases C, toxins, lipases and proteases are secreted by *P.aeruginosa*. Outer membrane proteins and membrane transporters are important drug targets in this pathogen due to their involvement in transport of antibiotics. Targets responsible for adhesion and motility are also of great interest in drug targeting.

Many multi-drug efflux systems are present in *P.aeruginosa* thus preventing the action of effective antibiotics. Thus targeting those genes responsible for inhibition of action of antibiotics would prevent the drug resistance property of this pathogen. This example thus illustrates the use of this approach to identify essential genes in pathogens that may be considered as drug targets with more confidence.

4   Conclusion

Our *in silico* approach of comparative metabolic pathway analysis resulted in the identification of potential drug targets. For the first time, the availability of complete genome sequences of many bacterial species is facilitating many computational approaches. The complete definition of all gene products by gene identification tools exemplified here is just the first step. The data presented here demonstrates that stepwise prioritization of genome open reading frames using simple biological criteria can be an effective way of rapidly reducing the number of genes of interest to an experimentally manageable number. This process is an efficient way for enriching potential target genes, and for identifying those that are critical for normal cell function. The generation of a comprehensive essential gene list will allow an accelerated genetic dissection of traits such as metabolic flexibility and inherent drug resistance that render *P. aeruginosa* such a tenacious pathogen. Such a strategy will enable us to locate critical pathways and steps in pathogenesis; to target these steps by designing new drugs; and to inhibit the infectious agent of interest with new antimicrobial agents.

References

1. Altschul, S.F., Thomas, L.M., Alejandro, A.S., Jinghui, Z., Zheng, Z., Webb, M., Lipman, D.J.: Gapped BLAST and PSI-BLAST: a new generation of protein database search programs. Nucleic Acids Res. 25, 3389–3402 (1997)
2. Asojo, O., Friedman, J., Adir, N., Belakhov, V., Shoham, Y., Baasov, T.: Crystal structures of KDOP synthase in its binary complexes with the substrate phosphoenolpyruvate and with a mechanism-based inhibitor. Biochemistry 40, 6326–6334 (2001)
3. Badger, J., Sauder, J.M., Adams, J.M., Antonysamy, S., Bain, K., Bergseid, M.G., Buchanan, S.G., Buchanan, M.D., Batiyenko, Y., Christopher, J.A., Emtage, S., Eroshkina, A., Feil, I., Furlong, E.B., Gajiwala, K.S., Gao, X., He, D., Hendle, J., Huber, A., Hoda, K., Kearins, P., Kissinger, C., Laubert, B., Lewis, H.A., Lin, J., Loomis, K., Lorimer, D., Louie, G., Maletic, M., Marsh, C.D., Miller, I., Molinari, J., Muller-Dieckmann, H.J., Newman, J.M., Noland, B.W., Pagarigan, B., Park, F., Peat, T.S., Post, K.W., Radojicic, S., Ramos, A., Romero, R., Rutter, M.E., Sanderson, W.E., Schwinn, K.D., Tresser, J., Winhoven, J., Wright, T.A., Wu, L., Xu, J., Harris, T.J.: Structural analysis of a set of proteins resulting from a bacterial genomics project. Proteins 60, 787–796 (2005)
4. Braun, V., Hantke, K., Koster, W.: Bacterial iron transport: mechanisms genetics, and regulation. Met. Ions. Biol. Syst. 35, 67–145 (1998)
5. Durham, D.R., Phibbs Jr., P.V.: Fractionation and characterization of the phosphoenolpyruvate: fructose 1-phosphotransferase system from *Pseudomonas aeruginosa*. J. Bacteriol. 149, 534–541 (1982)
6. Galperin, M.Y., Koonin, E.V.: Searching for drug targets in microbial genomes. Curr. Opin. Biotechnol. 10, 571–578 (1999)
7. Itaya, M.: An estimation of minimal genome size required for life. FEBS Lett. 362, 257–260 (1995)
8. Jacobs, M.A., Alwood, A., Thaipisuttikul, I., Spencer, D., Haugen, E., Ernst, S., Will, O., Kaul, R., Raymond, C., Levy, R., Chun-Rong, L., Guenthner, D., Bovee, D., Olson, M.V., Manoil, C.: Comprehensive transposon mutant library of *Pseudomonas aeruginosa*. Proc. Natl. Acad. Sci. USA 100, 14339–14344 (2003)
9. Kanehisa, M., Goto, S., Kawashima, S., Nakaya, A.: The KEGG databases at Genome Net. Nucleic Acids Res. 30, 42–46 (2002)
10. Khisimuzi, E., Mdluli, P.R., Witte, T.K., Adam, W.B., Erwin, A.L., et al.: Molecular validation of LpxC as an antibacterial drug target in *Pseudomonas aeruginosa*. Antimicrobial agents and Chemotherapy 50, 2178–2184 (2006)

11. Kobayashi, K., Ehrlich, S.D., Albertini, A., Amati, G., Andersen, K.K., Arnaud, M., Asai, K., Ashikaga, S., Aymerich, S., Bessieres, P., et al.: Essential *Bacillus subtilis* genes. Proc. Natl. Acad. Sci. USA 100, 4678–4683 (2003)
12. Koonin, E.V., Tatusov, R.L., Galperin, M.Y.: Beyond complete genomes: from sequence to structure and function. Curr. Opin. Struct. Biol. 8, 355–363 (1998)
13. Laskowski Roman, A., MacArthur, M.A., Smith, D.K., Jones, D.T., Gail Hutchinson, E., Morris, A.L., Moss, D.S., Thornton, J.M.: PROCHECK: a program to check the stereochemical quality of protein structures. J. Appl. Cryst. 26, 283–291 (1993)
14. Ma, Y., Stern, R.J., Scherman, M.S., Vissa, V.D., Yan, W., Jones, V.C., Zhang, F., Franzblau, S.G., Lewis, W.H., McNeil, M.R.: Drug Targeting *Mycobacterium tuberculosis* Cell Wall Synthesis: Genetics of dTDP-Rhamnose Synthetic Enzymes and Development of a Microtiter Plate-Based Screen for Inhibitors of Conversion of dTDP-Glucose to dTDP-Rhamnose. Antimicrob Agents Chemother. 45, 1407–1416 (2001)
15. Martina, M.O., Lu, C., Hancock, R.E.W., Abdelal, A.T.: Amino Acid-Mediated Induction of the Basic Amino Acid-Specific Outer Membrane Porin OprD from *Pseudomonas aeruginosa*. Journal of Bacteriology 181, 5426–5432 (1999)
16. Miesel, L., Greene, J., Black, T.A.: Genetic strategies for antibacterial drug discovery. Nat. Rev. Genet. 4, 442–456 (2003)
17. Nikaido, H.: Antibiotic resistance caused by gram-negative multidrug efflux pumps. Clin. Infect. Dis. 27, 32–41 (1998)
18. Rajashankar, R.K., Kniewel, R., Solorzano, V., Lima, C.D.: Crystal Structure of a Putative Peptidyl-Arginine Deiminase (to be published)
19. Sakharkar, K.R., Sakharkar, M.K., Chow, V.T.: A novel genomics approach for the identification of drug targets in pathogens with special reference to *Pseudomonas aeruginosa*. In Silico Biology. 4, 355–360 (2004)
20. Sali, A., Potterton, L., Yuan, F., van Vlijmen, H., Karplus, M.: Evaluation of comparative protein modeling by MODELLER. Proteins 23, 318–326 (1995)
21. Serino, L., Reimmann, C., Baur, H., Beyeler, M., Visca, P., Haas, D.: Structural genes for salicylate biosynthesis from chorismate in *Pseudomonas aeruginosa*. Mol. Gen. Genet. 249, 217–228 (1995)
22. Sheik, S.S., Sundararajan, P., Hussain, A.S.Z., Sekar, K.: Ramachandran plot on the web. Bioinformatics 18, 1548–1549 (2002)
23. Stover, K.C., Pham, X.Q., Erwin, A.L., Mizoguchi, S.D., Warrener, P., Hickey, M.J., Brinkman, F.S.L., Hufnagle, W.O., Kowalik, D.J., Lagrou, M., Garber, R.L., Goltry, L., Tolentino, E., Westbrock-Wadman, S., Yuan, Y., Brody, L.L., Coulter, S.N., Folger, K.R., Kas, A., Larbig, K., Lim, R., Smith, K., Spencer, D., Wong, G.K.-S., Wu, Z., Paulsen, I., Reizer, J., Saier, M.H., Hancock, R.E.W., Lory, S., Olson, M.V.: Complete genome sequence of *Pseudomonas aeruginosa* PAO1: an opportunistic pathogen. Nature 406, 959–964 (2000)
24. Strych, U., Huang, H.C., Krause, K.L., Benedik, M.J.: Characterization of the alanine racemases from *Pseudomonas aeruginosa* PAO1. Curr. Microbiol. 41, 290–294 (2000)
25. Tatusov, R.L., Koonin, E.V., Lipman, D.J.: A genomic perspective on protein families. Science 278, 631–637 (1997)
26. Whittington, D.A., Rusche, K.M., Shin, H., Fierke, C.A., Christianson, D.W.: Crystal Structure of LpxC, a Zinc-Dependent Deacetylase Essential for Endotoxin Biosynthesis. Proc. Natl. Acadl. Sci. USA 100, 8146–8150 (2003)

Understanding Prediction Systems for HLA-Binding Peptides and T-Cell Epitope Identification

Liwen You[1,2,3], Ping Zhang[3], Mikael Bodén[4], and Vladimir Brusic[3,5]

[1] School of Information Science, Computer and Electrical Engineering,
Halmstad University, Halmstad, Sweden
[2] Department of Theoretical Physics, Lund University, Lund, Sweden
[3] School of Land, Crop, and Food Sciences, and
[4] School of Information Technology and Electrical Engineering,
University of Queensland, Brisbane QLD, Australia
[5] Cancer Vaccine Center, Dana-Farber Cancer Institute, Boston MA, USA
liwen@thep.lu.se, p.zhang2@uq.edu.au, m.boden@uq.edu.au,
vladimir_brusic@dfci.harvard.edu

Abstract. Peptide binding to HLA molecules is a critical step in induction and regulation of T-cell mediated immune responses. Because of combinatorial complexity of immune responses, systematic studies require combination of computational methods and experimentation. Most of available computational predictions are based on discriminating binders from non-binders based on use of suitable prediction thresholds. We compared four state-of-the-art binding affinity prediction models and found that nonlinear models show better performance than linear models. A comprehensive analysis of HLA binders (A*0101, A*0201, A*0301, A*1101, A*2402, B*0702, B*0801 and B*1501) showed that non-linear predictors predict peptide binding affinity with high accuracy. The analysis of known T-cell epitopes of survivin and known HIV T-cell epitopes showed lack of correlation between binding affinity and immunogenicity of HLA-presented peptides. T-cell epitopes, therefore, can not be directly determined from binding affinities by simple selection of the highest affinity binders.

1 Introduction

Major histocompatibility complex (MHC) molecules present peptides, derived from antigens and host proteins, on the cell surface. The recognition of presented peptides by TCD8[+] cells is necessary for recognition of infected or pathologically mutated cells and induction of cellular immune responses and subsequent elimination of tumors and infected cells. Human MHC is known as the human leukocyte antigen (HLA). Antigen processing and presentation involves primarily three steps: proteasomal cleavage, translocation of cleaved fragments by transporter associated with antigen processing, and HLA-peptide binding. The HLA/peptide binding is by far the most discriminative step: natural prevalence of HLA-binding peptides is in the range of 0.1-5% for any given protein

J.C. Rajapakse, B. Schmidt, and G. Volkert (Eds.): PRIB 2007, LNBI 4774, pp. 337–348, 2007.
© Springer-Verlag Berlin Heidelberg 2007

of which some 20% remain functionally relevant [1,2]. Therefore, computational prediction and modeling of HLA/peptide binding can greatly facilitate peptide screening, with tremendous savings in time and experimental effort.

The HLA peptide binding prediction can be approached as simple classification problem of discriminating binders from non-binders. However, peptide binding is necessary but it does not guarantee an immune response. A binding affinity metric like inhibition concentration (IC_{50}) of a standard probe quantifies HLA/peptide binding. Given a large number of binding data recently made available [3] we have extended the approach by studying peptide binding as a regression problem.

Many different attempts have been made to predict MHC peptide binding. There are primarily three approaches: by structure modelling, data-driven using peptide sequences and their binding affinities, or by the combination thereof. Sequence-based approaches can be further categorized into motif/profile based methods [4,5,6,7,8,9] and machine learning methods which use Artificial Neural Networks (ANN) [10,11], Hidden Markov Models (HMM) [12], or Support Vector Machines (SVM) [13,14,15,16,17,18,19]. An example of combined method is the adaptive double threading [20]. The prediction methods have been compared for accuracy of classification (binders vs. non-binders) [3,21,22]. However different data sets are used to build models and evaluation data vary between different studies making it intrinsically difficult to compare predictor performance.

Recently, a comprehensive experimental relative binding affinity analysis of a complete overlapping peptide library derived from the tumor-associated antigen survivin [23] was reported for eight different types of HLA class I molecules. Also, a large data set of peptide binding affinities became available at the Immune Epitope Database and Analysis Resource (IEDB, www.immuneepitope.org) [3]. Combining these two data sets, we analysed the factors that affect accuracy of prediction models and explored the correlation of peptide binding and immunogenicity. The results of this study provide an improved understanding of the prediction systems design issues and their use for identification of HLA-binding peptides and T-cell epitopes.

2   Materials and Methods

Four different prediction methods were used in this study. Support Vector Regression (SVR) and epitope information were used to build a regression model to predict HLA-peptide binding affinities using the first of the data sets. The three models used at IEDB, namely ANN and two matrix methods were used as comparison predictors. We defined prediction performance criteria to enable fair comparisons between models. Specifically, we used match curve area and correlation coefficient to compare model performance.

2.1   Datasets

Survivin, a member of the apoptosis inhibitor protein family, is one of a limited number of shared tumor-associated antigens that is over-expressed in the majority of human cancers. There is an intense interest in using survivin as a target for therapeutic CTL response. Bachinsky et al. [23] used a high-throughput technique to identify peptides derived from survivin that bind eight HLA class I alleles: HLA-A*0101, HLA-A*0201, HLA-A*0301, HLA-A*2402, HLA-A*1101, HLA-B*0702, HLA-B*0801, HLA-B*1501. A library of 134 overlapping nonamers spanning the full length of the survivin protein (UniProt O15392 with 142 amino acids) was experimentally screened for peptides capable of binding each allele. Binding to each allele was reported as a percentage relative to a positive control peptide for that allele as values from 0 to >100%. An arbitrary cutoff of 30% of the control was used as a positive cutoff for experimental binders. Therewith, they identified nineteen HLA-A*0201, zero HLA-A*0101, seven HLA-A*0301, twelve HLA-A*1101, twenty-four HLA-A*2402, six HLA-B*0702, six HLA-B*0801 and eight HLA-B*1501 binding peptides.

Friedrichs et al. [24] collected a set of survivin-derived peptides, which can induce HLA restricted CTL responses. Two peptides reported as survivin-derived nonamer T-cell epitopes are HLA-A*0201 restricted $_{96}$LTLGEFLKL$_{104}$ and A*2402 restricted $_{20}$STFKNWPFL$_{28}$.

Another set of proteins that has been comprehensively studied in T-cell responses are HIV proteins. We analyzed all HIV protein T-cell epitopes available in the HIV molecular immunology database (www.hiv.lanl.gov/content /immunology). In addition, we analysed mutations of a small set of HLA-restricted CD8$^+$ T-cell epitopes.

Peters et al. [3] have made public a set of 48,828 quantitative peptide-binding affinity measurements relating to 48 different mammalian MHC class I alleles. They used this data to establish a set of predictions with one neural network method (IEDB ANN) and two matrix-based prediction methods (IEDB SMM and IEDB ARB) and compared them with other available online predictors. In this study, we only used eight nonamer datasets of the eight HLA alleles of interest in this study. The data set (which we denote as the IEDB data set) was downloaded from (mhcbindingpredictions.immuneepitope.org).The datasets used in this study were: A*0101 (1157 peptides), A*0201 (3089), A*0301 (2094), A*1101 (1985), A*2402 (197), B*0702 (1262), B*0801 (708), and B*1501 (978).

2.2 SVM Regression Model and Peptide Coding Using Extra Epitope Information

The SVM is firmly based on statistical learning theory. It can be used to solve both classification and regression problems by optimizing given generalization bounds. Its regression form (SVR) is based on a loss function that ignores errors within a certain distance of the true value (we use the ε-insensitive loss function). In SVMs data is implicitly projected into a high-dimensional feature space using a kernel function. We employed the Gaussian kernel, $K(x, z) = \exp(-||x - z||^2/\sigma^2)$, where x and z are two samples and σ is a kernel parameter. The Gaussian kernel requires peptides to be represented as numerical vectors. A sparse orthogonal coding was used to represent peptides, with each amino acid encoded by 20 bits (19 bits set to zero and 1 bit set to one). Hence, a nonamer is represented in a 180-dimensional space. The coding vector was extended by nine positions encoding the shape of the binding motif for each studied allele. The final coded peptide vector had 189 elements and we refer to it as the extended sparse coding.

2.3   IEDB Prediction Models

IEDB ANN, SMM and ARB prediction models are used by IEDB website and all three methods predict the quantitative binding affinity. The ANN is a nonlinear model and the other two generate scoring matrices. They have been used as benchmarking predictions [3] and we compared the SVR model performance with them.

2.4   Performance Evaluation Methods

To compare two classifiers discriminating binders vs. non-binders, area under ROC (Receiver operating characteristic curve; the AUC value) compares overall performance of classifiers and does not require a decision threshold to be determined. For regression, the correlation coefficient between predicted and true binding affinities were used. To assess potential epitopes/binders along protein sequences, we used "match curve" plots of the number of true binders in the top N ranked predicted affinities (y-axis) vs. N (x-axis). If most of true epitopes can be found within a short list of top ranked predicted binders, the prediction system is very useful for screening epitopes along protein sequences.

3   Study Design

This study has three parts designed to understand prediction systems for HLA-binding peptides, the relationship between binding affinities and known T-cell epitopes for survivin and selected HIV proteins, and the relationship of natural epitopes and their mutants.

3.1   SVR HLA-Binding Predictor

The Gaussian kernel and the extended sparse coding SVR were combined with the IEDB datasets to build a regression model for binding affinity prediction for each HLA allele. Since vast majority of IC_{50} values from the IEDB database for the eight HLA alleles are within 1 to 50,000 nM, we transformed the binding affinities to the range of 0 and 1 by using $1 - \log(\text{binding affinity})/\log(50000)$ as described in [25].

The model building was done using single-level five-fold cross-validation for each allele. We used the test data in each run to tune regularization and kernel parameters. We got five different regression models from the cross-validation process. We repeated the five-fold cross-validation five times. Regression performance was reported as the mean value of correlation coefficients from the five runs. In addition, we chose the models from a single cross-validation run which gave the best cross-validation performance. These five regression models were used as a committee for the corresponding single allele to get predicted binding affinity values for all yet unseen peptides.

3.2   Prediction of HLA-Binding Peptides in Survivin Protein

The comprehensive experimental relative binding affinity analysis of the complete survivin peptide set presented an opportunity to evaluate how different predictive models perform against an independent data set representing a complete protein. We applied our regression model on the survivin dataset and retrieved prediction results from the IEDB prediction servers to compare their performances. Comparison was based on the correlation coefficient calculated from the predicted binding affinities and the experimentally measured relative binding affinities.

3.3   Prediction of T-Cell Epitopes on HIV Proteins and Survivin Protein

We used the SVR model to predict binding affinities for known T-cell epitopes on HIV proteins and survivin within the eight alleles. When a known epitope is longer than a nonamer, we took the highest binding affinity of all possible nonamers within the epitope as its binding affinity. We also did one site mutagenesis on eight T-cell epitopes of HIV proteins and survivin in order to compare their affinities to corresponding ascendant epitopes and find mutation patterns.

4   Results

4.1   Cross-Validation Performance on the Eight HLA-Alleles

Table 1 shows the correlation coefficient (with standard deviation) of SVR models on the eight HLA-alleles using cross-validation. Most of correlation coefficient performances are satisfactory except for the B*0801 allele. Figure 1 shows the binding affinity distributions of data sets for HLA-A*0201, A*2402 and B*0801 and the horizontal line in each subplot denotes the binding affinity value, $\log_{10}(500)$. From the analysis of experimental data from [3], $\log_{10}(500)$ was taken as a threshold for binder and non-binder, which means that for a peptide with $\log_{10}(IC_{50})$ value less than $\log_{10}(500)$ it should be treated as a binder. Although the threshold, $\log_{10}(500)$, is arbitrary, it enables objective separation of binders from non-binders. From Figure 1, it is clear that there are very few samples (only 21) with binding affinities less than the threshold versus the total around 700 samples for B*0801 allele. For A*2402 alleles, there are only 197 samples in total, but the predictor performance is superior to that of B*0801. A possible explanation is that the dataset is more balanced. For the A*0201 allele the dataset is slightly unbalanced, however, there are still about 1000 samples with IC_{50} values below the binding threshold. This observation holds for other alleles. More samples with stronger binding affinities seem to imply better prediction performance.

Table 1. SVR cross-validation correlation coefficient (r) performance on HLA-alleles

Allele	A*0101	A*0201	A*0301	A*1101	A*2402	B*0702	B*0801	B*1501
Size	1157	3089	2094	1985	197	1262	708	978
Mean(r)	0.781	0.847	0.766	0.823	0.669	0.812	0.287	0.726
Std	0.005	0.002	0.004	0.002	0.003	0.004	0.073	0.008

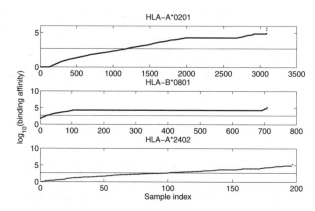

Fig. 1. Experimental binding affinity plots for A*0201, A*2402 and B*0801 data sets. The horizontal lines denote the binding affinity values, $\log_{10}(500)$.

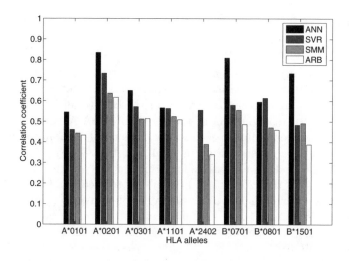

Fig. 2. Correlation coefficient performance comparison between IEDB ANN, SMM, ARB and SVR models on the whole survivin dataset

4.2 Prediction of Binding Affinity along Survivin Protein Sequence

For each of the eight HLA alleles we calculated the correlation coefficient between experimental binding values and predicted binding affinities. $\log_{10}(IC_{50})$ values of all nonamers in the survivin sequence were predicted using IEDB ANN, SMM and ARB prediction tools (tools.immuneepitope.org/analyze/html/mhc_binding.html). Figure 2 illustrates performance differences between models (HLA-A*2402 ANN method is not available from the IEDB web server). IEDB ANN is generally superior, followed by SVR. SMM and ARB show similar performance inferior to the other two.

For screening potential epitopes/binders in long protein sequences, it is useful to look at the match curve to assess how many peptides are typically required to test to identify all true epitopes/binders. Figure 3 shows the match curves for the eight alleles using the IEDB ANN and our SVR models using experimental data for survivin peptides. For A*0101 there are no binders according to the experimental settings in [23] and for A*2402 results for IEDB ANN are unavailable. The classification performance of ANN and SVR is very similar. For A*0201, A*0301 and B*0801, SVR is slightly better than the IEDB ANN. Most of experimentally determined binders are within a few numbers of the predicted top peptides. The worst performance by this measure is for the A*2402 allele, where 19 of 23 binders are within predicted top 65 peptides.

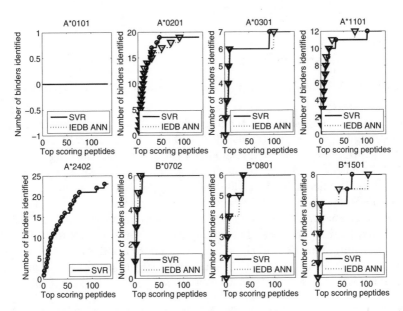

Fig. 3. Match curves for experimentally identified binders on survivin protein

4.3 Predictions of T-Cell Epitopes

Figure 4 displays the predicted binding affinities, $\log_{10}(IC_{50})$, using SVR for known HIV T-cell epitopes. The upper horizontal line indicates $\log_{10}(5000)$ and followed by $\log_{10}(500)$ and $\log_{10}(50)$. In [10], a peptide with $\log_{10}(IC_{50})$ less than $\log_{10}(5)$ is a very good binder; good binder with affinity between $\log_{10}(5)$ and $\log_{10}(50)$; intermediate binder between $\log_{10}(50)$ and $\log_{10}(500)$ and low affinity binder between $\log_{10}(500)$ and $\log_{10}(5000)$. In B*1501, all epitopes are low affinity binders. In B*0801 predictions are not informative, reflecting poor training set. For other alleles, T-cell epitopes show a broad range of binding affinities, most of which are between $\log_{10}(50)$ and $\log_{10}(5000)$.

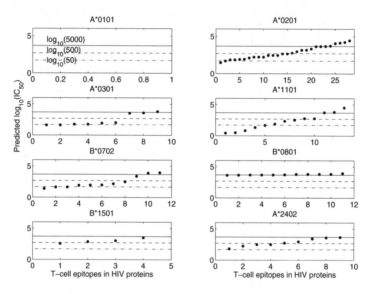

Fig. 4. SVR predicted binding affinities for known HIV T-cell epitopes

We compared the predicted binding affinities with experimental data for known survivin T-cell epitopes, shown in Table 2. Both known T-cell epitopes have moderate binding affinity. The 96-104 epitope is the third highest binder to A*0201 while 20-28 is the second highest binder to A*2402 of all survivin peptides [23]. Predicted binding affinities are approximately within 2-fold concentration of their experimental affinities indicating excellent correlation.

Table 3 shows natural epitopes and their *in silico* mutant versions with the highest predicted binding affinities. The affinity change varies from 2 to 10 fold. For A*0201, the mutant happens at the second position from T to M or L; for A*0101, it is at the ninth position from E to Y; for A*1101, at the seventh position from C to F; and for A*2402, at the second position from T to Y. This example illustrates

Table 2. Predicted binding affinities (IC_{50}) of known survivin T-cell epitopes

Allele Name	Start Position	End Position	Peptide	IC_{50} (SVR)	Experimental IC_{50} (from [23])
A*0201	96	104	LTLGEFLKL	893	430 nM
A*2402	20	28	STFKNWPFL	1290	740 nM

a possible vaccine engineering applications with modified peptides. In a successful case study of immunotherapy with modified survivin 96-104 peptide LMLGE-FLKL in a liver metasthasis of pancreatic cancer the result was a complete remission of metasthasis [26]. Table 4 shows *in silico* mutations of natural HIV epitopes and their mutants with the highest binding affinities.

Table 3. Mutant epitopes with the highest binding affinities vs. natural epitopes

Allele Name	Natural Epitope	IC_{50} (SVR)	Mutant Epitopes	IC_{50} (SVR)
A*0101	QFEELTLGE	33144	QFEELTLGY	2443
A*0201	STFKNWPFL	2077	SMFKNWPFL	891
A*0201	LTLGEFLKL	893	LLLGEFLKL	231
A*1101	LAQCFFCFK	21	LAQCFFFFK	3
A*2402	STFKNWPFL	1290	SYFKNWPFL	163

Table 4. Differences between natural HIV epitopes and their mutants resulting in improved binding affinity

Allele	P_5	P_4	P_3	P_2	P_1	$P_{1'}$	$P_{2'}$	$P_{3'}$	$P_{4'}$
A*0201	Q → Y	W/R → M G → L/P L → Q Y/F → L P/G → V I → Y	R → M T → K Q → R			A → I	G → F K → F		E → F G → F P → L Y → L A → Y
A*0301		Y → V D/Y → L C → P	D → M			Q → R			C → F
A*1101		Y → V	Q → R T → K	Q → F	A → R Q → V		W → A M → A	R → K A/T → F A → Y
A*2402	S → Y		R → Y			Q → V		G → A	A/P → F P → L
B*0702		V → P T → L E/F → M	R → Y T → K					G → A	G → L I → K K/W → F
B*0801	D → R	V/G → P G/P → L	Y → R			Q → R			G/Y/V → L C → F
B*1501		P → V	Y → R D → K			Q → R			

5   Conclusions

IEDB ANN model is the best among the four models, followed by the SVR model. The predictions by SMM and ARB models are inferior to them. The two non-linear methods produced more accurate predictions than two linear methods. We also found that, provided that training sets are representative, the more training samples result in a better prediction performance.

IEDB ANN and SVR models performed similarly in scanning potential binders when tested on survivin sequence. Both true epitopes were within top 2.5% of predicted binders. Therefore, *in silico* models can save significant experimental time and costs in screening potential targets. By analyzing predicted binding affinities of known HIV T-cell epitopes, we found that the range of binding affinities varies for different alleles; and the range of binding affinities include high, moderate, and low affinity. Most of the survivin epitopes are intermediate affinity binders compared to their one-site mutated descendants, some of which are high-affinity binders. The change of IC_{50} varies from 2 to 10 fold for mutated epitopes and most of them were at epitope anchors or auxiliary anchors. These phenomena indicate that high binding affinity binding and immunogenicity are not necessary correlated.

6 Discussion

Binding affinity alone is not sufficient to describe the interaction between HLA allele molecules and peptides. Other factors like dissociation rate or stability of each complex are also the determinants of the interaction. Classification of interaction into binders and non-binders only is not informative of immunogenic properties of peptides. Known survivin-derived T-cell epitopes are low affinity binders to their respective HLA molecules.

Most of its known T-cell epitopes of tumor antigen survivin are low affinity binders, which might offer an explanation for lack of response to antigens in cancer patients, self-tolerance, and subdominance [23]. It is unclear which epitopes within a given tumor-associated antigen should be selected to circumvent tolerance and hence serve as the best target in anti-tumor vaccination. One challenge for vaccine design is to enhance the immunogenicity of weak antigens and prevent silencing of active T-cell clones. One possible strategy is to optimize tumor-associated antigen epitope analogs for priming. The *in silico* mutation analysis demonstrated that the optimization should target mainly anchor or auxiliary anchor positions.

Acknowledgements. LY was supported by the National Research School in Genomics and Bioinformatics, Sweden. The authors acknowledge the support of the ImmunoGrid project, EC contract FP6-2004-IST-4, No. 028069.

References

1. Brusic, V., Zeleznikow, J.: Computational binding assays of antigenic peptides. Letters in Peptide Sci. 6, 313–324 (1999)
2. Yewdell, J.W.: Confronting complexity: real-world immunodominance in antiviral CD8+ T cell responses. Immunity 25, 533–543 (2006)

3. Peters, B., Bui, H.H., Frankild, S., Nielson, M., Lundegaard, C., Kostem, E., Basch, D., Lamberth, K., Harndahl, M., Fleri, W., Wilson, S.S., Sidney, J., Lund, O., Buus, S., Sette, A.: A community resource benchmarking predictions of peptide binding to MHC-I molecules. PLoS Comput. Biol. 2(6), 574–584 (2006)

4. Rammensee, H.G., Bachmann, J., Emmerich, N.P., Bachor, O.A., Stevanovic, S.: SYFPEITHI: database for MHC ligands and peptide motifs. Immunogenetics 50(3-4), 213–219 (1999)

5. Parker, K.C., Bednarek, M.A., Coligan, J.E.: Scheme for ranking potential HLA-A2 binding peptides based on independent binding of individual peptide side-chains. J. Immunol. 152(1), 163–175 (1994)

6. Udaka, K., Wiesmuller, K.H., Kienle, S., Jung, G., Tamamura, H., Yamagishi, H., Okumura, K., Walden, P., Suto, T., Kawasaki, T.: An automated prediction of MHC class I-binding peptides based on positional scanning with peptide libraries. Immunogenetics 51(10), 816–828 (2000)

7. Guan, P., Doytchinova, I.A., Zygouri, C., Flower, D.R.: MHCPred: bringing a quantitative dimension to the online prediction of MHC binding. Applied Bioinformatics 2(1), 63–66 (2003)

8. Peters, B., Sette, A.: Generating quantitative models describing the sequence specificity of biological processes with the stabilized matrix method. BMC Bioinformatics 6(132) (2005)

9. Bui, H.H., Sidney, J., Peters, B., Sathiamurthy, M., Sinichi, A., Purton, K.A., Mothe, B.R., Chisari, F.V., Watkins, D.I., Sette, A.: Automated generation and evaluation of specific MHC binding predictive tools: ARB matrix applications. Immunogenetics 57(5), 304–314 (2005)

10. Buus, S., Lauemoller, S.L., Worning, P., Kesmir, C., Frimurer, T., Corbet, S., Fomsgaard, A., Hilden, J., Holm, A., Brunak, S.: Sensitive quantitative predictions of peptide-MHC binding by a 'Query by Committee' artificial neural network approach. Tissue Antigens 62(5), 378–384 (2003)

11. Brusic, V., Bucci, K., Schonbach, C., Petrovsky, N., Zeleznikow, J., Kazura, J.W.: Efficient discovery of immune response targets by cyclical refinement of QSAR models of peptide binding. Journal of Molecular Graphics and Modelling 19(5), 405–411 (2001)

12. Mamitsuka, H.: Predicting peptides that bind to MHC molecules using supervised learning of hidden Markov models. Proteins 33(4), 460–474 (1998)

13. Dönnes, P., Elofsson, A.: Prediction of MHC class I binding peptides, using SVMHC. BMC Bioinformatics 3(25) (2002)

14. Zhao, Y., Pinilla, C., Valmori, D., Martin, R., Simon, R.: Application of support vector machines for T-cell epitopes prediction. Bioinformatics 19, 1978–1984 (2003)

15. Riedesel, H., Kolbeck, B., Schmetzer, O., Knapp, E.W.: Peptide binding at class I major histocompatibility complex scored with linear functions and support vector machines. Genome Informatics 15(1), 198–212 (2004)

16. Yang, Z.R., Johnson, F.C.: Prediction of T-cell epitopes using biosupport vector machines. J. Chem. Inf. Model 45(5), 1424–1428 (2005)

17. Bozic, I., Zhang, G.L., Brusic, V.: Predictive vaccinology: optimisation of predictions using support vector machine classifiers. In: Gallagher, M., Hogan, J.P., Maire, F. (eds.) IDEAL 2005. LNCS, vol. 3578, pp. 375–381. Springer, Heidelberg (2005)

18. Zhang, G.L., Bozic, I., Kwoh, C.K., August, J.T., Brusic, V.: Prediction of supertype-specific HLA class I binding peptides using support vector machines. Journal of Immunological Methods 320(1-2) (2007)

19. Cui, J., Han, L.Y., Lin, H.H., Zhang, H.L., Tang, Z.Q.: Prediction of MHC-binding peptides of flexible lengths from sequence-derived structural and physicochemical properties. Molecular Immunology 44(5), 866–877 (2007)
20. Jojic, N., Reyes-Gomez, M., Heckerman, D., Kadie, C., Schueler-Furman, O.: Learning MHC I-peptide binding. Bioinformatics 22(14), e227–235 (2006)
21. Yu, K., Petrovsky, N., Schonbach, C., Koh, J.Y., Brusic, V.: Methods for prediction of peptide binding to MHC molecules: a comparative study. Molecular Medicine 8(3), 137–148 (2002)
22. Trost, B., Bickis, M., Kusalik, A.: Strength in numbers: achieving greater accuracy in MHC-I binding prediction by combining the results from multiple prediction tools. Immunome Research 3, 5 (2007)
23. Bachinsky, M.M., Guillen, D.E., Patel, S.R., Singleton, J., Chen, C., Soltis, D.A., Tussey, L.G.: Mapping and binding analysis of peptides derived from the tumor-associated antigen survivin for eight HLA alleles. Cancer Immunity 5, 1–9 (2005)
24. Friedrichs, B., Siegel, S., Andersen, M.H., Schmitz, N., Zeis, M.: Survivin-derived peptide epitopes and their role for induction of antitumor immunity in hematological malignancies. Leukemia & Lymphoma 47(6), 978–985 (2006)
25. Nielsen, M., Lundegaard, C., Worning, P., Lauemøller, S.L., Lamberth, K., Buus, S., Brunak, S., Lund, O.: Reliable prediction of T-cell epitopes using neural networks with novel sequence representations. Protein Sci. 12(5), 1007–1017 (2003)
26. Wobser, M., Keikavoussi, P., Kunzmann, V., Weininger, M., Andersen, M.H., Becker, J.C.: Complete remission of liver metastasis of pancreatic cancer under vaccination with a HLA-A2 restricted peptide derived from the universal tumor antigen survivin. Cancer Immunology and Immunotherapy 55(10), 1294–1298 (2006)

Predicting Binding Peptides with Simultaneous Optimization of Entropy and Evolutionary Distance

Menaka Rajapakse[1,2] and Lin Feng[2]

[1] Institute for Infocomm Research, 21 Heng Mui Keng Terrace, Singapore 119613
[2] School of Computer Engineering, Nanyang Technological University, Block N4, Nanyang Avenue, Singapore 639798
menaka@i2r.a-star.edu.sg, asflin@ntu.edu.sg

Abstract. Identifying antigenic peptides that bind to Major Histocompatibility Complex (MHC) molecules plays a central role in determining T-cell epitopes suitable as vaccine targets. Prediction of the binding ability of antigenic peptides to MHC class II molecules is more complex that for class I. Class II molecules bind to peptides of different lengths and the core region that interacts with the binding site on the class II MHC molecule is located anywhere within the peptide. Obtaining an alignment for these binding sites is an important first step in determining the binding motif of MHC class II alleles. In this paper, we exploit entropy and evolutionary distance of the key binding positions (anchor positions) of an alignment in determining the best possible alignment for a given set of peptide data. Once an optimal alignment is found, a weight matrix representing the binding motif is estimated. The weight matrix designed is subsequently applied to predict MHC binding peptides.

1 Introduction

T cells play a key role as the mediators of immune response against diseases. These cells recognize viral antigens (short peptides) bound to Major Histocompatibility Complex (MHC) molecules through T cell receptors (TCR). Predicting such binding peptides assists in selecting epitopes for use in vaccine design. Prediction of MHC class II peptide binding is more difficult than that of class I [1]. This is due to the open-ended nature of MHC class II peptide binding groove which allows binding to a broader range of peptide lengths (approximately 11 to 22aa) [1,2]. While MHC class I binds to peptides of a narrow range (usually 8-10 aa), a core of nine aa within a peptide is sufficient to bind to MHC molecules of both classes [3]. However, often, the exact location of the binding core (motif) within a peptide longer than nine aa is unknown. Therefore, given a set of experimentally validated MHC class II binders of different length distribution, an accurate alignment of the binding cores must be first obtained before a motif can be determined. According to previous studies carried out on the structural features of MHC class II molecules indicate five binding sites, also known as *anchor positions* at positions 1,4,6,7 and 9 within a 9-mer peptide [4-6].

A peptide binding motif is represented either by a consensus sequence or as a quantitative matrix [7]. A widely used representation of a motif is the quantitative

J.C. Rajapakse, B. Schmidt, and G. Volkert (Eds.): PRIB 2007, LNBI 4774, pp. 349–355, 2007.

matrix. Each element in the matrix depicts a weight corresponding to the interaction between an amino acid and a position in the motif. Derivation of quantitative matrices based on experimentally derived position specific binding profiles is costly and time consuming. Hence, such matrices can not be easily updated as with machine-learning techniques when new data become available [8]. Other popular computational tools available for finding motifs in protein sequences are: MEME [9][23], Gibbs motif sampler [10] and Rankpep [11].

In this study, our aim is to obtain an optimal alignment of the binding cores for MHC class II, I-A^b molecule peptide sequence dataset. This is carried out with the help of an evolutionary algorithm [12] by simultaneously optimizing the relative entropy and the evolutionary distance of possible alignments. The obtained best alignment is then used to derive the quantitative matrix which will subsequently be used to predict binding peptides. Relative entropy, a measurement of uncertainty is often used to analyze sequence features and alignments, to measure sequence conservation. The evolutionary distance is measured using the BLOSUM62 substitution matrix, a matrix suitable for modeling evolutionary problems [13]. As anchor positions are known to influence peptide-MHC binding, a higher weightage is given to such positions during the estimation of evolutionary distance. In order to reduce the sequence redundancy in an alignment, we employed sequence clustering followed by sequence weighting.

2   Materials and Methods

2.1   Peptide Sequence Dataset

Peptide sequences and their binding affinities were obtained from SYFPEITHI [4], MHCPEP [17], AntiJen [20] and EPIMHC [21] databases. An independent test dataset was used to evaluate the predictive ability of the I-A^b mouse model. The extracted dataset consists of 251 unique binders with a length distribution ranging from 9 to 24 amino acid residues and 58 non-binders. Binder set was divided into two sets, training and test set so that there is no overlap between the two datasets. While training set consists of 167 binders, the testing dataset consists of 84 binders and 58 non-binders.

2.2   Peptide Sequence Clustering and Weighting

Sequence clustering and weighting is carried out according to [14]. A set of sequences with sequence identity greater or equal to 62% forms a cluster. Cluster assignment is followed by the sequences weighting. Sequence weighting reduces over-representation of sequences in an alignment. A peptide s of length k-mer in cluster, c is assigned a weight, $w_s = 1/n_c$, when n is the number of sequences in cluster c.

2.3   Pseudo-count Correction

Pseudo-count correction is carried out as given in [15], which uses the prior knowledge of amino acid relationships represented by substitution matrices. For a given column, pseudo-count frequencies, g_{al}, for amino acid a at position l of the alignment are calculated according to the following equation where f_{bl}, q_b, and q_{ab} represent observed frequency of amino acid b in position l, background frequency of amino acid b, and the target frequency implicit in the substitution matrix (the frequency by which amino acid a is aligned to amino acid b), respectively.

$$g_{al} = \sum_b \frac{f_{bl}}{q_b} q_{ab} = \sum_b f_{bl} q_{alb} \tag{1}$$

where q_{alb} is the conditional probability derived from the BLOSUM62 substitution matrix. The effective amino acid frequencies were then determined according to [15] by applying weight on pseudo-count correction as below:

$$g'_{al} = \frac{\alpha.f_{bl} + \beta.g_{al}}{\alpha + \beta} \tag{2}$$

Where α and β represent the effective sequence number and an arbitrary weight on the pseudo-count correction, respectively. Let the number of peptide clusters generated be C, and the value of $\alpha = C - 1$. An empirically determined suitable setting for $\beta = 10$ [15].

2.4 Identification of Binding Core of Peptides

The first step in designing a weight matrix is to obtain an accurate alignment of th binding cores that are distributed within experimentally determined binding peptides of varying length. Therefore, our goal here is to identify the starting position of the binding core in each peptide. Let S be a set of N peptide sequences, $S = \{s_1, s_2, \ldots s_i, \ldots s_N\}$. For a given alignment, let s_{il} denotes the ith peptide whose binding core starts at the lth position within the peptide. Let $\kappa = (k_1 k_2 \ldots k_9)$ represents the selected best nine aa length binding core in a peptide. Once all the starting positions are identified, an alignment is obtained for the binding peptides so that the weight matrix can be derived.

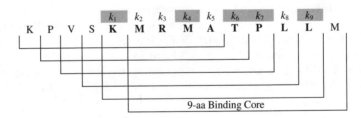

Fig. 1. An illustration of putative 9 aa binding cores within a peptide sequence, s_i. Highlighted positions indicate anchor positions within a putative binding core.

2.5 Generating an Optimal Alignment

We use the evolutionary approach described in [12] to optimize two objective functions associated with relative entropy (E) and evolutionary distance (D) of all alignments. Each individual in the evolving population represents possible starting positions of binding cores within each experimentally determined binding peptide in the training dataset. An individual is represented by a concatenated string of starting positions of peptides in an alignment. The bit size for representing each starting position is determined as below. Given a peptide of length r, the number of 9-mer peptides that can be derived from r is $p=r-9+1$. Hence, the starting positions are located between 0 and $p-1$ where each peptide is overlapped by a single amino-acid. The bit size, θ, is chosen such that $p < min\ (2^{\theta})$ whereby all 9-mer peptide positions in the peptide are taken care of.

Based on the starting positions embedded in an individual, an alignment is generated for the N peptides. The alignment is then use to estimate D and E for anchor positions as given by the Eq. (3) and Eq. (4) below. The evolutionary distance between two peptide sequences s_m and s_n at anchor positions $\kappa' = (k_1\ k_4\ k_6\ k_7\ k_9)$ of the κ binding core in the alignment is calculated as below, where $B(.)$ is the score estimates from the BLOSUM62 substitution matrix for s_m and s_n. Then D is estimated as:

$$D = \sum_{\substack{m=1 \\ n=m+1}}^{N} \sum_{j=k} W_j * B(s_{m,j}, s_{n,j}) \qquad (3)$$

Where W is a weighting factor; $W=w$ for $j= \kappa'$ and $W=1.0$ otherwise.

And E is estimated as:

$$E = -\sum_{j=k'} \sum_{a} g_{aj} \log \frac{g'_{aj}}{q_a} \qquad (4)$$

where g_{aj} is the frequency of amino acid a occupying at position j in the alignment, g'_{aj} is the frequency of pseudo-count and sequence weight corrected amino acid a at position j, and q_a is the background frequency of amino acid a. A number of different approaches are available for estimating background frequencies, also known as background model or null model: amino acid distribution in the SWISS-PROT database [16], a flat distribution where all amino acid frequencies are equal to 1/20, or an amino acid distribution estimated from a non-binder dataset.

Fitness of an alignment is scored according to Eq.(3) and Eq(4). Best population comprises of individuals that maximize Eq.(3) and minimize Eq.(4) simultaneously. The alignment, which scored the highest D and lowest E is then used to build the weight matrix, M, and subsequently for predicting binders in the testing datasets. Each position of the weight matrix, m_{aj} is calculated according to the equation given below.

$$m_{aj} = g_{aj} \log \frac{g'_{aj}}{q_a} \qquad (5)$$

3   Experiments and Results

The experiment in determining the weight matrix representation of I-A^b binding motif is carried out as follows. The values for the parameters β in Eq.(2) and w in Eq.(3) are chosen as 10.0 and 2.0, respectively.

During a single iteration of the evolutionary process, the values of the objective functions, D and E are estimated for the resulting alignments embodied in the individuals. A population of 1000 was evolved for 500 generations with the empirically determined values 0.9 and 0.0004 as the crossover and mutation probability. By using Eq. (5), the weight matrix, M is built with the best alignment, and subsequently used to test the peptides in the testing dataset. A peptide in the testing set is evaluated by scoring all possible 9 aa length binders within the peptide against the weight matrix. Of all the scores, the highest value obtained is assigned as the binding score of the tested peptide. Binding and non-binding status of peptides were determined using a threshold. The performance was measured by estimating Area under Receiver Operating Characteristics (AROC). Let the score estimated for the binding core κ in the peptide s_i be e_i. The binding status, binder (b) or non-binder (nb) is determined according to a threshold, t, as follows:

$$v_i = \begin{cases} b & if \ e_i \geq t \\ nb & if \ e_i < t \end{cases}$$

We obtained an ROC curve by evaluating sensitivity and specificity values for various thresholds as illustrated in Figure 2. The final AROC value estimated for the testing dataset is 0.79, a value considered as good prediction accuracy according to [22]. We also compared our results with MEME [23]. The same training dataset was submitted to the on-line web server http://meme.sdsc.edu/meme/meme.html, and the resulting log-odds matrix was used to measure the prediction accuracy. The AROC value estimated for the testing dataset is 0.71.

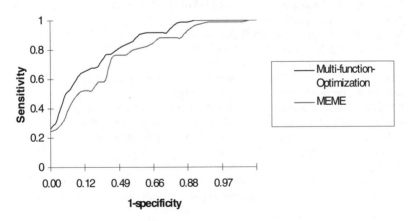

Fig. 2. The ROC plot illustrating the specificity and sensitivity values at different thresholds

4 Discussion and Future Directions

A weight matrix representing motif for MHC class II, I-A^b was derived by simultaneously optimizing entropy and evolutionary distance. The anchor positions of a putative binding core were given higher weightage during the calculation of evolutionary distance. In order to reduce the sequence redundancy in an alignment, we employed sequence clustering and weighting. The weight matrix developed was subsequently applied to discriminate binders from non-binders. The initial results are promising. Better predictive accuracy can be envisaged by incorporating structural properties as an additional objective function. Currently we are extending our investigations towards evaluating different background models, predictive accuracy of the proposed method on multiple alleles of HLA class I and class II molecules, and determining the applicability of multiple substitution matrices.

References

1. Reche, P.A., Glutting, J.P., Reinherz, E.L.: Prediction of MHC class I binding peptides using profile motifs. Hum. Immunology 63(9), 701–709 (2002)
2. Hammer, J., et al.: Precise prediction of major histocompatibility complex class II – peptide interaction based on peptide side chain scanning. J. Exp. Medicine 180(6), 2353–2358 (1994)
3. Rammensee, H., et al.: MHC ligands and peptide motifs: first listing. Immunogenetics 41(4), 178–228 (1995)
4. Rammensee, H., et al.: SYFPEITHI:database for MHC ligands and peptide motifs. Immunogenetics 50, 213–219 (1999)
5. Stern, L.J., et al.: Crystal structure of the human class II MHC protein HLA-DR1 complexed with an inluenza virus peptide. Nature 368, 215–221 (1994)
6. Dessen, A., et al.: X-ray crystal structure of HLA-DR4 (DRA*0101, DRB1*0401) complexed with a peptide from human collagen II. Immunity 7, 473–481 (1997)
7. Mamitsuka, H.: Predicting peptides that bind to MHC molecules using supervised learning of hidden Markov models. Proteins 33(4), 460–474 (1998)
8. Nielsen, M., et al.: Improved Prediction of MHC class I and class II epitopes using a novel Gibbs Sampling Approach. Bioinformatics 20(9), 1388–1397 (2004)
9. http://meme.scdc.edu/meme/website/meme.html
10. Neuwald, A.F., et al.: Gibbs motif sampling: detection of bacterial outer membrane protein repeats. Protein Science 4, 1618–1632 (1995)
11. Reche, P.A., et al.: Enhancement to the RANKPEP resource for the prediction of peptide binding to MHC molecules using profiles. Immunogenetics 56, 405–419 (2004)
12. Deb, K., et al.: A Fast and Elitist Multiobjective Genetic Algorithm. IEEE Trans. on Evolutionary Computation 6, 182–197 (2002)
13. Heinkoff, S., Heinkoff, J.: Amino acid substitution matrices from protein blocks. Proc. Natl. Acad.Sci. USA 89, 10915–10919 (1992)
14. Hobohm, U., et al.: Selection of representative protein datasets. Protein Sci. 1, 409–417 (1992)
15. Altschul, S.F., et al.: Gapped BLAST and PSI-BLAST: a new generation of protein database search programs. Nucleic Acids Research 25, 3389–3402 (1997)
16. Bairoch, A., Apweiler, R.: The SWISS-PROT protein sequence database and its supplement TrEMBL. Nucleic Acids Research 28, 45–48 (2000)
17. Brusic, V., Rudy, G., Harrison, L.C.: MHCPEP, a database of MHC-binding peptides: update 1997. Nucleic Acids Res. 26, 368–371 (1998)
18. Bhasin, M., Singh, H., Raghava, G.P.S.: MHCBN, a comprehensive database of MHC binding and non-binding peptides. Bioinformatics 19, 665–666 (2003)

19. Nielsen, M., Lundegaard, C., Worning, P., Hvid, C.S., Lamberth, K., Buus, S., Brunak, S., Lund, O.: Improved prediction of MHC class I and class II epitopes using a novel Gibbs sampling approach. Bioinformatics 20(9), 1388–1397 (2004)
20. Blythe, M.J., Doytchinova, I.A., Flower, D.R.: JenPep: a database of quantitative functional peptide data for immunology. Bioinformatics 18, 434–439 (2002)
21. Reche, P.A., Zhang, H., Glutting, J.P., Reinherz, E.L.: EPIMHC: a curated database of MHC-binding peptides for customized computational vaccinology. Bioinformatics 21, 2140–2141 (2005)
22. Swets, J.A.: Measuring the accuracy of diagnostic systems. Science 240, 1285–1293 (1988)
23. Bailey, T.L., Elkan, C.: Fitting a mixture model by expectation maximization to discover motifs in biopolymers. In: Proceedings of the Second International Conference on Intelligent Systems for Molecular Biology, pp. 28–36. AAAI Press, Menlo Park, California (1994)

3D Automated Nuclear Morphometric Analysis Using Active Meshes

Alexandre Dufour[1,3], JooHyun Lee[2], Nicole Vincent[3],
Regis Grailhe[2], and Auguste Genovesio[1]

[1] Image Mining Group, Institut Pasteur Korea
[2] Dynamic Imaging Platform, Institut Pasteur Korea
[3] Intelligent Perception Systems (SIP-CRIP5) team, Paris Descartes University
{alexandre.dufour,agenoves}@ip-korea.org

Abstract. Recent advances in bioimaging have allowed to observe biological phenomena in three dimensions in a precise and automated fashion. However, the analysis of depth-stacks acquired in fluorescence microscopy constitutes a challenging task and motivates the development of robust methods. Automated computational schemes to process 3D multi-cell images from High Content Screening (HCS) experiments are part of the next generation methods for drug discovery. Working toward this goal, we propose a fully automated framework which allows fast segmentation and 3D morphometric analysis of cell nuclei. The method is based on deformable models called Active Meshes, featuring automated initialization, robustness to noise, real-time 3D visualization of the objects during their analysis and precise geometrical shape measurements thanks to a parametric representation of each object. The framework has been tested on a low throughput microscope (classically found in research facilities) and on a fully automated imaging platform (used in screening facilities). We also propose shape descriptors and evaluate their robustness and independence on fluorescent beads and on two cell lines.

1 Introduction and Related Efforts

The combination of microscopy and robotics enables to perform 2D visual cell based experiments in parallel and in a fully automated fashion. As a consequence, the exponential increase of images to analyze has motivated the development of fully automated frameworks. However, 2-dimensionality has some limitations, in particular for objects that are heterogeneous along the depth axis such as cell nuclei. Much more information can be obtained by acquiring depth-stacks of images, which allows to analyze the entire 3D structure of cellular or sub-cellular compartments [1].

The cell nuclear morphology constitutes a good start for such a study. A large array of biological functions is accompanied by major changes in the geometry of the nucleus [2]. Determining exactly how geometric characteristics relate to cellular function requires accurate 3D morphological information.

In addition to quantitative measurements, visual observation is also a key aspect of scene interpretation and understanding. Yet, visualizing a 3D scene

J.C. Rajapakse, B. Schmidt, and G. Volkert (Eds.): PRIB 2007, LNBI 4774, pp. 356–367, 2007.

during its analysis remains a challenging task. Most methods employ a 3D reconstruction algorithm (e.g. the Marching Cubes [3]) to produce an intuitive rendering of the scene. These algorithms are time-consuming and suffer from surface approximation errors, therefore real-time visualization remains an issue.

The analysis of 3D fluorescent stacks is not trivial. Indeed, fluorescent images generally suffer from many disturbances induced by the imaging protocol (medium autofluorescence, acquisition noise etc.). However, one of these disturbance factors, namely the convolution with the microscope PSF, has a different impact in 2D and 3D. The PSF is not constant along the depth, and has a much stronger blurring effect on slices below and over the focus plane, yielding very fuzzy boundaries along the depth axis (cf. Fig. 1), causing most algorithms to fail detecting the edges correctly in 3D.

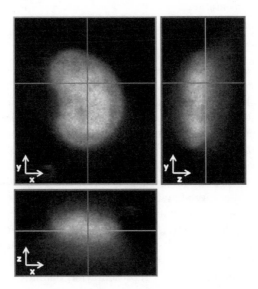

Fig. 1. Axis-based view of a 3D image of size 100 × 115 × 60 pixels and resolution 0.28 × 0.28 × 0.5 μm. Center: XY plane view. Right: YZ plane view. Bottom: XZ plane view. The YZ and XZ planes emphasize the blurring effect of the microscope PSF on the lower and higher Z planes of the volume.

In this context, deformable models (also known as "active contours") have shown to be efficient thanks to their handiness and robustness to noise [4]. The idea is to deform an initial contour under the influence of various forces until it fits the target structure. These forces are usually computed from the minimization of a so-called *energy* functional describing the characteristics of the structure, and defined to be minimal when the model coincides with the target. Deformable models also offer a semantic interpretation of each object, allowing independent and precise object measurements (e.g. size, shape, resemblance to a reference model etc.) rather than global image measurements.

Two main families of deformable models can be distinguished, depending on the mathematical representation of the contour: in explicit models (known as *snakes* in 2D [5]), the boundary is represented by a parametric function, and in implicit models, the contour is defined as the zero level of a higher dimensional scalar function (called *level set* function) [6]. Each family has advantages and drawbacks, the choice thus mostly depends on the application. We briefly summarize the main advantages and drawbacks of both approaches in table 1.

Table 1. Brief comparison of the different advantages of 3D explicit and implicit deformable models. We focus on the aspects that concern our applicative context.

	Explicit	Implicit
Topology handling	−	+
Implementation	−	+
Memory consumption	+	−
Real-time visualization	+	−
Shape description	+	−

Implicit models handle contour splitting and merging implicitly, thus they are well suited to segment an unknown number of objects with a single contour. They are easy to implement in any dimension, however they manipulate a heavy data structure (of the size of the image), easily reaching hundreds of megabytes in 3D. Biological applications can be found in [7][8], however visualization is achieved using a 3D reconstruction algorithm, hence real-time visualization is not possible. Finally, geometrical measurements in a voxel-type structure is dependent on the resolution and thus yields approximation errors.

Explicit models perform faster, but are complex to implement in 3D. More and more methods therefore work directly with the discrete form of the surface (often called *polygonal mesh*) consisting of a set of connected points forming a closed polygonal manifold [9]. This representation enables the introduction of geometric rules that can handle surface splitting and merging [10]. Also, geometrical measures can be computed directly from the mesh in a simpler and more precise manner [11]. Since polygonal meshes rely on the same data structure as conventional computer graphic cards, 3D rendering is available with no additional time-cost, allowing real-time visualization. More popular in medical imaging [12], this approach has been recently applied to automated cell segmentation in fluorescence microscopy (the Active Mesh framework [13]).

In this paper, we propose a fully automated framework for nuclear shape segmentation and analysis based on the Active Mesh framework and propose a set of shape descriptors that can be used to discriminate different phenotypes of a given cell line, showing how this framework is suitable for 3D HCS applications. In section 2, we describe the biological experiment and present the analysis framework. Then we evaluate the method as well as the shape descriptors in section 3. Section 4 concludes the paper and discusses pending applications for the proposed framework.

2   Material and Methods

2.1   Biological and Imaging Protocol

A first experiment was conducted on two cell lines: HEK-293 (Human Embryonic Kidney) and Hela (Henrietta Lack), and a second was performed on $10\mu m$ fluorescent beads (96-Whatman without skirt, Evotec, Germany). All cells were grown on 96-well optical bottom plates, black (Greiner) under same culture conditions (DMEM with 10% FBS). Nuclei were labeled using DNA-specific DRAQ5 fluorescent dye (Biostatus, UK) following the instructions of the manufacturer.

Images were acquired at room temperature using $633nm$ excitation wavelength with $650nm$ long pass emission filter. The Z-stacks were obtained, for HEK-293 nuclei, on a confocal line-scanning microscope equipped with a oil-immersed plan apochromat 63x lens of NA 1.4 (LSM 5 Live, Zeiss, Germany), and for both Hela nuclei and fluorescent beads, on an automated Nipkow-disk confocal microscope (Opera, Evotec, Germany) equipped with a water-immersed plan apochromat 40x lens of NA 0.9 (Olympus, Japan).

2.2   Quantitative Analysis Method

In this section we describe the principal components of the nuclei analysis work flow, from segmentation to quantitative analysis. We start by describing the characteristics of the core segmentation method (the Active Mesh model), and then present each step of the final analysis work flow.

Definition of an Active Mesh. An active mesh [13] is a three-dimensional discrete surface defined by a list of vertices forming a closed set of oriented triangles, such that the mesh boundary represents at all times the contour of a volumetric object. The deformation of the mesh is driven by that of its vertices, which evolve in a real-coordinates space bounded by the image (i.e. the vertices are not fixed on the image grid). To avoid excessive complexity in the manifold structure, a regular sampling is imposed, such that all connected mesh vertices remain within an arbitrary distance interval $[d_{min}, d_{max}]$ from each other. Therefore, as the mesh grows or shrinks, vertices are respectively added or deleted automatically in order to maintain homogeneous edge lengths over the surface. To speed up computation, a multi-resolution approach is chosen, such that the distance interval varies during the evolution: the initial surface has a coarse resolution (vertices are far from each other). Then, as the surface approaches to the solution, d_{min} and d_{max} are progressively reduced, causing a global refinement of the mesh, and so until a suitable resolution is reached. This scheme allows fast and efficient sub-resolution segmentation.

Energy minimization. In our method, we choose to minimize the well-known Mumford-Shah piecewise-smooth functional (or *reduced* Mumford-Shah functional) [14]. This functional reads

$$F(\Gamma, c_1, \cdots, c_n) = \lambda \sum_{i=1}^{n} \left[\int_{R_i} |u_0 - c_i|^2 d\omega \right] + \mu \int_{\Gamma} ds \qquad (1)$$

and states that the target regions R_i, described by their mean intensity c_i, should resemble to the original image u_0 (first term), while the boundary set Γ between the regions should be minimal to avoid over-segmentation (second term). λ and μ are non-negative weighting parameters, and $d\omega$ and ds are the elementary volume and surface respectively. This functional has shown to be efficient for cell and nucleus segmentation in both 2D and 3D fluorescence imaging [15][8], since the target entities are fully stained and have very few corners and cusps. One region R_{out} represents the image background, and every other region $R_{i>0}$ represents an object that will be segmented by a specific mesh. The boundary set Γ thus corresponds to the set of meshes that evolve in the image domain, and the equation above can be rewritten as follows:

$$F(\mathcal{M}_1, \cdots, \mathcal{M}_n, c_{out}, c_1, \cdots, c_n) = \lambda \int_{R_{out}} |u_0 - c_{out}|^2 d\omega +$$
$$\sum_{i=1}^{n} \left[\lambda \int_{R_i} |u_0 - c_i|^2 d\omega + \mu \int_{\mathcal{M}_i} ds \right], \quad (2)$$

where R_{out} denotes the background component of the image with mean intensity c_{out}, and c_i is the mean intensity inside the mesh $\mathcal{M}_i$ segmenting the object i. The minimization is done using a steepest gradient-descent method using the Euler-Lagrange equations (see details in [13] and [16]). The final algorithm complexity is $O(N)$ per iteration, where N is the total number of vertices forming the n meshes. The number of iterations depends on the model initialization, as we shall discuss below.

Initialization. Due to the non-convexity of the energy functional in Eq. 2, convergence is only guaranteed to a local minima. Therefore, deformable models perform better and faster when they are initialized close to the solution. To avoid manual initialization, we propose the following automatic scheme:

- a. Blur the original stack with a Gaussian filter,
- b. Threshold the blurred stack using a 2-class K-Means algorithm,
- c. Extract the connected components (number and average diameter),
- d. Eliminate the objects partially visible (i.e. on the image edge),
- e. Initialize each surface by a coarse 3D reconstruction of each component,
- f. Evolve all surfaces simultaneously on the original (non-blurred) stack.

The 3D reconstruction involved in step (e) utilizes the *Marching Tetrahedra* algorithm [17]. This algorithm has the interesting property of using the same data structure as an active mesh. Hence, no data conversion is necessary, and the surface can be directly used as an initialization, that will hence be very close from the target boundary. Although 3D reconstruction algorithms are time-consuming at fine resolution, a coarse (i.e. fast) reconstruction is sufficient in our case since the model handles refinement automatically during the segmentation.

Visualization. Since each active mesh utilizes the same data structure as current graphic cards (typically a set of connected vertices), the rendering is straightforward and performed on the graphic card parallely to the main computation, yielding no additional time cost. This feature first allows real-time visual monitoring of the analysis, for instance to tweak the algorithm parameters. Secondly, it allows the method to save the 3D scene corresponding to each stack in a database, in order to provide off-line visual feedback after the analysis.

2.3   Statistical Analysis

In order to describe the nuclei shapes as best as possible, we wish to find a set of independent measures in order to compute robust statistics on the objects. In the following, we compute the following criteria from the final mesh: *Surface, Volume, LongAxis, Roughness, RadiusCV, HullDiff*. While *Surface* and *Volume* are quite self-explanatory, the other criteria are less obvious and detailed below.

- The *LongAxis* measure is the longest distance between two mesh vertices, eventually serving as an object elongation indicator.
- The *Roughness* is a measure outlining the local vibrations of the surface membrane. This measure should be low for convex objects and higher when the surface exhibits local concavities. To compute this value, we start by defining a local curvature measure for each mesh vertex v as the dot product between the outer normal $\overrightarrow{N_v}$ (of unit length) and the barycentric normal $\overrightarrow{B_v}$ linking v to the center of its neighbor vertices in the mesh (see figure 2). If the vectors have opposite directions (i.e. negative dot product), the surface is locally convex. If the vectors have same directions (i.e. positive dot product), the surface exhibits a local concavity at the given vertex. Finally, the roughness measure is defined as the standard deviation of all the local curvature values. Reference value is 0 for a sphere.
- The *RadiusCV* measure describes how different the object shape is from a sphere. This measure is obtained for each mesh by computing the standard deviation of the distances between the mass center and each vertex, normalized by the mean radius (definition of the coefficient of variation). Reference value is 0 for a sphere.

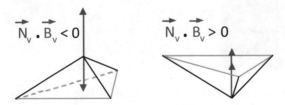

Fig. 2. Description of the roughness measure at a given vertex v. In case of a local convexity (left), the outer normal (red) and barycentric normal (blue) have a negative dot product. In case of a local concavity (right), the dot product is positive.

- The *HullDiff* measure is the difference percentage between the volume of the object and that of its convex hull (i.e. the smallest convex surface that can contain it). This measure will be useful to discriminate bean-shaped objects for instance. Reference value is 0 for a convex object.

3   Experiments and Results

3.1   Segmentation

The automated segmentation and shape measurement protocol was first tested on a set of Z-stacks of HEK-293 cells acquired one by one on a Zeiss LSM 5 *Live* microscope, yielding 22 stacks of size $512 \times 512 \times 60$ voxels and spatial resolution $0.28 \times 0.28 \times 0.5$ μm, totalizing 121 nuclei. Then, the method was applied on Hela cells using a automated imaging platform (Evotec Opera). We used 20 wells of a 96-well plate, and acquired in each well one Z-stack of size $688 \times 520 \times 31$ voxels and spatial resolution $0.327 \times 0.327 \times 0.75$ μm, totalizing 201 nuclei. The computation time ranged from 20 to 40 seconds per stack for all experiments, depending on the number of objects. This time includes: stack loading into memory, initialization (see section 2.2), segmentation and shape measurements of the detected objects.

Figures 3 and 4 present results for the HEK-293 and Hela cells experiments respectively. Left images show a maximum intensity projection (MIP) of one of the Z-stacks. Middle images show a snapshot of the 3D scene taken right after initialization. One can clearly see that cells touching the image edge have been automatically removed, and that the coarse 3D reconstruction using the Marching Tetrahedra are fast and efficient estimates of the nuclei surfaces. Right images show a similar snapshot at the end of the segmentation.

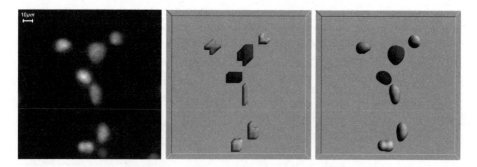

Fig. 3. Segmentation of a HEK-293 cell nuclei Z-stack (size $512 \times 512 \times 60$). Left: maximum intensity projection of the original stack. Middle: snapshot after initialization (coarse 3D reconstruction). Right: snapshot after segmentation.

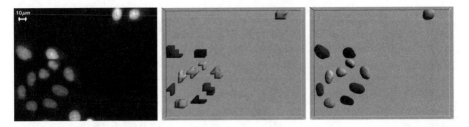

Fig. 4. Segmentation of a Hela cell nuclei Z-stack (size $688 \times 520 \times 31$). Left: maximum intensity projection of the original stack. Middle: snapshot after initialization (coarse 3D reconstruction). Right: snapshot after segmentation.

3.2 Shape Analysis

The validation contains two steps. First, we check that our shape measures are consistent on fluorescent beads. Then, we check their independence in order to keep a compact set of non-redundant shape descriptors.

Validation on Fluorescent Beads. We have conducted a screening experiment on fluorescent beads following the protocol described in section 2.1. Expected values and average measures over 100 beads are given in table 2. Although all measures are close from the expected values, detected objects seem generally bigger than the real objects (e.g. the *LongAxis* measure is 14% higher). This is due to the growing effect of the microscope PSF along the Z axis. This effect decreases as the objects size increases, therefore this error is expected very low for our real experiments, where nuclei are bigger than the beads.

Table 2. Evaluation of shape descriptors on 10 μm fluorescent beads. Measured values are averaged over 100 beads. Coefficients of variation below 1 indicate low-variance populations.

	Surface	Volume	LongAxis	Roughness	RadiusCV	HullDiff
Expected	314.1	523.5	10	0	0	0
Measured	326.9	546.2	11.4	0.02	0.09	0.002
Coef. Var.	0.230	0.015	0.016	0.058	0.032	0.165

Dispersion. We evaluate the dispersion of each measure by computing its coefficient of variation (CV) on each population, i.e. the standard deviation-to-mean ratio. Results are shown in table 3. All coefficients are below 1, implying stable measures, nonetheless, some measures have a higher value than others. For instance, the *HullDiff* measure has a CV around 0.5 for both populations, therefore care should be taken in its interpretation in a shape comparison context. Same remark applies to the *Volume* measure in the HEK-293 case.

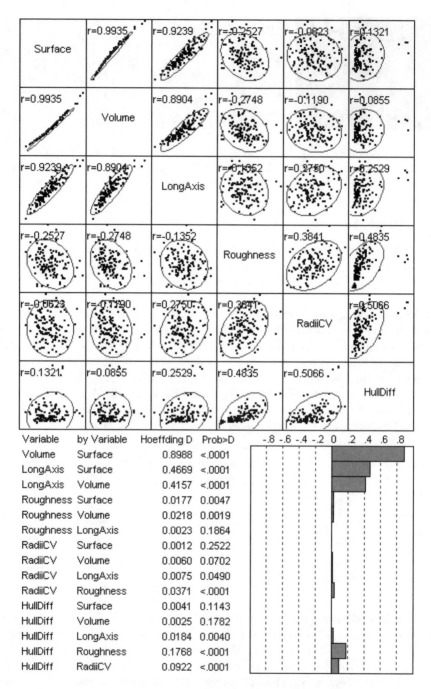

Fig. 5. Statistics on the HEK-293 cell line. Correlation (top) and Hoeffding's D (bottom) measures are given for the criteria presented in section 2.3. D values range from −0.5 to 1, 1 indicating complete dependence. Red ellipses cover 90% of the population.

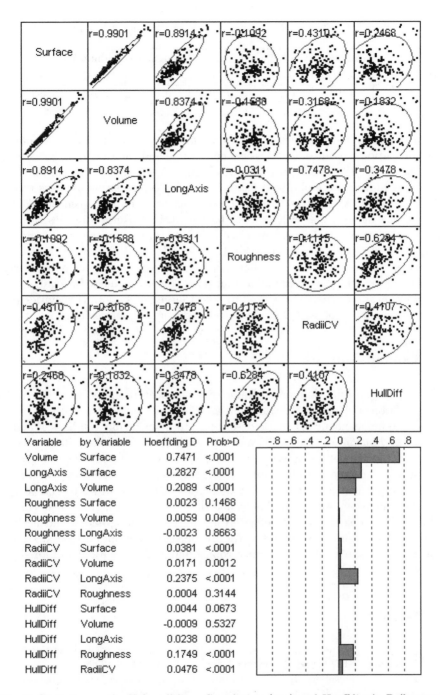

Fig. 6. Statistics on the Hela cell line. Correlation (top) and Hoeffding's D (bottom) measures are given for the criteria presented in section 2.3. D values range from -0.5 and 1, 1 indicating complete dependence. Red ellipses cover 90% of the population.

Table 3. Coefficient of variation of each measure on each population

	Surface	Volume	LongAxis	Roughness	RadiusCV	HullDiff
HEK nuclei	0.280	0.425	0.145	0.108	0.281	0.544
Hela nuclei	0.196	0.278	0.135	0.128	0.264	0.455

Robustness. Finally, we determine the robustness of our criteria by computing two correlation measures: the classical correlation and the Hoeffding measure of dependence D [18]. Results are given in Figures 5 (HEK cell line) and 6 (Hela cell line). Figures were obtained using the multiple correlation analysis tool of JMP software (SAS Institute, 1994). The strong correlation between the *Volume* and *Surface* measures, as well as with the *LongAxis* measure, coincides with the fact that these three measures are closely linked for any convex object. Another interesting observation is the relation between the *HullDiff* and the *Roughness* measures. This is due to the fact that a surface concavity at a given point creates a volume gap with the convex hull at that point. However, due to its local nature, the *Roughness* measure is not suited to detect large but smooth concavities such as for bean-shaped objects, for which *HullDiff* is much more efficient.

4   Conclusion

In this paper, a fully automated framework has been proposed for efficient 3D segmentation and morphometric analysis of cell nuclei, in live cells. We have found five independent 3D shape descriptors to describe our cell lines: *Volume, LongAxis, Roughness, RadiusCV*, and *HullDiff*. These measures will be used to study changes in cell phenotypes under challenging conditions. The method is robust and particularly well adapted to 3D fluorescence microscopy. We further plan to implement a larger array of shape descriptors, in order to enable better discrimination. Although this is not the case of nuclei, it is clear that in some applications the objects of interest may be touching and would need to be separated before their analysis. We are thus working on automated separation algorithms, in order to provide a robust and generic analysis tool for shape analysis in 3D HCS.

References

1. Vonesch, C., Aguet, F., Vonesch, J.L., Unser, M.: The colored revolution of bioimaging. IEEE Signal Processing Magazine 23(3), 20–31 (2006)
2. Leman, E., Getzenberg, R.: Nuclear matrix protein as biomarkers in prostate cancer. Journal of Cell Biochemistry 86(2), 213–223 (2002)
3. Lorensen, W., Cline, H.: Marching cubes: a high resolution 3D surface construction algorithm. In: SIGGRAPH'87. 14th annual conference on Computer graphics and interactive techniques, pp. 163–169. ACM Press, New York (1987)

4. Zimmer, C., Zhang, B., Dufour, A., Thebaud, A., Berlemont, S., Meas-Yedid, V., Olivo-Marin, J.C.: On the digital trail of mobile cells. Signal Processing Magazine 23(3), 54–62 (2006)
5. Kass, M., Witkin, A., Terzopoulos, D.: Snakes: Active contour models. International Journal of Computer Vision 1, 321–331 (1988)
6. Sethian, J.A.: Level set methods and fast marching methods, 2nd edn. Cambridge University Press, Cambridge (1999)
7. Malpica, N., de Solorzano, C.O.: Automated Nuclear Segmentation in Fluorescence Microscopy. In: Science, Technology and Education of Microscopy: an Overview. Microscopy Book Series, Formatex, vol. 2, pp. 614–621 (2002)
8. Dufour, A., Shinin, V., Tajbaksh, S., Guillen, N., Olivo-Marin, J., Zimmer, C.: Segmenting and tracking fluorescent cells in dynamic 3d microscopy with coupled active surfaces. IEEE Transactions on Image Processing 14(9), 1396–1410 (2005)
9. Delingette, H.: General object reconstruction based on simplex meshes. International Journal on Computer Vision 32, 111–146 (1999)
10. Lachaud, J., Montanvert, A.: Deformable meshes with automated topology changes for coarse-to-fine three-dimensional surface extraction. Medical Image Analysis 3(2), 187–207 (1999)
11. Zhang, C., Chen, T.: Efficient feature extraction for 2D/3D objects in mesh representation. In: International Conference on Image Processing, Thessaloniki, pp. 935–938 (2001)
12. Zhukov, L., Bao, Z., Guskov, I., Wood, J., Breen, D.: Dynamic deformable models for 3D MRI heart segmentation. In: SPIE Medical Imaging, vol. 4684, pp. 1398–1405 (2002)
13. Dufour, A., Vincent, N., Genovesio, A.: 3D Mumford-Shah based active mesh. In: Martínez-Trinidad, J.F., Carrasco Ochoa, J.A., Kittler, J. (eds.) CIARP 2006. LNCS, vol. 4225, pp. 208–217. Springer, Heidelberg (2006)
14. Mumford, D., Shah, J.: Optimal approximations by piecewise smooth functions and associated variational problems. Comm. Pure App. Math. 42, 577–684 (1989)
15. Zhang, B., Zimmer, C., Olivo-Marin, J.C.: Tracking fluorescent cells with coupled geometric active contours. In: International Symposium on Biomedical Imaging, Arlington, pp. 476–479 (2004)
16. Zimmer, C., Olivo-Marin, J.C.: Coupled Parametric Active Contours. IEEE Transactions on Pattern Analysis and Machine Intelligence 27(11), 1838–1842 (2005)
17. Gueziec, A., Hummel, R.: Exploiting triangulated surface extraction using tetrahedral decomposition. IEEE Transactions on Visualization and Computer Graphics 1(4), 328–342 (1995)
18. Hoeffding, W.: A class of statistics with asymptotically normal distribution. The Annals of Mathematical Statistics 19(3), 293–325 (1948)

Time-Frequency Method Based Activation Detection in Functional MRI Time-Series

Arun Kumar[1,2] and Jagath C. Rajapakse[1,3]

[1] School of Computing, NTU, Singapore
[2] School of EEE, Singapore Polytechnic, Singapore
[3] Biological Engineering Division, MIT, Cambridge, USA

Abstract. A time-frequency method based on Cohen's class of distribution is proposed for analysis of functional magnetic resonance imaging (fMRI) data and to detect activation in the brain regions. The Rihaczek-Margenau distribution among the various distributions of Cohen's class produces the least amount of cross products and is used here for calculating the spectrum of fMRI time-series. This method also does not suffer from the time and frequency resolution trade-off which is inherent in short-term Fourier transform (STFT). Other than detecting activation, the time-frequency analysis is also capable of providing us with more details about the non-stationarity in fMRI data, which can be used for clustering the data into various brain states. The results of brain activation detection with this techniques are presented here and are compared with other prevalent techniques.

1 Introduction

Human brain is a complex organ anatomically and more so in terms of its functionality. A number of signal and image based techniques have been used to understand the functionality of the brain. Functional Magnetic Resonance Imaging (fMRI) is one such imaging technique which is also non-invasive. It effectively captures the changes in the Blood Oxygenation Level Dependent (BOLD) contrast, allowing the evaluation of brain activity due to external stimuli [1]. Usually, fMRI data consists of time-series emanating from each brain voxel, collected over the periods of activation and rest. The low signal-to-noise ratio (SNR) of fMRI data makes detection of the activation-related signal changes difficult; hence most of the data is collected from periodic stimulation alternating with the rest condition. The temporal dynamics of the activation response, which is delayed and is relatively slow compared to actual brain activity, is another problem that must be dealt with during analysis. Most of the present methods rely on exclusive modeling of the hemodynamic response function to detect this delayed activation [2]. The most extensively used fMR data analysis techniques are variants of general linear model based on t-test, F-test, correlation coefficients (between observed responses and stimulus function) or multiple linear regression. These techniques require accurate knowledge of stimulus function [3].

J.C. Rajapakse, B. Schmidt, and G. Volkert (Eds.): PRIB 2007, LNBI 4774, pp. 368–377, 2007.

The frequency-domain analysis methods are able to overcome some of these shortcomings, as they do not require exclusive modeling of the hemodynamic response function and also the accurate knowledge of stimulus signal is not necessary. Spectral analysis methods have also been found to obtain better statistical estimators for short data segments [4]. Fourier transform has been used traditionally to calculate the signal spectrum but in the process of calculating the spectrum, it looses the signal time information. Short-time Fourier transform (STFT), or spectrogram, is able to provide the time-based spectrum, but suffers from the time and frequency resolution trade-off condition. A number of techniques have been developed to overcome this shortcoming of STFT. The first one of these time-frequency techniques were Wigner-Ville distribution based on the autocorrelation function. This technique although overcomes the time and frequency resolution trade-off condition, it suffers from the presence of cross terms in the spectrum [5]. Various distributions in the Cohen's class aim at reducing the cross-talk, while keeping the advantages of Wigner-Ville distribution [6]. The Rihaczek-Margenau distribution of Cohen's class was found to produce the least amount of cross products and is used here for calculating the spectrum of fMRI time-series and hence for activation detection.

2   Methods

2.1   Functional MRI Time Series

Functional magnetic resonance imaging data consists of a series of three-dimensional brain scans taken at regular intervals during an experiment. This complete set of spatio-temporal data can be considered as a functional image of the brain. A functional image F can hence be defined as $F : \Omega \times \Theta \rightarrow \{0, 1 \ldots 32767\}$, where $\Omega \in N^3$ denotes the three-dimensional spatial domain of image voxels and Θ represents the scanning times. As these scans are acquired at regular intervals of time, $\Theta = \{\Delta, 2\Delta \ldots n\Delta\}$ where Δ denotes the scanning interval and n the total number of brain scans. A functional time-series is further defined as the functional image at a particular voxel over the experimental duration. Consider an fMRI experiment with series of brain scans taken at regular intervals; let the stimulus signal be denoted by and the change in the BOLD signal at a voxel i or the mean-corrected fMRI time-series be represented by.

The Wigner-Ville and other distributions of Cohen's class use the approach of calculating power spectrum from autocorrelation function, as is used in calculation of power spectral density (psd) [7]. In the standard autocorrelation function, summation is carried out over time as shown in Eq. (1), resulting in the autocorrelation function $r_i(\tau)$which is a function of lag/time-shift τ only.

$$r_i(\tau) = \sum_{t \in \Theta} y_{i,t} y_{i,t+\tau} \qquad (1)$$

The Wigner-Ville and other distributions of Cohen's class use a variation of the autocorrelation function where time remains in the result. In this case also, the comparison of waveform with itself is carried over all lag values, but instead of integrating over time, comparison is done over all possible values of time. This results in so-called instantaneous autocorrelation $R_i(t, \tau)$ of the fMRI data as shown in Eq. (2).

$$R_i(t, \tau) = y_{i,t+\tau} y_{i,t-\tau}^* \tag{2}$$

where τ is the time lag and * represents the complex conjugate of the signal. The instantaneous autocorrelation function retains both the lag and time value. The Fourier transform of $R_i(t, \tau)$ is taken along the τ dimension, hence the result is a function of both time and frequency. The relationship for determining the time-frequency distribution of Cohen's class from the instantaneous autocorrelation function is as given in Eq. (3).

$$\rho(t, f) = \sum_{\tau=0}^{(n-1)\Delta} R_i(t, \tau) G(t, \tau) e^{-j2\pi f\tau}; \forall t \in \Theta, f \in \mathbf{f}_i \tag{3}$$

where function $G(t, \tau)$ is based on a two-dimensional filter (filter for the auto-correlation function) and this filter is what distinguishes various distributions within the Cohen's class, as described by Semmlow [6]. The expression for filter $G(t, \tau)$ is given by the Eq. (4-5) below for both the Wigner-Ville and Rihaczek-Margenau distributions of Cohen's class.

$$G(t, \tau) = \int_{-\infty}^{\infty} g(v, \tau) e^{j\pi vt} dv \tag{4}$$

$$g(v, \tau) = \begin{cases} 1 & \text{; for Wigner-Ville distribution} \\ e^{jv\tau/2} & \text{; for Rihaczek-Margenau distribution} \end{cases} \tag{5}$$

Wigner-Ville distribution can be considered as a special class of Cohen's distribution as it does not apply a filter i.e. $g(v, \tau) = 1$ and hence is simplified as in Eq. (6).

$$W(t, \mathbf{f}_i) = \text{FFT}(R_i(t, \tau)) \tag{6}$$

The main problem with Wigner-Ville distribution is the presence of cross products in the spectrum and various distributions in the Cohen's class aim to reduce the amplitude of these cross products. All the transformations in Cohen's class of distribution produce better results when applied to a modified version of the waveform termed the Analytic signal, which is a complex version of the original signal. As the analytic signal does not contain negative frequencies, its use reduces the number of cross products. The approach based on Hilbert transform of signal, as described by Semmlow [6], is being used here to derive the analytic function. The Rihaczek-Margenau distribution of Cohen's class was found to produce the least amount of cross products and is used here for calculating the spectrum of fMRI time-series and hence for activation detection [5]. The original spectrum for Rihaczek-Margenau distribution is three dimensional as it is the time-frequency distribution with amplitude of spectral components as the third dimension. For activation detection, spectral magnitude at stimulus frequency can be obtained by summing together the spectral magnitude at this frequency over all time-points. The comparison of magnitude at stimulus frequency hence is used to select the set of activated voxels. The complete three dimensional time-frequency information belonging to these activated voxels can then be used for clustering and in describing the various brain states. The time-frequency data can also be used to study the variation of characteristics of hemodynamic response over different brain regions.

3   Results and Discussion

The proposed approach was tested on both synthetic and real functional MRI time-series and the activation detection results were compared with the statistical parametric mapping (SPM) method.

3.1   Synthetic Data

A two-dimensional dataset with 64×64 pixels per image scan was generated for the synthetic functional time-series, with 5 cycles of eight rest samples followed by eight task samples. The duration between two scans was taken to be two seconds (RT = 2s) and the box-car time-series was designed for activated pixels while inactive pixels remain unchanged over time. The response of the activated pixels was then generated by convolving the box-car time-series with a gamma hemodynamic response function (lag = 5s and dispersion = 6s). Independent and identically distributed (i.i.d.) Gaussian random noises was then added to the time-series of both activated and inactive pixels. Pixel intensities of image scans are given by the synthetic functional time-series (see Fig. 1). The signal-to-noise ratio (SNR) is defined as $SNR = h^2/\sigma^2$, where h is the amplitude of the box-car time-series, and σ is the standard deviation of the noise. Two different values of SNR=1.2 and SNR=2.0 were used in generation of two sets of synthetic data.

The results obtained for the SPM (F-test) and those obtained from Rihaczek-Margenau distribution based spectrum, for synthetic images with SNR=1.2 and SNR=2.0 are as shown in the Fig. 2 and Fig. 3 respecivey.

The ROC curves obtained for two sets of synthetic data with SNR=1.2 and SNR=2.0 are shown in Fig. 4(a) and ROC curves for various number of epochs considered in the synthetic data are shown in Fig. 4(b). Both the plots indicate better performance for Rihaczek-Margenau (Cohen's class of distribution) spectrum based activation detection, as compared to the SPM.

3.2   Functional MRI Data

FMR images analyzed in this section were obtained on a 3.0 Tesla Medspec 30/100 scanner (Bruker Medizintechnik GmbH, Ettlingen, Germany) at the MRI Centre of the Max-Planck-Institute of Cognitive Neuroscience. A visual-stimulation experiment using a FLASH Protocol was carried out to obtain these

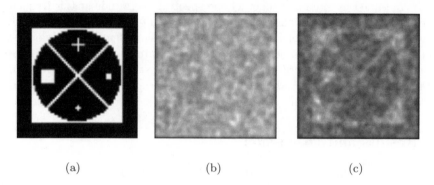

Fig. 1. Synthetic functional images (SNR=1.2) with i.i.d. Gaussian noises (a) Actual activation (b) representative scan of rest state (40th scan) (c) representative scan of stimulus state (80th scan).

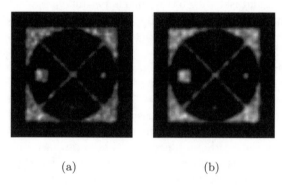

Fig. 2. Illustration of the detected activation on synthetic data with independent noise (SNR=1.2) based on (a) SPM (F-test) (b) Rihaczek-Margenau distribution spectrum

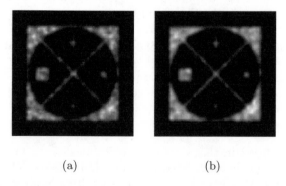

Fig. 3. Illustration of the detected activation on synthetic data with independent noise (SNR=2.0) based on (a) SPM (F-test) (b) Rihaczek-Margenau distribution spectrum

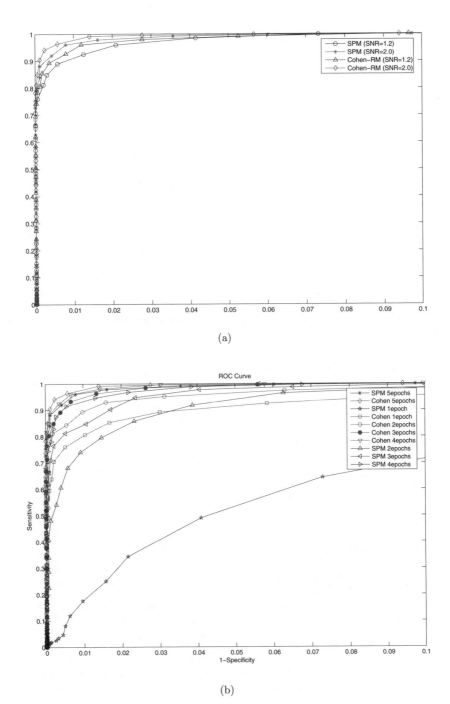

Fig. 4. ROC curves for synthetic data comparing the Rihaczek-Margenau distribution and SPM (*F*-test) technique considering (a) multiple epochs with varying SNR (SNR=1.2 and SNR=2.0) (b) varying number of epochs with constant SNR.

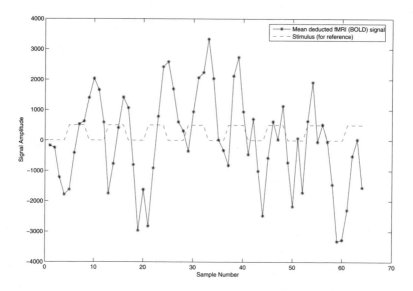

Fig. 5. Illustration of noisy fMRI data from a single voxel with respect to the stimulus signal

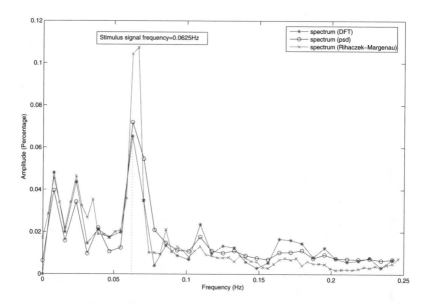

Fig. 6. Comparison of spectrum based on psd, discrete Fourier transform and Rihaczek-Margenau distribution of Cohen's class for the noisy voxel data (Fig. 5)

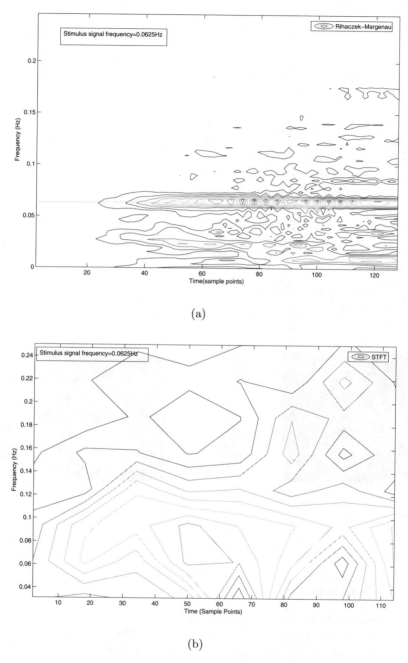

(a)

(b)

Fig. 7. Contour plot of noisy voxel data of Fig. 5 based on (a) Rihaczek-Margenau distribution (b) Short time Fourier transform (window size of 16 samples and overlap of 8)

images (TR=80.5ms; TE = 40ms; matrix = 128 × 64; The image matrices were zero-filled to obtain 128 × 128 images with a spatial resolution of 1.953 × 1.953 mm; slice thickness = 5mm and 2mm gap). In all the experiments, on and off stimuli were presented at a rate of RT = 5.16s per sample. Each stimulation period had four successive stimulated, ON, scans followed by four rest scans, i.e., stimulation OFF scans. Further details of this experiment can be found in [8].

The spectrum as obtained from Rihaczek-Margenau distribution for a noisy voxel data (Fig. 5) is compared with the spectrum obtained from the DFT and psd in the Fig. 6. It is clear from the figure that the spectrum obtained with Rihaczek-Margenau distribution has comparatively higher magnitude at stimulus frequency and much lower magnitude at noise frequencies as compared to the psd and DFT spectrum.

The original spectrum for Rihaczek-Margenau distribution is three dimensional as it is the time-frequency distribution with amplitude of spectral components as the third dimension (Fig. 7) and the two-dimensional plot (Fig. 6) has been obtained by summing together the spectral magnitude for a given frequency at all time-points. The comparison of magnitude at stimulus frequency hence is used to select the set of activated voxels.

Activation detection results for the multiple cycle visual task fMRI data for Rihaczek-Margenau distribution and SPM (F-test) are shown in the Fig. 8.

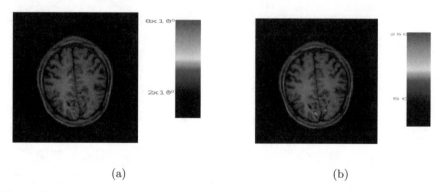

(a)   (b)

Fig. 8. Illustration of activation detection using (a) Rihaczek-Margenau distribution (b) SPM (F-test)

4   Conclusion

A time-frequency method based on Cohen's class of distribution for spectrum calculation and detection of activated voxels in fMRI data was proposed and presented in this paper. The technique is found to produce more accurate activation maps as compared to the existing techniques. The technique is robust to uncorrelated noise and is also able to detect voxels with delayed activation. Our technique does not require prior information of the HRF, also the precise information of stimulus signal is not required. The information from the time-frequency analysis has not been exploited fully as yet in the present paper and will be used in studying the variation of characteristics of hemodynamic response over different brain regions and also in describing the various brain states in our future work. Our future work also includes incorporation of variation of structures and tissues into the detection of brain activation in multi-modality frameworks [9].

References

1. Farckowiak, R.S.J., Friston, K.J., Frith, C.D., Dolan, R.J., Mazziota, J.C.: Human Brain Function. Academic Press, USA (1997)
2. Friston, K.J., Jezzard, P., Jezzard, T.R.: Analysis of functional MRI time-series. Human Brain Mapping 1, 153–171 (1994)
3. Friston, K.J., Holmes, A.P., Worsley, K.J., et al.: Statistical Parametric Maps in Functional Imaging: A General Linear Approach. Human Brain Mapping 2, 189–210 (1995)
4. Mitra, P.P., Pesaran, B.: Analysis of Dynamic Brain Imaging Data. Biophysical Journal 76, 691–708 (1999)
5. Cohen, L.: Time-Frequency Analysis. Prentice Hall, Englewood Cliffs, NJ (1995)
6. Semmlow, J.L.: Biosignal and biomedical image processing. Marcel Dekker Inc., USA (2004)
7. Kumar, A., Rajapakse, J.C.: Power spectral based detection of brain activation. Neural Computing and Applications (accepted March 2, 2007) (in print)
8. Rajapakse, J.C., Kruggel, F., Maisog, J.M., Cramon, D.Y.: Modeling Hemodynamic Response for Analysis of Functional MRI Time-Series. Human Brain Mapping 6, 283–300 (1998)
9. Zhou, J., Rajapakse, J.C.: Segmentation of subcortical brain structures using fuzzy templates. Neuroimage 28(4), 927–936 (2005)

High Performance Classification of Two Imagery Tasks in the Cue-Based Brain Computer Interface

Omid Dehzangi, Mansoor Zolghadri Jahromi, and Shahram Taheri

School of Computer Engineering,
Nanyang Technological University, Nanyang Avenue, Singapore
Omid0002@ntu.edu.sg,
{Zjahromi,Taheri}@cse.shirazu.ac.ir

Abstract. Translation of human intentions into control signals for a computer, so called Brain-Computer Interface (BCI), has been a growing research field during the last years. In this way, classification of mental tasks is under investigation in the BCI society as a basic research. In this paper, a Weighted Distance Nearest Neighbor (WDNN) classifier is presented to improve the classification rate between the left and right imagery tasks in which a weight is assigned to each stored instance. The specified weight of each instance is then used for calculating the distance of a test pattern to that instance. We propose an iterative learning algorithm to specify the weights of training instances such that the error rate of the classifier on training data is minimized. ElectroEncephaloGram (EEG) signals are caught from four familiar subjects with the cue-based BCI. The proposed WDNN classifier is applied to the band power and fractal dimension features, which are extracted from EEG signals to classify mental tasks. Results show that our proposed method performs better in some subjects in comparison with the LDA and SVM, as well-known classifiers in the BCI field.

Keywords: Nearest Neighbor, Weighted distance, Brain-Computer Interface, EEG.

1 Introduction

Classification of mental imagery tasks is used to help amyotrophic lateral sclerosis (ALS) patients to enable them to communicate with their environment [1]. A bright view to the future of this research is to help ALS patients by enabling them to move their limbs with their thoughts. Limb movement can be done by Functional Electrical Stimulation (FES) [2], which is controlled by the BCI system. This interesting application is in its primary stages mainly due to low classification rate even between two imagery tasks in some subjects.

The research in the BCI field can be categorized into *synchronous* [1] and *asynchronous* [3] methods. Most articles focus on the synchronous BCI which is so called cue-based BCI. In this way, Boostani *et al.* [4] applied Adaboost classifier on the fractal dimension features (extracted from the EEG signals) and showed that this

J.C. Rajapakse, B. Schmidt, and G. Volkert (Eds.): PRIB 2007, LNBI 4774, pp. 378–390, 2007.
© Springer-Verlag Berlin Heidelberg 2007

combination has a good prediction ability. In a comprehensive study, Boostani *et al.* [5] employed genetic algorithm on different features and used three different classifiers on the weighted features to show that choosing the band power and fractal dimension as features (by genetic weighting) can significantly improve the performance of cue-based BCI system. The Graz-BCI research group has employed discriminative features based on second order statistics such as band power [1], adaptive autoregressive coefficients [6], and wavelet coefficients [7] with well-known classifiers containing Fisher's Linear Discriminant Analysis (FLDA) [8], Finite Impulse Response Multi-Layer Perceptrons (FIRMLP) [9], Linear Vector Quantization (LVQ) [10], Hidden Markov Models (HMM) [1], and Distinction Sensitive Learning Vector Quantization (DSLVQ) [11] to improve the classification rate between the various movement in imagery tasks. Deriche *et al.* [12] selected the best feature combination among variance, AR coefficients, wavelet coefficients, and fractal dimension by modified mutual information method. They showed that a combination of the aforementioned features is more efficient than each of them individually.

As a simple but efficient supervised learning algorithm, the nearest neighbor classifier has been used successfully on pattern classification problems [13], [14]. However, this method fails to perform satisfactorily in cases that different classes are overlapped in some regions of feature space. Another problem is the noisy training instances that can degrade the performance of this classifier in the generalization phase.

The basic NN uses all training data in the generalization phase. It also considers all the stored instances with the same importance for classification, but the instances are different in being representative of their typical classes.

Recently, many improving techniques have been proposed and added to the nearest neighbor algorithm such as editing, condensing, learning, and weighting [15] for overcoming to its drawbacks. Moreover, there has been considerable research interest in learning mechanisms to locally adapt the distance metrics [16], [17]. Wang et al. [18], [19] have shown that by including a local weight and introducing a simple adaptive distance measure the performance of the NN improves significantly. In this paper a novel learning algorithm is presented which is used to assign a weight to each stored instance, which is then contributed in distance measure, with the goal of improvement in generalization ability of the basic NN. Our proposed learning method is used to adjust the weights of instances in the training set. The basic component of the learning algorithm is an optimization procedure that finds the best operating point of a classifier (i.e., resulting in minimal error rate of the classifier on train data). The proposed scheme achieves two desirable goals at the same time. The classification rate is improved by adjusting a weight for each instance and considering it while calculating distance measure. Our experiments show that the proposed WDNN algorithm can make a robust and accurate classifier system that improves the performance of the cue-based BCI.

The rest of this paper is organized as follows. In section 2, subjects and the method of data acquisition are described. In section3, features are illustrated. In section 4, the proposed WDNN and our proposed method of learning the weights of training instances are described. In section 5, the experimental results are presented and in section 6, conclusion is discussed.

2 Subjects and Data Acquisition

Four subjects (L1, O3, O8, and G8), familiar with the Graz-BCI, participated in this study. Subjects are ranged from 25 to 35 years old. Each subject sat in a armchair about 1.5 meters in front of the computer screen. Three bipolar EEG-channels were recorded from 6 Ag/AgCl electrodes placed 2.5 cm anterior and 2.5 cm posterior to the standardized positions C3, Cz and C4 (international 10-20 system). The EEG was filtered between 0.5 and 50 Hz and recorded with a sample frequency of 128 Hz.

The training in Graz-BCI paradigm is consisted of a repetitive process of triggered movement imagery trials. Each trial lasted 8 seconds and started with the presentation of a blank screen. A short acoustical warning tone was presented at second 2 and a fixation cross appeared in the middle of the screen. At the same time, the trigger was set from 0 to 1 for 500 milliseconds. From second 3 to second 7, the subjects performed left or right hand motor imagery according to an arrow (cue) on the screen. An arrow pointing either to the left or to the right indicated the imagination of a left hand or right hand movement. The order of appearance of the arrows was randomized and at second 7 the screen content was erased. The trial finished with the presentation of a randomly selected inter-trial period (up to 2 seconds) beginning at second 8. Figure 1. shows the timing scheme. Three sessions were recorded for each subject on 3 different days. Each session consisted of 3 runs with 40 trials each.

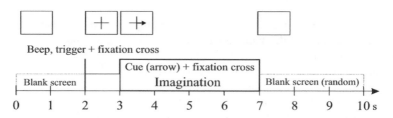

Fig. 1. Training paradigm

3 Feature Extraction

The goal of feature extraction is to find an informative representation of the data that simplifies the detection of brain patterns. The signal features should encode the commands sent by the user. Band power and fractal dimension features are used in this paper. These are briefly described in the following sections.

3.1 Band Power (BP)

The EEG contains different specific frequency bands, that is standard alpha (10-12Hz) and beta (16-24Hz) bands, which are particularly important in classifying different brain states, especially for discriminating imagery tasks. For this study, band power features were calculated by applying a Butterworth filter (order 5), squaring of the samples and then averaging of subsequent samples (1 s average with 250 ms overlap).

3.2   Fractal Dimension (FD)

BP and AAR features are based on the second order statistics of the signal and thus they describe the spectral information in the data. FD, however, captures nonlinear dynamics in the signal. Although all features here try to capture the underlying neurophysiological patterns in the signal, FD has a direct relationship with the entropy of the signal, which in turn is related to information content of the signal. FD is a measure of complexity of a signal. More fluctuation in the attractor shape is reflected by a higher value of FD. There are several methods to calculate the FD [20]. In this study we employed Higuchi's method [21], which is described as follows: Consider a signal containing N samples $\{x(1), x(2),...,x(N)\}$. Construct k new time series $x_m^{\ k}$ (embedded space) as:

$$x_m^{\ k} = \left\{ x(m), x(m+k), x(m+2k),...,x\left(m+\left[\frac{N-m}{k}\right]k\right) \right\} \quad for \quad m=1, 2,..., k. \tag{1}$$

where m indicates the initial time value, and k represents the discrete time interval between points. For each of the k time series $x_m^{\ k}$, the length $L_m(k)$ is computed by:

$$L_m(k) = \frac{\sum_{i=1}^{\left[\frac{N-m}{k}\right]} |x(m+ik) - x(m+(i-1)k)|(N-1)}{\left[\frac{N-m}{k}\right]k} \tag{2}$$

where N is the total length of the data sequence x and $(N-1)/[(N-m)/k]k$, is a normalization factor. An average lengyth of every sub-sequence is computed as the mean of the k lengths $L_m(k)$. This procedure is repeated for the different values of k ($k = 1,2, ...,k_{max}$), that k_{max} varies for each k. There is no analytical formula for determining the value of k, therefore, it has to be found experimentally. An average length for each k is obtained which may be expressed as proportional to k^{-D}, where D is the signal's FD. In order to find the best value of k, from the log-log plot of $log(L(k))$ versus $log(1/k)$, one obtains the slope of the least-squares linear best fit. The FD of the signal, D, is then calculated as:

$$D = [\log L(k)] / \log(1/k) \tag{3}$$

4   Weighted Distance Nearest Neighbor (WDNN)

We briefly describe the NN rule to introduce the notation. For an M-class problem, assume that a set of training examples of the form $\{(X_i, C_i) \mid i = 1,..., N\}$ is given. Where, X_i is a n-dimensional vector of attributes $X_i = [x_{i1}, x_{i2}, ...,x_{in}]^T$ and $C_i \in [1,2, ...,M]$ defines the corresponding class label. To identify the NN of a query pattern Q, a distance function has to be defined to measure the distance between two patterns. Euclidean distance has conventionally been used to measure the distance (i.e., dissimilarity) between two patterns X_i and X_j:

$$d(X_i,X_j) = \sqrt{\sum_{k=1}^{n}\left(x_{ik} - x_{jk}\right)^2} \qquad (4)$$

Assuming that each attribute of the problem is normalized to the interval [0,1], we can equivalently work with the following similarity measure (instead of Euclidean dissimilarity measure), which normalizes the similarity between two instances X_i and X_j to a real number in the interval [0,1]:

$$\mu(X_i,X_j) = 1 - \frac{d(X_i,X_j)}{\sqrt{n}} \qquad (5)$$

With basic NN rule, the query pattern Q is classified by the class most similar training pattern X_p in the training set. This can be formally stated as:

$$p = \underset{1 \le i \le N}{\operatorname{argmax}} \left\{\mu(Q,X_i)\right\} \qquad (6)$$

The NN rule assumes that all classifiers (i.e., stored instances) are equally reliable and uses equation (6) to find the NN of a query pattern. This paper is based on the idea that some of the stored instances are more reliable classifiers than others. We accomplish this by assigning a weight w_k to each instance X_k. The weights of the training instances are used in the test phase to find the NN of a query pattern:

$$p = \underset{1 \le j \le N}{\operatorname{argmax}} \left\{w_j \times \mu(Q,X_j)\right\} \qquad (7)$$

We refer to this classifier as WDNN. Alternatively, the scheme can be viewed as a form of adaptive distance measure for NN that allow the distance measure to vary as a function of instances in the training set. In the next section, we present an algorithm that finds the best operating point in 2-class problems. This algorithm will be used as the basic component of the proposed scheme in section 4.2 to learn the weights of training instances in a WDNN classifier.

4.1   Learning the Best Operating Point in 2-Class Problems

A discrete classifier such as a classification tree only produces a class label for an input pattern. For a 2-class problem (with positive and negative class labels), given a test set of P positive and N negative labeled patterns, a classifier of this type generates a 2×2 confusion matrix (shown in Fig.2) representing the performance of the classifier. The accuracy of the classifier is defined as:

$$Accuracy = \frac{TP + TN}{P + N} = \frac{TP - FP}{P + N} + \frac{N}{P + N} \qquad (8)$$

Many classifiers, such as Bayesian classifier or neural networks naturally assign a score $S(X_t)$ to each input pattern X_t (i.e., scoring classifiers). For example, naive Bayes classifiers output posterior probability distribution over classes. In this case, the score of a pattern for our 2-class problem can be defined as:

$$S(X_t) = \frac{pr(n, X_t)}{pr(p, X_t)} = \frac{pr(n, X_t)}{1 - pr(n, X_t)} \qquad (9)$$

Where $pr(p, X_t)$ and $pr(n, X_t)$ denote the estimated probabilities that the pattern X_t is of positive and negative class, respectively. With the above definition, the score is a numeric value (in the range 0 to ∞) expressing the degree that X_t is thought to be of negative class.

A scoring classifier can be converted to a discrete classifier by specifying a threshold on score. A pattern is classified as negative if its score is greater than the specified threshold and positive otherwise. In this way, the accuracy corresponding to each specified threshold can be calculated using (8).

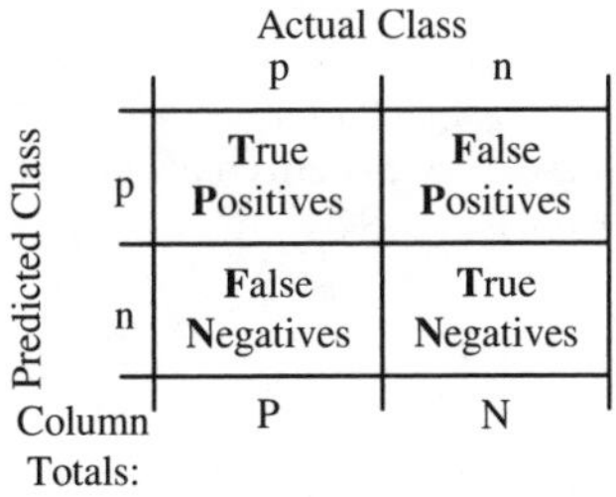

		Actual Class	
		p	n
Predicted Class	p	True Positives	False Positives
	n	False Negatives	True Negatives
Column Totals:		P	N

Fig. 2. Confusion matrix for a discrete classifier

Having the relation between a threshold and corresponding accuracy of the classifier, the best threshold can be easily found by varying the threshold from 0 to ∞. Actually, it is sufficient to consider those thresholds such that classification of an instance changes from negative to positive. Based on this idea, an efficient algorithm for calculating the best threshold is given in [8, 17]. For this purpose, the patterns are ranked in ascending order of their scores (i.e., $S(X_1) < S(X_2) < \ldots < S(X_{P+N})$). Considering any threshold between $S(X_K)$ and $S(X_{K+1})$, the first K patterns will be classified as positive and the remaining $P+N-K$ patterns as negative. In this way, a maximum of $P+N+1$ different thresholds should be examined to find the best threshold. The first threshold classifies everything as negative and the last threshold classifies everything as positive. The rest of the thresholds are chosen in the middle of two non-equal successive scores $S(X_K)$, $S(X_{K+1})$ in the list such that $S(X_K) \neq S(X_{K+1})$. The best threshold is simply the one that maximizes the accuracy (8) of the classifier. An algorithm to find the best threshold is given in Table 1. This algorithm receives a set of patterns and their scores as input and returns the best threshold (i.e. giving maximum classification rate) as output.

The important point is that the value of the best threshold (i.e., *best-th*) calculated using the algorithm of Table 1 can be used as the weight for positive class. That is, instead of classifying a pattern X_t as positive if $S(X_t) < best\text{-}th$, we can equivalently classify the pattern as positive if $best\text{-}th \times pr(p, X_t) > pr(n, X_t)$.

Table 1. Algorithm for finding the best threshold

Inputs: patterns X_t, scores $S(X_t)$
Output: the value of best threshold (*best-th*)
  current = number of misclassified patterns corresponding to the threshold of *th* = 0 (i.e.,
  classifying everything as $\overline{class\ T}$)
  optimum = *current*
  best-th = 0
  rank the patterns in ascending order of their scores
  {assume that X_k and X_{k+1} are two successive patterns in the list}
  for each different threshold *th* = (*Score*(X_k)+*Score*(X_{k+1}))/2
    current = number of misclassified patterns corresponding to the specified threshold
    (i.e., all patterns X_t having *Score*(X_t) < *th* are classified as *class T*)
    if *current* < *optimum* **then**
      optimum = *current*
      best-th = *th*
    end if
  end for
  {assume that *last* is the score of last pattern in the list and τ is a small positive number}
  current = number of misclassified patterns corresponding to *th* = *last* + τ (i.e.,
  classifying everything as *class T*)
  if *current* < *optimum* **then**
    optimum = *current*
    best-th = th
  end if
  return *best-th*

4.2   Learning Weights of Training Instances

For an M-class problem, assume that a training set Γ consisting of N labeled training patterns (i.e. $\Gamma=\{X_j, j=1, 2, ..., N\}$) is available. In this section, we propose an efficient algorithm that attempts to maximize the classification accuracy of the WDNN classifier on training data by learning the weights of instances in the training set.

In its basic form, the proposed algorithm is a hill-climbing search method. The algorithm starts with an initial solution to the problem (i.e., $\{w_k = 1, k =1, 2, ..., N\}$), and sequentially improves the solution by finding a neighbor solution that is better than the current one. A neighbor solution is different from the current solution in the value of just one parameter (i.e., the weight of one instance). Without the algorithm proposed in this section, many neighbor solutions should be examined (i.e., making the search process slow) to find a solution that is better than the current solution. This is due to the complexity of the problem and the fact that the optimization parameters (instance weights) are continuous.

The algorithm given below can provide neighbor solutions that are at least as good as the current solution. This algorithm, which is actually an extended version of the algorithm given in Table 1, finds the optimal weight of an instance assuming that the weights of all other instances are given and fixed. Note that, by optimizing the weight of one instance, the algorithm is indeed providing a neighbor solution that is better than (or at least as good as) the current solution.

We illustrate this algorithm to find the optimal weight of $X_t \in \Gamma$ assuming that the weight of all other instances in Γ are given and fixed. Further, assume that X_k is a training instance of class T, where $T \in \{1, 2, ..., M\}$. The optimal weight of X_k can be found using the following steps.

1. $I = \{\}$
2. Classify all training examples using $w_k = \infty$ (i.e., a very large positive number)
3. Classify all training examples using $w_k = 0.0$
4. Add to I those training examples that are classified correctly only in one of the previous steps (step 2 or 3).
5. Calculate the score of each training example $X_t \in I$ using the following measure.

$$S(X_t) = \frac{\max\limits_{1 \le j \le N} \left\{ w_j \cdot \mu(X_t, X_j) | X_j \in \Gamma \right\}}{\mu(X_t, X_k)} \quad (10)$$

6. The algorithm of Table 1 can now be used to find the best threshold (i.e. the best weight w_k of instance X_k). For this purpose, a 2-class situation is formed by considering class T as positive class and a class by merging all other classes as negative class.

The purpose of steps 1 to 4 is to identify those training examples (by collecting them in the set I) that their correct classification depends directly on the value of w_k.

Overall, the search for locally optimum solution starts with an initial solution to the problem and sequentially improving the solution using a hill-climbing approach. The above 6-step algorithm can be used to find all neighbor solutions (i.e., being different just in the value of one parameter) that are at least as good as the current solution. The usual hill-climbing search can be performed by sequentially replacing the current solution with the best neighbor solution. Finding all neighbor solutions can take a long time when the number of parameters (i.e., instance weights) is large. To speed up the search, in our implementation, the value of just one parameter is updated in each step of the hill-climbing search. That is, the search for locally optimum solution is conducted by optimizing each instance in turn assuming that the order of the instances to be optimized is fixed. The search terminates if no neighbor solution could be found that is better than the current one (or after a certain number of iterations). Obviously, with this modification, the algorithm is sensitive to the order of instances considered for optimization. In simulation results reported in section 5, we fix the ordering by sorting the instances based on their similarity to their nearest enemy instance.

$$\mu_{enemy}(X_k) = \max\limits_{1 \le j \le N} \left\{ \mu(X_k, X_j) | class(X_j) \ne class(X_k) \right\} \quad (11)$$

Where, $\mu_{enemy}(X_k)$ is the similarity of instance X_k with nearest instance of enemy class, and $class(X_k)$ is used to denote the true class of X_j. The instances in the training set are ranked in the descending order of measure (11) and optimized in that order.

When finding the best weight of an instance, it can happen that the set I is empty after step 4 of the above algorithm. This actually indicates that the classification rate on the training data cannot be improved by setting the weight of this instance. In other words, the instance can be pruned (i.e. by setting its weight to zero) without affecting the classification rate. Note that an instance can also be pruned if the value of *best-th* returned by the algorithm of Table 1 is zero. One key feature of the proposed scheme is pruning redundant instances visited during the learning process. The final set of instances usually contains much fewer instances contributed in the classification process. This feature is very useful since, the proposed algorithm can be considered as a serious instance reduction technique for the NN classifier.

5 Experimental Results

In our experiments, in order to compare our proposed method with the well-known classifiers in the BCI field, Support Vector Machine (SVM) [24] and Linear Discriminant Analysis (LDA) [25] are employed. The classification results are produced in the way by which one of the features (i.e. BP or FD) is applied to one of the classifiers (i.e. FLDA or SVM or WDNN), which referred to as the combinations of single feature classifier.

At first, EEG signals from four trained subjects (L1, O3, O8, and G8) were recorded. Out of 360 trials recorded for each subject, 240 trials were used in the train and the rest in the test phase. FD and BP features were extracted from the signals and basic NN, WDNN, SVM, and FLDA were used as the classifiers. In the train phase, significant features were selected using average accuracy on the validation set by ten times ten folds cross validation (10CV). In this way, a classifier is trained with the features of all trials in each 250 ms through the paradigm. Our paradigm is 8sec.; therefore, we have 32 different feature sets for all trials. To choose the best feature set, we calculate the error rate of the classifier by 10CV on training data. Classifiers, trained with the best feature set for each subject, were then used to classify test feature vectors.

Results of test data for th   e subjects with all of the classifiers and features are shown in Table 2. It can be seen that the WDNN has a good compatibility with the FD feature in comparison with the BP. Results of combination of FD and WDNN, in three cases out of four, led to the better results in comparison with FLDA, SVM, and standard NN. Also, in the case O8, combination of WDNN and BP shows a supremacy compare to the other considered classifiers. Incidentally, in Table 3, number of stored instances in NN and WDNN are compared which shows a drastic instance reduction performed while generalization property is increased dramatically. The reason is that in the NN method all trained samples (240) are considered for measuring the class label of an unknown pattern but in the WDNN method, a lot of trained samples have zero weights, therefore, they are removed automatically from the training set.

In Fig. 3, the average error rates of each method on every four subjects are shown. For each classification method, minimum and maximum error rates on the whole cases under investigation is also depicted. In Fig. 3(a), the obtained results are shown in case of using BP features for classification and Fig. 3(b) displays the same statistics in case of using FD features.

Table 2. Test error rates for different subjects using different features and classifiers

Subject	Feature	Basic NN	SVM	FLDA	WDNN
L_1	FD	32.16	27.38	28.57	24.34
	BP	28.65	23.81	20.24	18.67
O_3	FD	23.98	19.8	9.59	17.39
	BP	30.65	26.03	21.92	12.83
O_8	FD	22.43	21.90	17.14	16.48
	BP	25.72	21.90	22.86	12.82
G_8	FD	23.26	16.43	16.43	22.37
	BP	26.29	21.43	22.86	18.86

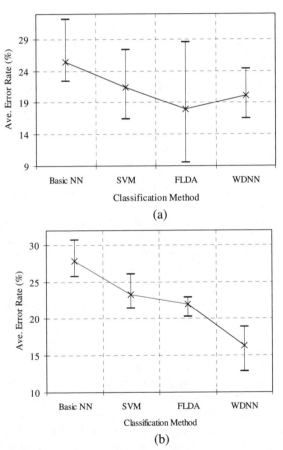

(a)

(b)

Fig. 3. Average error rates on four subjects obtained by applying compared methods and the maximum and minimum error rates achieved by them are shown in (a) for BP features and in (b) for FD

Table 3. Number of training instances stored in the test phase for NN and WDNN methods

Subjects	Feature	Basic NN	WDNN
O_3	FD	240	29
	BP	240	34
O_8	FD	240	30
	BP	240	20
L_1	FD	240	28
	BP	240	34
G_8	FD	240	17
	BP	240	19

As it can be seen in Fig. 3(a), in case of using BP features, WDNN performs better than Basic NN and SVM, but FLDA is the best method in this case. Note that WDNN is still more reliable, since the deviation of error rate is less than FLDA method. In Fig. 3(b), in case of using FD features, it can be seen that the proposed WDNN method has the highest classification accuracies among the compared classifiers. Then, it can predict between the left and right imagery tasks with less error rates. Note that in this case, maximum error rate resulted by applying WDNN is still less than minimum error rates occurred by the other methods presented in this article.

6   Conclusion

In this paper, an innovative approach has been proposed in order to improve the classification rate on the left and right imagery tasks in the cue-based BCI. NN is a traditional method but it is still a powerful method in various applications. In this research, a new version of NN is developed. The noisy instances degrade the performance of the nearest neighbor. NN also considers the importance of all the stored instances the same, but in fact, their relevancy is not the same. This paper presents a novel learning algorithm which is used to assign a local weight to each stored instance, which is then contributed in distance measure, with the goal of improvement in generalization ability of the basic NN. The learning algorithm optimizes the instance weights based on the classification accuracy. The presented scheme achieves two purposes at the same time. The classification rate is improved by adjusting a weight for each instance and considering it while calculating distance measure. It is also faster in predicting between the left and the right tasks for a new subject introduced to it. Since, majority of training instances are removed by assigning a zero weight to them and will not contributed in classification task. In order to evaluate our method, it has been applied on the BP and FD features of two imagery tasks of four subjects (L1, O3, O8, and G8) participated in this study. The results showed that the proposed method is effective to achieve higher accuracy to choose between the left and right tasks.

References

1. Pfurtscheller, G., Neuper, C.: Motor imagery and direct brain-computer communication. In: Proc. IEEE, pp. 1123–1134. IEEE Computer Society Press, Los Alamitos (2001)
2. Pfurtscheller, G., Lopes, S.: Event related desynchronization: Hand book of electroenceph. And clininical Neurophisiology. Revised edition, vol. 6. Elsevier, Amsterdam (1999)
3. Bozorgzadeh, Z., Birch, G.E., Mason, S.G.: The LF-ASD brain computer interface: online identification of imagined finger flexions in the spontaneous EEG of able-bodied subjects. In: IEEE Intern. Conf. Acous. Speech proc., vol. 6, pp. 2385–2388 (2000)
4. Boostani, R., Moradi, M.H.: A new approach in the BCI research based on fractal dimension as feature and Adaboost as classifier. Journ. Neural Eng. 1, 4 (2004)
5. Boostani, R., Graimann, B., Moradi, M.H., Pfurtscheller, G.: A Comparison Approach toward Finding the Best Feature and Classifier in Cue-Based BCI. Journ. Medic. Bio. Eng. Comp. 6 (2007)
6. Schlögl, A., Flotzinger, D., Pfurtscheller, G.: Adaptive Autoregressive Modeling used for Single-trial EEG Classification. Biomediz. Tech. 42, 162–167 (1997)
7. Graimann, B., Huggins, J.E., Levine, S.P., Pfurtscheller, G.: Toward a direct brain interface based on human subdural recordings and wavelet-packet analysis. IEEE Trans. Biomed. Eng. 51, 954–962 (2004)
8. Pfurtscheller, G., Neuper, C., Schlogl, A., Lugger, K.: Separability of EEG signals recorded during right and left motor imagery using adaptive autoregressive parameters. IEEE Trans. Rehab. Eng. 6, 316–328 (1998)
9. Haselsteiner, E., Pfurtscheller, G.: Using time-dependent neural networks for EEG classification. IEEE Trans. on Rehab. Eng. 8, 457–463 (2000)
10. Kalcher, J., Flotzinger, D., Pfurtscheller, G.: A New Approach to a Brain-Computer-Interface (BCI) based on Learning Vector Quantization (LVQ3). In: Proceed. Ann. Intern. Conf. IEEE, vol. 4, pp. 1658–1659 (1992)
11. Flotzinger, D., Pregenzer, M., Pfurtscheller, G.: Feature selection with distinction sensitive learning vector quantisation and genetic algorithms. IEEE Intern. Conf. Comp. Intell. 6, 3448–3451 (1994)
12. Deriche, M., Al-Ani, A.: A new algorithm for EEG feature selection using mutual information. In: IEEE Intern. Conf. Acous. Speech, Signal Proc. ICASSP, vol. 2, pp. 1057–1060 (2001)
13. Cover, T.M., Hart, P.E.: Nearest Neighbor Pattern Classification. IEEE Trans. on Info. Theo. 13, 21–27 (1967)
14. Dasarathy, B.V.: Nearest Neighbor (NN) Norms: NN Pattern Classification Techniques. IEEE Computer Society Press, Los Ala. CA (1991)
15. Goldberger, J., Roweis, S., Hinton, G., Salakhutdinov, R.: Neighbourhood component analysis. Neur. Info. Proc. Sys (NIPS) 17, 513–520 (2004)
16. Weinberger, K., Blitzer, J., Saul, L.: Distance metric learning for large margin nearest neighbor classification. In: Weiss, B.S., Platt, J. (eds.) Advances in Neural Information Processing Systems, p. 18 (2005)
17. Wang, J., Neskovic, P., Cooper, L.N.: Neighbourhood selection in the k-nearest neighbor rule using statistical confidence. Patt. Recog. 39, 417–423 (2006)
18. Wang, J., Neskovic, P., Cooper, L.N.: Improving nearest neighbor rule with a simple adaptive distance measure. Pattern Recognition Letters 28, 207–213 (2007)
19. Xiao-Yuan, J., David, Z., Yuan-Yan, T.: An improved LDA Approach. IEEE Trans. Sys. Man Cyber. 34, 5 (2004)

20. Esteller, R.: Detection of Seizure Onset in Epileptic Patients from Intracranial EEG Signals. Ph. D. thesis, School of Electrical and Computer Engineering Georgia Institute of Technology (2000)
21. Higuchi, T.: Approach to an Irregular Time Series on the Basis of Fractal Theory. Physica D 31, 277–283 (1988)
22. Fawcett, T.: ROC Graphs: Notes and Practical Considerations for Researchers, Technical Report HPL-2003-4, HP Labs (2003)
23. Lachiche, N., Flach, P.: Improving accuracy and cost of two-class and multi-class probabilistic classifiers using ROC curves. In: 20th Intern. Conf. Machine Learning (ICML'03), pp. 416–423 (2003)
24. Vapnic, V.N.: Statistical learning theory. John Wiley and Sons, New York (1998)
25. Fukunaga, K.: Introduction to Statistical Pattern Classification. Academic Press, San Diego, Calif. (1999)

Human Brain Anatomical Connectivity Analysis Using Sequential Sampling and Resampling

Bo Zheng[1] and Jagath C. Rajapakse[1,2]

[1] BioInformatics Research Center, School of Computer Engineering,
Nanyang Technological University, 50 Nanyang Avenue,
Singapore 639798
[2] Singapore-MIT Alliance, N2-B2C-15, 50 Nanyang Avenue, Singapore

Abstract. Diffusion Tensor MR Imaging (DTI) provides non-invasive approach to track white matter (WM) trajectories within human brain *in vivo*, and thereby facilitates studies of anatomical connectivity between sub-cortical and cortical regions. This paper presents a probabilistic fiber tracking framework, which aims to address the two problems in earlier approaches: first, it does not adopt fractional anisotropy (FA) as the stopping criteria so that the exploration of cortico-cortical connectivity is feasible; secondly, fiber tracking process is regularized so that trajectory with low curvature means high belief of connection between two voxels.

1 Introduction

Diffusion tensor MR Imaging (DTI) fiber tracking algorithms provide potential for non-invasive reconstructing of white matter (WM) trajectories of human brain *in vivo*, as well as assessing changes due to disease, such as multiple sclerosis, amyotrophic lateral sclerosis (ALS), stroke, schizophrenia, reading disability, etc [1]. In early work, one of the most popular approaches is a simple deterministic line propagation technique whereby a single trajectory is propagated bidirectionally from a manually defined seed point by moving in a direction parallel to principle diffusion direction (PDD) [2,3,4,5,6,7]. However, there are three major limitations to such a deterministic approach. First, they assume that PDD is the tangent vector of underlying dominant WM trajectory orientation in each voxel and estimated from measured DTI data that is discrete, coarsely sampled noisy, and voxel averaged. Hence, there is uncertainty caused by the noise and artifacts present in any MR scan and incomplete modeling of diffusion signal [8] associated with PDD. Furthermore, the reliability of reconstructed WM trajectory cannot be estimated using such deterministic approaches. Secondly, they cannot handle a voxel which contains more than one WM trajectories with different orientations since only one trajectory within each voxel is presumed by single tensor model. Especially, at millimeter-scale resolution of DTI the number of such voxels may be considerate given widespread divergence and convergence of WM trajectories [9,10]. And such voxels manifest by the form of oblate diffusion tensor (pancake-like shape) [11]. By examining DTI data of single subject, about 9.97% of within brain voxels possibly does not contain one WM trajectory. Third, most of them

J.C. Rajapakse, B. Schmidt, and G. Volkert (Eds.): PRIB 2007, LNBI 4774, pp. 391–400, 2007.

adopt fractional anisotropy (FA) as the stopping criteria. When FA is smaller than a certain value, fiber tracking process terminates making exploration of cortico-cortical connectivity impossible since FA near sub-cortical and cortical regions is very small. Therefore, these deterministic fiber tracking approaches work best on datasets with high FA and without voxel containing multiple WM trajectories. In addition, they cannot reveal sufficient and reliable information for analysis of human brain anatomical connectivity.

More recently, several probabilistic fiber tracking approaches [8,12,13,14] have been proposed to address the three limitations to deterministic approaches by modelling uncertainty associated with PDD. Often, a probability density function (PDF) derived from either raw diffusion weighted images [8,14] or estimated diffusion tensor [12,13] for each voxel, which estimates the probability of the orientation of local trajectories in all possible directions, is proposed. Then, simple line propagation is carried out using derived PDF of seed point as well as those voxels that the reconstructed trajectory passes through. This process is repeated in Monte-Carlo fashion to generate adequate number of WM trajectories to represent PDF of trajectories starting from seed point. Finally, statistical connectivity map is produced from reconstructed trajectories to estimate probability of connection from seed point to any other voxels. In [8,13,14], such a map is estimated using Probabilistic Index of Connectivity (PICo), which is calculated as the number of trajectory passes through the region divided by the total number of trajectory reconstructed. This index is reasonable estimation of belief that there is a connection between two regions. And it is also adopted in this paper. Further normalization on this index is applied making comparison of connectivity starting from different regions feasible.

By studying previous probabilistic approaches, it was realized that there is still one problem in them, high curvature of reconstructed trajectories. It is believed that trajectory with low curvature means high belief of connection between two voxels. The factors that determine curvature of reconstructed 3D piecewise trajectory are turning angel (i.e. inverse cosine of dot product of previous direction and current direction vectors) and step size at each propagation step. In deterministic approach, low curvature is achieved by terminating fiber tracking process when turning angel is too high at the expense of coverage area of reconstructed trajectories. In probabilistic approach [8,12,13], such a priori information is not incorporated into PDF of local trajectory orientation. Therefore, most of reconstructed trajectories are irregular. In [14], Friman et al. considered the trajectory as first-order Markov Chain to minimize the probability of occurrence of large turning angel. However, step size was not taken into consideration. Furthermore, first-order Markov Chain is insufficient to minimize curvature, especially when trajectory passes through regions with low FA.

Inspired by work [14], a new probabilistic fiber tracking approach is presented in this paper. The approach models the trajectory as high order Markov Chain, which means that current direction is determined by several previous directions. At each propagation step, unlike previous approaches, both the current direction and step size are sampled. This new approach was applied to young normal DTI data to study trajectories passing through splenium of corpus callosum, then compared with [13].

2   Method

The reconstructed WM trajectory x is a piecewise continuous 3D space curve which is considered as a sequence of space vector $\{x_k, k = 1, 2, ..., N\}$. The reconstruction process can be modelled as sequential sampling using equation (1)

$$x_{k+1} = x_k + \alpha_k \nu_k \qquad (1)$$

where ν_k and α_k are the direction vector and step size of kth sampling step, respectively.

WM trajectory x is modelled as m^{th} order Markov chain to regularize the reconstructed trajectory, that is, (ν_k, α_k) depend on previous m sampling steps. In this paper, Monte Carlo estimation of anatomical connectivity from region A to region B $\theta_{AB} = E_\pi h_{AB}(x)$ is of interest, where x is a random WM trajectory sampled from the distribution $\pi(x)$ can be expressed as:

$$\pi(x) = P(\nu_1, \alpha_1 | y, \hat{D}) \prod_{l=2}^{m} P(\nu_l, \alpha_l | \nu_l, ..., \nu_{l-1}, y, \hat{D})$$

$$\prod_{k=m+1}^{N} P(\nu_k, \alpha_k | \nu_{k-1}, ..., \nu_{k-m}, y, \hat{D}) \qquad (2)$$

where $\hat{D}$ is estimated diffusion tensor using multivariate linear regression, and y is logarithm of raw diffusion weighted images (DWI).

In [8,14], posterior PDF of PDD ν_k and α_k is derived from DWI using two-tensor model, which assumes two fiber trajectories with different orientation present within each voxel. Although problem of incomplete modelling in single tensor model is partially resolved, more data acquisitions as well as computation effort are required since two-tensor model involves more latent parameters. Considering limited acquisition time and limitations inherent to MR Imaging scanner (i.e. millimeter-scale resolution of DTI data), single tensor model is a reasonable compromise to reflect the averaged diffusion coefficient over a voxel in any direction of space. Therefore, posterior PDF of PDD ν_k and α_k is derived from estimated diffusion tensor $\hat{D}$. Any additional trajectory within a voxel is simply considered as uncertainty. And ν_k's posterior distribution becomes disperser when multiple trajectories present within a voxel. Furthermore, instead of constant step size, it is adaptable so that the lower the uncertainty is, the larger it is.

The anatomical connectivity from region A to region B is measured using Probabilistic Index of Connectivity in [8,13,14] by first sampling a large number of trajectories starting from region A, and calculating the proportion of trajectories that pass through region B. This measure indicates the belief that there is connection from region A to region B. Note that this index is not symmetric. To compute connectivity from region B to A, sampling trajectories starting from region B needs to be performed. This index of connectivity is not comparable suppose the starting region is different. Therefore, we proposed a coefficient associated with starting region, which measures the true relative number of trajectories passing through the region. And if the starting region is gray matter (GM), the volume of the region is a good candidate. Here, we assume the neurons are uniformly distributed in the region. If the starting region is WM, average FA is chosen. Currently, only comparison of connectivity starting from homogeneous brain tissue is possible. The connectivity function $h_{AB}(x)$ is:

$$h_{AB}(x) = \begin{cases} c_A & \text{If x passes through region B} \\ 0 & \text{otherwise} \end{cases} \tag{3}$$

2.1   Diffusion Tensor Estimation

DT-MR Imaging consists of acquiring DWI $I_i, i = 1, 2, \ldots, K; K \geq 6$, which measures a single scalar apparent diffusion constant (ADC) along different diffusion-sensitizing directions $g_i, i = 1, 2, \ldots, K; K \geq 6$. In DT-MR Imaging, diffusion tensor D that characterizes anisotropic water diffusion within a macroscopic voxel is estimated from the set of at least 6 DWIs with non-collinear and non-coplanar diffusion-sensitizing directions, which are uniformly distributed on a unit sphere surface, plus the non-diffusion weighted image I_0 (i.e. $b = 0$) using equation (4) [16,17] via multivariate linear regression (equation (5)):

$$I_i = I_0 \exp(-b g_i^T D g_i) \tag{4}$$

$$Y = X\beta + \varepsilon, \varepsilon \; N(0, \sigma^2) \tag{5}$$

where

$$Y_i = \ln I_i - \ln I_0$$

$$X = -b \begin{pmatrix} g_{1x}^2 & g_{1y}^2 & g_{1z}^2 & 2g_{1x}g_{1y} & 2g_{1x}g_{1z} & 2g_{1y}g_{1z} \\ g_{2x}^2 & g_{2y}^2 & g_{2z}^2 & 2g_{2x}g_{2y} & 2g_{2x}g_{2z} & 2g_{2y}g_{2z} \\ \vdots & \vdots & \vdots & \vdots & \vdots & \vdots \\ g_{Kx}^2 & g_{Ky}^2 & g_{Kz}^2 & 2g_{Kx}g_{Ky} & 2g_{Kx}g_{Kz} & 2g_{Ky}g_{Kz} \end{pmatrix}$$

$$\beta = [D_{xx} \quad D_{yy} \quad D_{zz} \quad D_{xy} \quad D_{xz} \quad D_{yz}]^T$$

where b-value renders the amount of diffusion weighting.

Then eigenvalues, eigenvectors, and FA were determined. A mask was also generated based on FA map to prevent fiber tracking outside brain.

2.2   Estimation of Joint Posterior Distribution

This probability relates estimated diffusion tensor $\hat{D}$ to WM trajectory local direction ν_k and α_k. Assume that ν_k and α_k are conditionally independent, the joint posterior distribution was decomposed into two distributions (equation (6)). The key point is to introduce the *a priori* knowledge of low curvature of trajectories into posterior PDF. By applying Bayes' theorem, such *a priori* knowledge was incorporated in the form of prior distribution $P(\nu_k|\nu_{k-1}, \ldots, \nu_{k-m})$ (equation (7))

$$P(\nu_k, \alpha_k|\nu_{k-1} \ldots \nu_{k-m}, Y, \hat{D}) = P(\alpha_k|\nu_k, \nu_{k-1} \ldots \nu_{k-m}, Y, \hat{D})$$
$$P(\nu_k|\nu_{k-1} \ldots \nu_{k-m}, Y, \hat{D}) \quad (6)$$

$$P(\nu_k|\nu_{k-1} \ldots \nu_{k-m}, Y, \hat{D}) = \frac{P(Y|\nu_k \ldots \nu_{k-m}, \hat{D})P(\nu_k|\nu_{k-1} \ldots \nu_{k-m}, Y)}{P(Y|\nu_{k-1} \ldots \nu_{k-m}, \hat{D})} \quad (7)$$

Since single tensor model was used here, there are no other latent parameters besides estimated diffusion tensor $\hat{D}$. Assume Y depends on current propagation direction only; the equation (7) can be further simplified into:

$$P(\nu_k|\nu_{k-1} \ldots \nu_{k-m}, Y, \hat{D}) = \frac{P(Y|\nu_k, \hat{D})P(\nu_k|\nu_{k-1} \ldots \nu_{k-m})}{\int_{\nu_l} P(Y|\nu_l, \hat{D})P(\nu_l|\nu_{l-1} \ldots \nu_{l-m})} \quad (8)$$

It is impossible to evaluate integral term of equation (8) in continuous domain. Hence, ν_k is discretized into a large number of samples by uniformly sampling the sphere to transform integration into summation. The multivariate linear regression model assumes Gaussian distributed noise. The likelihood of Y $P(Y|\nu_k, \hat{D})$ was modelled as:

$$P(Y|\nu_k, \hat{D}) = \prod_{i=1}^{K} \frac{1}{\hat{\sigma}\sqrt{2\pi}} \exp(-\frac{(Y_i + bg_i^T R(\nu_k)\hat{D}R(\nu_k)^T g_i)^2}{2\hat{\sigma}^2}) \quad (9)$$

where $R(\nu_k)$ is a rotation matrix that rotates PDD to ν_k; the noise variance σ^2 was estimated using equation (10).

$$\hat{\sigma}^2 = \frac{\sum_{i=1}^{K}(Y_i + bg_i^T \hat{D}g_i)^2}{K - 6} \quad (10)$$

The prior distribution $P(\nu_k|\nu_{k-1} \ldots \nu_{k-m})$ follows:

$$P(\nu_k|\nu_{k-1} \ldots \nu_{k-m}) \propto \begin{cases} \langle \nu_k, \frac{x_k - x_{k-m}}{\|x_k - x_{k-m}\|} \rangle & \langle \nu_k, \frac{x_k - x_{k-m}}{\|x_k - x_{k-m}\|} \rangle > 0 \\ 0 & \text{otherwise} \end{cases} \quad (11)$$

This prior distribution gives the direction made up of low curvature trajectory high probability. In addition, the constraint in equation (11) is to avoid backward tracking.

Distribution for step size $P(\alpha_k|\nu_k, \nu_{k-1} \ldots \nu_{k-m}, \hat{D})$ follows Gaussian distribution with mean $cFA_k\langle\nu_k, \frac{x_k - x_{k-m}}{\|x_k - x_{k-m}\|}\rangle + b$ (where b and c are constant) and standard deviation σ. Therefore, when trajectory passes through regions with high FA and turning angle is small, the probability of α_k being large is high.

2.3 Sampling and Re-sampling Trajectory

The simple line propagation approach used for sampling trajectory is called Fiber Assignment by Continuous Tracking (FACT) described in our previous work [18]. In brief, starting from user-defined seed voxel, fiber trajectory is reconstructed from the diffusion tensor by propagating forward and backward, following the PDD. As given in equation (1), suppose the current point is x_k, the next point x_{k+1} along the path is calculated by adding the normalized PDD ν_k multiplied by the step size α_k, where (ν_k, α_k) is randomly generated using its joint posterior distribution. The tracking process is terminated when the net change in direction within a single voxel exceeded $\pm\pi/2$ [13], or the boundary of brain is reached. This is to compare our results with [13], since different stopping criterion may give different results. Note that, diffusion tensor D is discrete, but continuous tensor field is required in propagation process. A statistical interpolation approach proposed by [8] is adopted to interpolate diffusion tensor field. In this statistical framework, diffusion tensor at point x_k is assigned one of its nearest neighbors tensor values. The probability of picking one neighbor is inversely proportional to the distance between point x_k and center of the neighbor. To estimate the connectivity from starting point to other regions, re-sampling of generated WM trajectories is performed to give Monte Carlo estimation as well as its standard error.

3 Results

The proposed probabilistic fiber tracking approach was carried out on real DTI data. DTI data consists of 15 DWIs and one non-diffusion weighted image. Images were obtained from one healthy volunteer on a Philips 3T MRI scanner using pulsed-gradient echo planar sequence with the following parameters: field of view (FOV) $= 230mm$; $TR = 3700ms$; $TE = 56ms$; 256×256 acquisition matrix; slice thickness $= 3mm$; b factor $= 800smm^{-2}$. All scans were approved by ethics committee of National Neuroscience Institute, Singapore.

To validate the results, the corpus callosum, which is the largest fiber bundle interconnecting the two cerebral hemispheres, was chosen to be studied, since its topography has been well defined in literature (e.g. [19,20,21]). The seed point (Fig. 1) was placed on the midline in splenium (posterior part) of corpus callosum and 1000 trajectories were generated for first-order, third-order and fifth-order Markov Chain, respectively. We also defined a region of interest (ROI) (Fig. 1) composed of cuneus, superior occipital gyrus, middle occipital gyrus to study the anatomical connectivity from splenium of corpus callosum to occipital lobe. As suggested by previous qualitative and quantitative studies [14,19,20,21], it

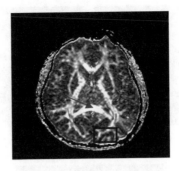

Fig. 1. Fractional Anisotropy map at middle axial slice. The red arrow shows the seed point; the blue rectangle shows region of interest.

is expected the connectivity is nearly 1 as most of fiber trajectories through splenium span out to occipital lobe, and a few, named tapetum, extend laterally on left side of human brain into the inferior temporal lobe.

In most previous probabilistic fiber tracking approaches, constant step size was used. And only direction vector was sampled at each propagation step. Compared to one of previous approach [13] (referred to as GJM method), the proposed method which incorporates adaptive step size gives less average number of sampling steps (table 1). Hence, the adaptive step size ensures effective sequential sampling of fiber trajectory. Furthermore, GJM method did not involve any *a priori* information of direction vector which controls the curvature of sampled fiber trajectories. Consequently, probability of violation of stopping criterion due to noise or partial volume effect before reaching ROI is higher than proposed method. In conclusion, *a priori* information helps reduce curvature of sampled fiber trajectories and fiber tracking process pass through noisy or branching regions. This is also confirmed by average curvature and connectivity shown in table 1. Since many sampled fiber trajectories by GJM method stops before they enter into ROI, the connectivity index is much lower than proposed method. In addition, high order Markov Chain does not necessarily give lower curvature. That curvature of sampled trajectories modelled by third and fifth order Markov Chain is higher than those modelled by first order Markov

Table 1. Comparison of characteristics of generated fiber trajectories by different method

Method	Step	Curvature	Connectivity
GJM Method	186.21 ± 84.49	0.1208 ± 0.0914	0.4807 ± 0.0153
1st Markov Chain	128.42 ± 53.33	0.1048 ± 0.0429	0.9074 ± 0.0085
3rd Markov Chain	148.33 ± 64.15	0.1148 ± 0.0439	0.9761 ± 0.0049
5th Markov Chain	148.93 ± 60.57	0.1127 ± 0.0411	0.9812 ± 0.0044

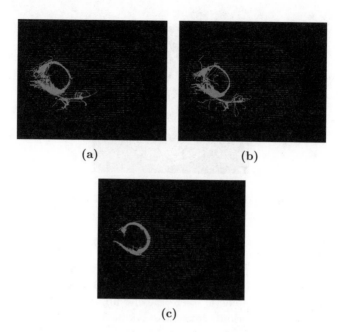

Fig. 2. Top view of generated fiber trajectories (a) 3rd order Markov Chain; (b) 5th order Markov Chain; (c) GJM Method

Chain, since high order Markov Chain makes violation of stopping criterion more difficult and trajectories near sub-cortical and cortical regions are more curved in nature.

Fig. 2 shows top view of sampled fiber trajectories by proposed method (Fig. 2(a) and (b)) and GJM method (Fig. 2(c)). Given same number of samples, the proposed method discovered more branches than GJM method. The proposed method found trajectories that project into inferior temporal lobe (i.e. tapetum), which were not found by GJM method. Hence, the convergence rate of PDF in proposed method is possibly faster than PDF in GJM method. And fewer samples are required to estimate the anatomical connectivity. In conclusion, the proposed method gives better estimation of PDF of fiber trajectory as well as anatomical connectivity.

Although high order Markov Chain is very effective in fiber tracking process, higher order does not mean better propagation results. The order of Markov Chain depends on the length of trajectory as well as its shape. If the trajectory is not so long and is very straight, lower order is sufficient to regularize the trajectory. Furthermore, if the trajectory is curved, high order may possibly distort its shape. To study fiber bundles (consist of thousands of trajectories), Monte-Carlo simulation with different order of Markov Chain can be carried out to reveal different levels of information. For fiber trajectories passing splenium of corpus callosum, 3 is an appropriate order. There is no significant difference of characteristics of sampled fiber trajectories between third and fifth order.

4 Conclusion

A new probabilistic fiber tracking approach allowing effective regularization of reconstructed trajectories as well as exploration of cortico-cortical anatomical connectivity was proposed. The advantage of our approach over previous ones is that both turning angel and step size are taken into consideration to regularize reconstructed trajectories. Furthermore, there is no need to specialize threshold for FA or turning angel, which is usually difficult to determine, to terminate fiber tracking process. This helps fiber tracking process passes through noisy or branching regions and propagates into sub-cortical and cortical regions so that estimation of cortico-cortical anatomical connectivity is feasible.

Future work may include development of technique to fuse anatomical connectivity with function connectivity derived from fMRI.

References

1. Mori, S., van Zijl, P.C.M.: Fiber tracking: principles and strategies - a technical review. NMR Biomed. 15, 468–480 (2002)
2. Mori, S., Crain, B.J., Chacko, V.P., van Zijl, P.C.M.: Three dimensional tracking of axonal projections in the brain by magnetic resonance imaging. Ann. Neurol. 45, 265–269A (1999)
3. Xue, R., van Zijl, P.C.M., Crain, B.J., Solaiyappan, M., Mori, S.: In vivo three-dimensional reconstruction of rat brain axonal projections by diffusion tesnor imaging. Magn. Reson. Med. 42, 1123–1127 (1999)
4. Jones, D.K., Simmons, A., Williams, S.C., Horsfield, M.A.: Non-invasive assessment of axonal fiber connectivity in the human brain via diffusion tensor MRI. Magn. Reson. Med. 42, 37–41 (1999)
5. Basser, P.J., Pajevic, S., Pierpaoli, C., Duda, J., Aldroubi, A.: In vivo fiber tractography using DT-MRI data. Magn. Reson. Med. 44, 625–632 (2000)
6. Lazar, M., Weinstein, D., Hasan, K., Alexander, A.L.: Axon tractography with tensorlines. In: Proceedings of the 8th Annual Meeting of ISMRM, Denver, pp. 482–482 (2000)
7. Conturo, T.E., Lori, N.F., Cull, T.S., Akbudak, E., Snyder, A.Z., Shimony, J.S., McKinstry, R.C., Burton, H., Raichle, M.E.: Tracking neuronal fiber pathways in the living human brain. Proc. Natl. Acad. Sci. USA 96, 10422–10427 (1999)
8. Behrens, T.E.J., Woolrich, M.W., Jenkinson, M., Johansen-Berg, H., Nunes, R.G., Clare, S., Matthews, P.M., Brady, J.M., Smith, S.M.: Characterization and Propagation of Uncertainty in Diffusion-Weighted MR Imaging. Magn. Reson. Med. 50, 1077–1088 (2003)
9. Makris, N., Worth, A.J., Sorensen, A.G., Papadimitriou, G.M., Wu, O., Reese, T.G., Wedeen, V., Davis, T., Stakes, J.W., Caviness, V.S., Kaplan, E., Rosen, B., Pandya, D.N., Kennedy, D.N.: Morphometry of in vivo human white matter association pathways with diffusion-weighted magnetic resonance imaging. Ann. Neurol. 42, 951–962 (1997)
10. Makris, N., Meyer, J.W., Bates, J.F., Yeterian, E.H., Kennedy, D.N., Caviness, V.S.: MRI-based topographic parcellation of human cerebral white matter and nuclei II. Rationale and applications with systematics of cerebral connectivity. Neuroimage 9, 18–45 (1999)

11. Wiegell, M.R., Larsson, H.B.W., Wedeen, V.J.: Fiber crossing in human brain depicted with diffusion tensor MR imaging. Radiology 217, 897–903 (2000)
12. Hagmann, P., Thiran, J.P., Jonasson, L., Vandergheynst, P., Clarke, S., Maeder, P., Meuli, R.: DTI mapping of human brain connectivity: statistical fiber tracking and virtual dissection. NeuroImage 19, 545–554 (2003)
13. Parker, G.J.M., Haroon, H.A., Claudia, A.M.: Wheeler-Kingshott: A Framework for a Streamline-Based Probabilistic Index of Connectivity (PICo) Using a Structural Interpretation of MRI Diffusion Measurements. J. Magn. Reson. 18, 242–254 (2003)
14. Friman, O., Farneback, G., Westin, C.-F.: A Bayesian Approach for Stochastic White Matter Tractography. IEEE Trans. Med. Imag. 25, 965–978 (2006)
15. Mattiello, J., Basser, P.J., Lebihan, D.: Analytical expression for the b matrix in NMR diffusion imaging and spectroscopy. J. Magn. Reson. A 108, 131–141 (1994)
16. Basser, P.J., Mattiello, J., Lebihan, D.: Diagonal and off-diagonal components of the self-diffusion tensor: their relation to and estimation from the NMR spin-echo signal. In: 11th Annual Meeting of the SMRM, Berlin, p. 1222 (1992)
17. Basser, P.J., Mattiello, J., Lebihan, D.: Estimation of the effective self-diffusion tensor from the NMR spin-echo. J. Magn. Reson. B 103, 247–254 (1994)
18. Zheng, B., Rajapakse, J.C.: Effect of diffusion weighting and number of sensitizing directions on fiber tracking in DTI. Neural Information Processing, 102–109 (2006)
19. Parker, G.J.M., Wheeler-Kingshott, C.A.M., Barker, G.J.: Estimating Distributed Anatomical Connectivity Using Fast Marching Methods and Diffusion Tensor Imaging. IEEE Trans. on Med. Imag. 21, 505–512 (2002)
20. Hofer, S., Frahm, J.: Topography of the human corpus callosum revisited–Comprehensive fiber tractography using diffusion tensor magnetic resonance imaging. NeuroImage 32, 989–994 (2006)
21. Jones, D.K., Griffin, L.D., Alexander, D.C., Catani, M., Horsfield, M.A., Howard, R., Williams, S.C.R.: Spatial Normalization and Averaging of Diffusion Tensor MRI Data Sets. NeuroImage 17, 592–617 (2002)

Classification of CT Brain Images of Head Trauma

Tianxia Gong[1], Ruizhe Liu[1], Chew Lim Tan[1], Neda Farzad[2], Cheng Kiang Lee[3],
Boon Chuan Pang[3], Qi Tian[4], Suisheng Tang[4], and Zhuo Zhang[4]

[1] Department of Computer Science, School of Computing, National University of Singapore,
3 Science Drive 2, Singapore 117543
{gong_tianxia,liurz,tancl}@comp.nus.edu.sg
[2] Department of Learning, Management, Informatics & Ethics (LIME),
Karolinska Institute Berzelius v. 3, Stockholm, Sweden 17177
neda.farzad.182@student.ki.se
[3] National Neuroscience Institute, Tan Tock Seng Hospital, 11 Jalan Tan Tock Seng,
Singapore 308433
{cheng_kiang_lee,boon_chuan_pang}@nni.com.sg
[4] Insitute of Infocomm Research, 21 Heng Mui Keng Terrace, Singapore, 119613
{tian,suisheng,zzhang}@i2r.a-star.edu.sg

Abstract. A method for automatic classification of computed tomography (CT) brain images of different head trauma types is presented in this paper. The method has three major steps: 1. The images are first segmented to find potential hemorrhage regions using ellipse fitting, background removal and wavelet decomposition technique; 2. For each region, features (such as area, major axis length, etc.) are extracted; 3. Each extracted feature is classified using machine learning algorithm; the images are then classified based on its component regions' classification. The automatic medical image classification will be useful in building a content-based medical image retrieval system.

1 Introduction

Due to the advances of multi-slice Computed Tomography (CT) Scan with up to 64 slices per scan, a huge amount of CT images are produced in modern hospitals. Today, CT scan images are in the standard DICOM (Digital Imaging and Communications in Medicine) format which incorporates textual information together with the images. Display and retrieval of CT scan images are via PACS (Picture Archives and Communication System) hardware [1]. However with such standards and hardware, the CT scan images currently can only be retrieved using patient names or identity card numbers. To retrieve an image pertaining to a particular anomaly without the patient name is literally like looking for a needle in a haystack. In the domain of CT brain images, very often doctors already overloaded with day-to-day medical consultation simply could not remember patients' names when they need to refer to cases of certain type of brain trauma seen before and as such valuable information are lost in the sea of raw image pixels.

However, if the CT brain images are automatically classified according to trauma types and incorporated to the medical image search system, then the system with search

J.C. Rajapakse, B. Schmidt, and G. Volkert (Eds.): PRIB 2007, LNBI 4774, pp. 401–408, 2007.
© Springer-Verlag Berlin Heidelberg 2007

functions not just by patients' names but by trauma types provides solution to the problem. In this paper, we propose a method to classify CT brain images of head trauma automatically and quickly, so that it facilitates the building of such search systems.

Head trauma has the following major types [2]: epidural hemorrhage[1] (EDH), acute subdural hemorrhage (SDH_Acute), chronic subdural hemorrhage (SDH_Chronic), intracerebral hemorrhage (ICH), intraventricular hemorrhage (IVH) and subarachnoid hemorrhage (SAH). In this paper, we focus on classification of EDH, SDH_Acute and ICH, for they are the dominant types in most head trauma cases. Our images are from CT brain scans performed in the two-year period of 2003 and 2005 as a result of hospital admission for mild head injured patients in National Neuroscience Institute, Tan Tock Seng Hospital [3]. Some of the mild head injuries were later found to be insignificant with no hemorrhage detected. Such cases are treated as belonging to the "normal" class in our training data.

The rest of the paper is organized as follows. In section 2, we will present our method of automatic classification which basically consists of three phases: namely, pre-processing, feature extraction and classification. In section 3, we will discuss our experimental results involving machine learning and validation. Finally, section 4 concludes the paper with our future works.

2   A Method for Automatic Classification of CT Brain Images

Our proposed method to automatically classify CT brain images consists of three phases: preprocessing, feature extraction and classification. In the preprocessing phase, we segment the hemorrhage regions from the CT brain image using ellipse fitting [4], background removal and wavelet decomposition technique [5, 6]. The segmented result is a binary image with potential hemorrhage regions in white and the others in black. Then for each of the potential hemorrhage regions, we extract information about size, shape and position, and create a feature vector accordingly. Lastly, we use a machine learning algorithm to classify the potential hemorrhage regions into different hemorrhage types or normal regions according to the extracted features. The CT brain images are then classified according to the classification of its potential hemorrhage regions.

2.1 Preprocessing

Preprocessing algorithm consists of 4 steps. Step one removes the skull and fits an ellipse to the skull to construct an "interior region", which is the brain inside the skull. Step two removes the gray matter. Step three uses a wavelet decomposition to reduce noise and set a threshold automatically to identify the hemorrhage regions. The last step generates a binary image containing the hemorrhage regions in white and the others in black.

Step 0: Input CT brain image in JPEG format of dimension 512×512. (Figure 1)

---

[1] The terms "hemorrhage" and "hematoma" are often used interchangeably. In this paper, we use "hemorrhage" for consistency.

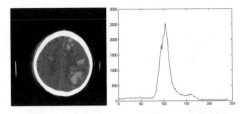

Fig. 1. Image *I*, Original raw input image; left: image; right: intensity distribution[2]

Step 1: Remove the skull and segment the "interior region"

The skull is in white color, whose intensity is above 250 in a gray scale map. Hence, we simply treat those pixels with intensity 250 and above as the skull. The interior region refers to the brain content inside the skull. Since most traumas are diagnosed according to blood clots or edema inside the skull, it is important to segment the interior region. Firstly, we do a boundary detection based on the skull removed. The boundary contains points with intensity above 250, which belong to the skull. Note that there are two other regions that are also in white color. These two regions belong to the CT scan device. However, since they are much smaller than the skull, they can be removed by doing a simple area comparison. Next, we do an ellipse fitting on the boundary points, and compute the center (X_c, Y_c), the major axis, the minor axis, and the parameters of the ellipse. There are 6 parameters, a,b,c,d,e and f, and thus the ellipse has an equation of the following form:

$$ax^2 + bxy + cy^2 + dx + ey + f = 0$$

Hence, a point $[x, y]$ given to the equation that has a result less than zero is inside the ellipse. Finally, we segment the interior points based on the following rules on the original image *I*.

1. The point should be inside the ellipse;
2. The point should be set apart from the center of the ellipse with a distance less than 80% of the average of the major and minor axes of the ellipse.
3. Its intensity is between 10 and 250.
   We denote the interior image to be T_0. (Figure 2)

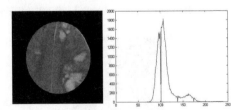

Fig. 2. Image T_0, interior region; left: image; right: intensity distribution

---

[2] The intensity below 10 (background) and above 250 (skull) are not shown in the histogram so that the intensity of inner part of the brain is shown in more detail.

Step 2: Remove the gray matter

Most parts of the content inside the skull are the gray matter. In the histogram of intensity on T_0, the peak corresponds to the gray matter. (Figure 2) Hence, a simple subtraction off the peak intensity from T_0 will give us an image with the gray matter removed. We call it T_1. (Figure 3)

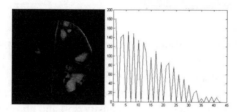

Fig. 3. Image T_1, gray matter removed; left: the image; right: intensity distribution

Step 3: Reduce noise

There is much noise as white dots or tiny fragments produced in T_1, because we subtract only a single intensity value from various parts of the gray matter. A second level 2D *Biorthogonal* wavelet transform is used to reduce the noise [5, 6]. We finally get the image with reduced noise but more distinguishable diseased parts. We denote the resultant image to be T_2. (Figure 4)

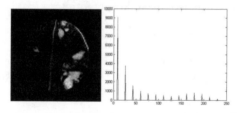

Fig. 4. Image T_2, noise reduced; left: image; right: intensity distribution

Step 4: Generate a binary image of hemorrhage

After the preprocessing, we can define thresholds according to the intensity distribution of image T_2. We set the hemorrhage threshold to be the median of the peaks obtained from the wavelet transform. Finally we get a binary image T_3. (Figure 5)

Fig. 5. Image T_3, each white pixel group represents a possible hemorrhage region of the image

2.2 Feature Extraction

As human doctors use size, shape and position of the potential hemorrhage region to classify them, we need quantifiable features that describe the size, shape and position for antomatic classification. For each potential hemorrhage region, we use the Matlab function *regionprops* [7] to extract the area, major and minor axis lengths, eccentricity, solidity and extent. Also, features for the skull and the background are extracted from the labeled skull and background regions [8]. These features describe the size, shape and position of the potential hemorrhage region; therefore, they are useful for classification. The class of each feature vector is one of the following values: EDH, SDH_ Acute, ICH and normal. All features are described in Table 1.

Table 1. Features extracted from each region

	Name	Description[7]	Example
1	Area	The actual number of pixels in the region.	607
2	Major axis length	The length (in pixels) of the major axis of the ellipse that has the same second-moments as the region.	38.1135
3	Minor axis length	The length (in pixels) of the minor axis of the ellipse that has the same second-moments as the region.	23.5155
4	Eccentricity	The eccentricity of the ellipse that has the same second-moments as the region. The eccentricity is the ratio of the distance between the foci of the ellipse and its major axis length.	0.7295
5	Solidity	The proportion of the pixels in the convex hull that are also in the region. Computed as Area/ConvexArea[3].	0.8772
6	Extent	The proportion of the pixels in the bounding box that are also in the region. Computed as the area divided by area of the bounding box.	0.6485
7	Skull	Whether the region is adjacent to skull or not.	false
8	Background	Whether the region is adjacent to background or not.	true

2.3 Classification

According to the features extracted in section 2.2, we classify the regions into five categories: EDH, SDH_ Acute, ICH, other and normal, where the first three classes refer to the three types of hemorrhages we focus on, the other refers to the remaining types of hemorrhage, and normal means that the region is not a hemorrhage. For example, the potential hemorrhage regions of Figure 5 classified as ICH are shown in Figure 6.

As there may be more than one type of hemorrhage present in a brain CT image, the class for each image cannot have only one of the class values as the regions have. Instead, the class for each image is a boolean vector <EDH, SDH_ Acute, ICH, normal>, where each boolean value indicates the presence of certain type of hemorrhage.

---

[3] The number of pixels in convex image, which is the convex hull, with all pixels within the hull filled in.

Fig. 6. Each white pixel group represents an ICH region of the image

The class of the image is classified according the classifications of its regions. If the regions are classified as some type(s) of hemorrhage (EDH, SDH_Acute, or ICH), the image is also classified as the same type(s) of the hemorrhage(s). Otherwise if all regions are classified as normal, the image itself is also classified as normal.

3 Experimental Results

We obtained 35 CT brain images (15 EDH, 9 SDH_Acute, 6 ICH and 5 normal) belonging to 12 patients from the National Neuroscience Institute, Tan Tock Seng Hospital, Singapore. After preprocessing, we obtained 818 potential hemorrhage regions (15 EDH, 19 SDH_Acute, 47 ICH and 737 normal).

3.1 Classification of Potential Hemorrhage Regions

We used J48 classifier, a decision tree classifier based on C4.5 [9], from WEKA [10] to train and test the region features. 10-fold cross validation was used. The average accuracy (correctly classified regions / all regions) is 93.0%. As there are many more normal class cases than the other classes, the data is highly imbalanced, which causes high accuracy for normal class and relatively lower accuracy for other classes. The detailed testing results for each class are reported as shown in Table 2.

Table 2. Detailed testing results for each class

	EDH	SDH_Acute	ICH	normal
Precision	60.0%	53.8%	60.0%	95.9%
Recall	60.0%	36.8%	44.7%	98.2%

The decision tree obtained from J48 is shown in Figure 7. The knowledge represented by the decision tree is actually very close to the doctor's knowledge in classifying potential hemorrhage regions. For example, if the region's area is less than or equal to 2891 pixels ($6.89cm^2$) and greater than 91 pixels ($0.22cm^2$), and the eccentricity is less than or equal to 0.9426 (the greater the eccentricity is, the elongated is the region), and the region is not adjacent to skull, then the region is ICH. This is also a typical rule for doctors to recognize ICH manually.

```
Area <= 2891
|    Eccentricity <= 0.9426
|    |    skull = false
|    |    |    Area <= 91
|    |    |    |    Extent <= 0.6485: normal
|    |    |    |    Extent > 0.6485: ICH
|    |    |    Area > 91: ICH
|    |    skull = true
|    |    |    Area <= 1263: normal
|    |    |    Area > 1263
|    |    |    |    Eccentricity <= 0.9322: EDH
|    |    |    |    Eccentricity > 0.9322: SDH_Acute
|    Eccentricity > 0.9426: normal
Area > 2891
|    Eccentricity <= 0.8579: ICH
|    Eccentricity > 0.8579
|    |    Area <= 7185
|    |    |    Extent <= 0.1852
|    |    |    |    MajorAxisLength <= 274.1822: EDH
|    |    |    |    MajorAxisLength > 274.1822: SDH_Acute
|    |    |    Extent > 0.1852: SDH_Acute
|    |    Area > 7185: EDH
```

Fig. 7. Decision tree obtained from the training data using the J48 classifier

3.2 Classification of Images

The classification of the image is considered as: 1. correct, if the predicted class(es) and the actual class(es) are exactly the same; 2. partially correct, if the actual class(es) is/are included in the prediction, but other class(es) is/are also predicted; 3. incorrect, if the predicted class(es) is different from the actual class. Among the 35 images, 18 are classified correctly, 6 are classified partially correctly, and 11 are classified incorrectly.

4   Conclusion

In this paper, we propose a method to classify CT brain images of head trauma automatically and quickly. The method consists of three phases: preprocessing, feature extraction and classification. In the preprocessing phase, we segment the hemorrhage regions from the CT brain image using ellipse fitting, background removal and wavelet decomposition technique. The segmented result is a binary image with potential hemorrhage regions in white and the rest in black. Then for each of the potential hemorrhage regions, we extract information about its size, shape and relative location, and create a feature vector. Lastly, we use machine learning algorithms to classify the potential hemorrhage regions into different hemorrhage types or normal regions according to the extracted features. The CT brain images are then classified according to the classification of its potential hemorrhage regions.

The fast and scalable automatic medical image classification can help to build a medical image search system according to the syndrome types (in our case of CT brain images, the syndrome types are head trauma types) and facilitate doctors' research on certain syndrome as well as education for medical profession.

In our future work, we will extend the classification types to include other head traumas. We will also explore other machine learning algorithms and compare their classification results. Finally, we will do text mining to extract further information from the text of neuroradiologists' report to find more features for classification.

Acknowledgement

This project is supported in part by the National University of Singapore under Academic Research Fund grant no. R252-000-290-112.

References

1. Huang, H.K.: PACS: Basic Principles and Applications. Wiley-Liss Inc., Canada (1999)
2. Downie, A.: Tutorial: CT in Head Trauma, http://www.radiology.co.uk/srs-x/tutors/cttrauma/tutor.htm
3. Pang, B.C., Yin, H.: Analysis of clinical criterion for "talk and deteriorate" following minor head injury using different data mining tools. Journal of Neurotrauma, National Neurotrauma Society, Gainesville, Florida, USA (in press)
4. Fitzgibbon, A., Pilu, M., Fisher, R.B.: Direct Least Square Fitting of Ellipses. IEEE Pattern Analysis and Machine Intelligence 21(5) (May 1999)
5. Daubechies, I.: Ten Lectures on Wavelets. SIAM: Society for Industrial and Applied Mathematics (June 1, 1992)
6. Misiti, M., Misiti, Y., Oppenheim, G., Poggi, J.M.: Wavelet Toolbox 4 User's Guide. Matlab Wavelet Toolbox User's Guide, Version 4 (2007)
7. Image Processing Toolbox User's Guide. Matlab Image Processing Toolbox User's Guide Version 3 (2002)
8. Cosic, D., Loncaric, S.: Rule-Based Labeling of CT Head Image. In: Keravnou, E.T., Baud, R.H., Garbay, C., Wyatt, J.C. (eds.) AIME 1997. LNCS, vol. 1211, pp. 453–456. Springer, Heidelberg (1997)
9. Quilan, J.R.: C4.5: Programs for machine learning. Morgan Kaufmann, San Francisco (1993)
10. Holmes, G., Donkin, A., Witten, I.H.: WEKA: a machine learning workbench. In: Proceedings Second Australia and New Zealand Conference on Intelligent Information Systems, Brisbane, Australia, pp. 357–361 (1994)

Author Index

Printing: Mercedes-Druck, Berlin
Binding: Stein+Lehmann, Berlin

Lecture Notes in Bioinformatics

Vol. 3594: J.C. Setubal, S. Verjovski-Almeida (Eds.), Advances in Bioinformatics and Computational Biology. XIV, 258 pages. 2005.

Vol. 3500: S. Miyano, J. Mesirov, S. Kasif, S. Istrail, P. Pevzner, M. Waterman (Eds.), Research in Computational Molecular Biology. XVII, 632 pages. 2005.

Vol. 3388: J. Lagergren (Ed.), Comparative Genomics. VII, 133 pages. 2005.

Vol. 3380: C. Priami (Ed.), Transactions on Computational Systems Biology I. IX, 111 pages. 2005.

Vol. 3370: A. Konagaya, K. Satou (Eds.), Grid Computing in Life Science. X, 188 pages. 2005.

Vol. 3318: E. Eskin, C. Workman (Eds.), Regulatory Genomics. VII, 115 pages. 2005.

Vol. 3240: I. Jonassen, J. Kim (Eds.), Algorithms in Bioinformatics. IX, 476 pages. 2004.

Vol. 3082: V. Danos, V. Schachter (Eds.), Computational Methods in Systems Biology. IX, 280 pages. 2005.

Vol. 2994: E. Rahm (Ed.), Data Integration in the Life Sciences. X, 221 pages. 2004.

Vol. 2983: S. Istrail, M.S. Waterman, A. Clark (Eds.), Computational Methods for SNPs and Haplotype Inference. IX, 153 pages. 2004.

Vol. 2812: G. Benson, R.D.M. Page (Eds.), Algorithms in Bioinformatics. X, 528 pages. 2003.

Vol. 2666: C. Guerra, S. Istrail (Eds.), Mathematical Methods for Protein Structure Analysis and Design. XI, 157 pages. 2003.